Analysis and Design for Networked Teleoperation System

Changchun Hua · Yana Yang ·
Xian Yang · Xinping Guan

Analysis and Design for Networked Teleoperation System

Springer

Changchun Hua
Institute of Electrical Engineering
Yanshan University
Qinhuangdao, Hebei, China

Xian Yang
Institute of Electrical Engineering
Yanshan University
Qinhuangdao, Hebei, China

Yana Yang
Yanshan University
Qinhuangdao, China

Xinping Guan
Department of Automation
Shanghai Jiao Tong University
Shanghai, China

ISBN 978-981-13-7938-3 ISBN 978-981-13-7936-9 (eBook)
https://doi.org/10.1007/978-981-13-7936-9

This Springer imprint is published by the registered company Springer Nature Singapore Pte Ltd.
The registered company address is: 152 Beach Road, #21-01/04 Gateway East, Singapore 189721, Singapore

Preface

Recently, teleoperation system, which is integrated with the capacities of control theory, computer technology, and communications, has attracted a considerable amount of interest around the world due to its theory and application. To date, teleoperation system has embraced widespread applications, such as in nuclear accident rescue, undersea operation, space exploration, and robotic telesurgery. Attracted by its tremendous potential, many researchers from academic and industry communities have already devoted huge efforts to the research of teleoperation system. Unfortunately, there are still many restrictions for the teleoperation system in practical application: (i) In order to facilitate the operation of a specific object (such as dangerous goods, etc.), a communication network is required to ensure real-time and efficient transmission of data, and then a networked teleoperation system is generated, meaning that information communication between master and slave is inevitably influenced by limited communications bandwidth; (ii) Teleoperation system applied in reality has strong nonlinearity, the system model and parameters are always uncertain, and the system model is disturbed by the task environment easily; (iii) Due to the extension of applied range and the increasement of task complexity, the performance demands (such as rapidity, stability, accuracy, overshoot, etc.) on the teleoperation control are increasing.

This book is devoted to present cutting-edge results on stability analysis and control scheme designs for networked teleoperation systems. It highlights new research on commonly encountered nonlinear teleoperation systems, including the stability analysis of teleoperation systems with asymmetric time-varying delays, stability analysis of teleoperation systems with interval time delays, and so on. Moreover, the book presents several high-performance control scheme designs for teleoperation systems when the velocity is available and unavailable, and for systems with nonlinear input. The results presented here mark a substantial contribution to nonlinear teleoperation system theory, robotic control theory, and networked control system theory. As such, the book will be of interest to university researchers, R&D engineers, and graduate students in control theory and control engineering who wish to learn about the core principles, methods, algorithms, and applications of networked teleoperation systems, robotic systems, and nonlinear control systems.

The readers of our book are assumed to have some background in robot control and nonlinear systems control. Although this book is primarily intended for students and practitioners of nonlinear teleoperation system control theory, it is also a valuable reference for those in fields such as communication engineering. Moreover, we believe that this book should be suitable for certain advanced courses or seminars.

In the introduction chapter, a motivation on the research and its history and an overview of recent development of nonlinear teleoperation system are provided.

Then, this book will present four parts:

Part I: Some development of stability analysis results for nonlinear teleoperation system is introduced. In Chap. 2, based on the designed P+d (Proportion plus damping) controller, the stability of teleoperation system with asymmetrical time-varying delays by using proper Lyapunov-Krasovskii functional is analyzed. In Chap. 3, the convergence problem of teleoperation system with asymmetrical time-varying delays and nonpassive human operator inserted force is addressed. Furthermore, Chap. 4 and Chap. 5 consider the stability problem of teleoperation system with asymmetrical interval time-varying delays and quantized information feedback, respectively.

Part II: Some new results on nonlinear teleoperation control design are presented. In Chap. 6, a new class of controller with delayed torque feedback is developed to provide better transparency performance. In Chap. 7, compared with the previous control results with asymptotic convergence, the exponential convergence results are guaranteed by designing a new control approach. Moreover, Chap. 8 and Chap. 9 address the finite-time control and prescribed performance control problems for teleoperation system, and thus the control performances on rapidity, robustness, and accuracy can be guaranteed simultaneously.

Part III: In Part two, different kinds of velocity observers are designed for nonlinear teleoperation system. In Chap. 10, a simple but effective high-gain observer is given to obtain the unavailable velocity, and then the exponent stability of the master-slave is guaranteed. In Chap. 11, the TSM (Terminal Sliding Mode) based finite-time velocity observer is proposed by applying strong approximation abilities of the NN (Neural Networks). Compared with the general velocity observers, the finite-time observer designed in this chapter can estimate unknown velocity signals in finite time. Moreover, higher precision and strong robust performances can be achieved with the developed finite-time velocity observer.

Part IV: The last Part of this book is dedicated to teleoperation system with actuator nonlinear input. In Chap. 12, for actuator saturated and time-varying delayed teleoperation system, a saturated controller is represented as a convex combination of actual feedback and an auxiliary counterpart in the space surrounded by two pairs of intersected parallel surfaces. Furthermore, the estimate of domain of attraction (DOA) is explored, and an optimization problem is proposed to maximize the estimate of DOA. In Chap. 13, the output-feedback based controller design problem is investigated for networked teleoperation system with actuator saturation. An adaptive SP+Sd-type (saturated proportion plus saturated damping) controller is designed by considering the unknown gravity term, and then the global asymptotic stability of the bilateral teleoperation system with time-varying delays and bounded inputs can be guaranteed.

The support from the National Natural Science Foundation of China (61825304, 61751309, 61673335, 61703361, 61603328), Basic Research Program of Hebei Province (F2016203467), Program for the Outstanding Young Innovative Talent of China is gratefully acknowledged.

Qinhuangdao, China — Changchun Hua
Qinhuangdao, China — Yana Yang
Qinhuangdao, China — Xian Yang
Shanghai, China — Xinping Guan
January 2019

Contents

Symbols and Acronyms

R	Field of real numbers
R^n	n-dimensional real Euclidean space
$R^{n\times m}$	Space of $n \times m$ real matrices
R_+	n-dimensional positive real Euclidean space
I	Identity matrix
O	Zero matrix
A	System matrix
A^{-1}	Inverse of matrix A
A^T	Transpose of matrix A
$A \geq 0$	Symmetric positive semi-definite
$A > 0$	Symmetric positive definite
$A \leq 0$	Symmetric negative semi-definite
$A < 0$	Symmetric negative definite
lim	Limit
min	Minimum
max	Maximum
LMI	Linear matrix inequality
SMC	Sliding mode control
SISO	Single-input single-output
MIMO	Multiple-input multiple-output
$\mathrm{tr}(A)$	Trace of matrix A
$\det(A)$	Determinant of matrix A
$\mathrm{rank}(A)$	Rank of matrix A
$0_{n\times m}$	Zero matrix of dimension $n \times m$
$\lambda(A)$	Eigenvalue of matrix A
$\lambda_{\min}(A)$	Minimum eigenvalue of matrix A
$\lambda_{\max}(A)$	Maximum eigenvalue of matrix A
$\mathrm{sign}(x)$	sign of x
sat()	Saturation function with unit limits
$\|x\|$	Absolute value (or modulus) of x

$\|x\|$	Euclidean norm
$\forall$	For all
$\in$	Belong to
$\rightarrow$	Tend to, or mapping to (case sensitive)
$\sum$	Sum
sup	Supremum
inf	Infimum
$*$	Symmetric terms in a symmetric matrix
$\mathrm{diag}\{X_1, X_2, \cdots, X_m\}$	Diagonal matrix with X_i as its ith diagonal element
L_2 norm	$\|f\|_2 = \int_0^\infty \|f(t)\|^2 dt, f : R_+ \rightarrow R^n$
L_∞ norm	$\|f\|_\infty = \sup_{t\in[0,\infty)} \|f(t)\|, f : R_+ \rightarrow R^n$
L_2 space	$\{f : R_+ \rightarrow R^n : \|f\|_2 < \infty\}, f : R_+ \rightarrow R^n$
L_∞ space	$\{f : R_+ \rightarrow R^n : \|f\|_\infty < \infty\}, f : R_+ \rightarrow R^n$

Chapter 1
Introduction

1.1 Background

In recent years, with the rapid development of computer technology, electric communication technology and control technology, teleoperation system integrating these technologies has gained more and more attention [1–3, 8]. The teleoperation system allows human operators to operate in dangerous or hostile environments, and allows human-sensing and manipulative capabilities in remote environment to be extended [4–7]. In general cases, a typical teleoperation system is composed of human operator, master robot, communication channel, slave robot, and environment (see Fig. 1.1). For a teleoperation system the bilateral control supplies force feedback information to human operator through a master robot. On the other hand, a slave robot applies to interact with unknown environments. Both the master and slave robots are coupled via a communication channel, where the position velocity and force information are transferred. For the command of the human operator, the slave robot situated at a remote location tracks the motion of the master robot. In order to improve the task performance, the force feedback from the slave robot to the master robot is needed. If the information flows in both directions between master and slave, the teleoperator system is said to be controlled bilaterally. Bilateral teleoperation is a challenging topic of current control technologies with a number of traditional and potential applications, such as space and undersea exploration, handling of hazardous objects (see Fig. 1.2), and robotic telesurgery (see [8] and references therein).

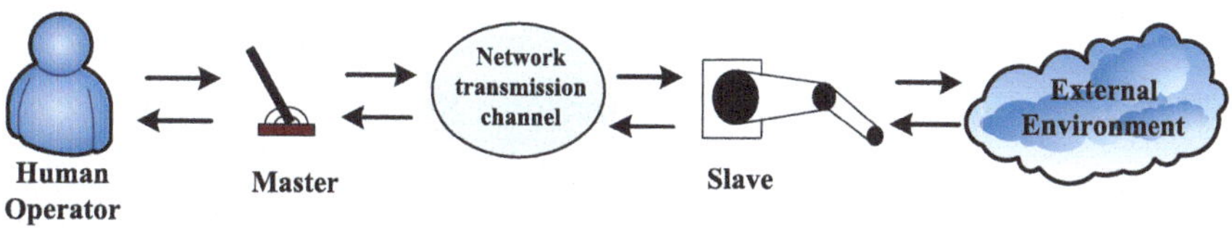

Fig. 1.1 Networked teleoperation system

C. Hua et al., *Analysis and Design for Networked Teleoperation System*,
https://doi.org/10.1007/978-981-13-7936-9_1

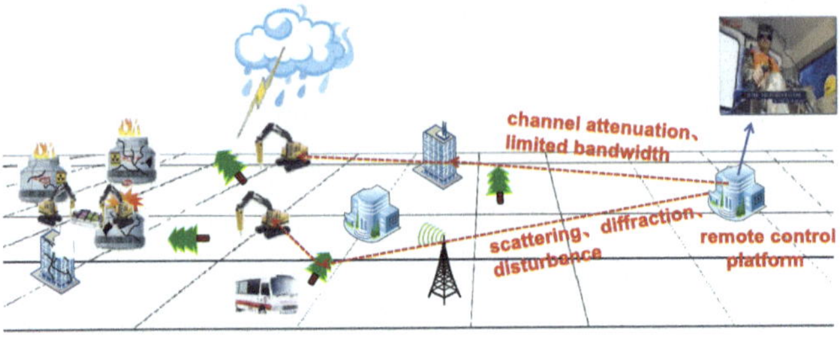

Fig. 1.2 Teleoperation system in nuclear accident

Different from the single robot control system, from a control theoretic point of view the main goals of a teleoperation system are twofold [9–12].

Stability: Maintain stability of the closed-loop system irrespective of the behavior of the operator or the environment.

Telepresence: Provide the human operator with a sense of telepresence, with the latter regarded as transparency of the system between the environment and the operator.

It is worth noticing that the stability and transparency of the teleoperation system are often conflicting in practical applications. The system will inevitably compromise the transparency of the system when we overemphasize the system stability. Similarly, the system will inevitably result in instability of the teleoperation system when the transparency is overemphasized. Therefore, in most cases, controller should be designed in order to balance the stability and transparency of the system, simultaneously.

1.2 Methods and Issues

A typical teleoperation system is composed of two robotic subsystems a master and a slave that exchange signals (positions, velocities and/or forces) [13, 14]. In general cases, the slave tries to mimic the behavior of the master, and the master also will take into account the force feedback from the slave when planning the motion trajectories. By referring to a large number of literature on teleoperation control, the dynamical model of general teleoperation system is presented in following subsection.

1.2.1 Dynamical Model of Teleoperation System

This subsection presents the dynamical model of nonlinear teleoperator together with some properties needed for the stability analysis of the controllers covered in this book.

Notations. The time argument of all functions in the following contents will be omitted, e.g., $x \triangleq x(t)$, expect for those which appear with time delay, e.g., $x(t - d(t))$. The argument of signals inside integral will be omitted, it is supposed to be equal to the variable on the differential, unless otherwise noted, e.g., $\int_0^t x d\sigma \triangleq \int_0^t x(\sigma) d\sigma$.

In this book, the local and remote manipulators are modeled as a pair of n-Degree of Freedom (DOF) serial links. Their corresponding nonlinear dynamics, together with the human operator and environment torques are given by

$$\begin{cases} M_m(q_m)\ddot{q}_m + C_m(q_m, \dot{q}_m)\dot{q}_m + G_m(q_m) = \tau_m + F_h \\ M_s(q_s)\ddot{q}_s + C_s(q_s, \dot{q}_s)\dot{q}_s + G_s(q_s) = \tau_s - F_e \end{cases} \tag{1.1}$$

where $q_m(t), q_s(t) \in R^n$ are the vectors of the joint displacements; $\dot{q}_m(t)$, $\dot{q}_s(t) \in R^n$ are the vectors of joint velocities; $\ddot{q}_m(t), \ddot{q}_s(t) \in R^n$ are the vectors of joint accelerations; $M_m(q_m), M_s(q_s) \in R^{n \times n}$ are the positive definite inertia matrices; $C_m(q_m, \dot{q}_m)$, $C_s(q_s, \dot{q}_s) \in R^{n \times n}$ are the matrices of Coriolis and centrifugal effects, defined using the Christoffel; $G_m(q_m), G_s(q_s) \in R^n$ are the gravitational torques; $F_h, F_e \in R^n$ are human operator force (torque) and environmental force (torque), respectively; $\tau_m, \tau_s \in R^n$ are applied control torques. In the following chapters, it is assumed that the manipulators are composed by actuated revolute joints. Moreover, the friction torques and the external disturbances will be considered according to the actual situations.

It is well known that these dynamical models have the following important properties [15–18]

Property 1.1 *The inertia matrix $M_i(q_i)$ is a symmetric positive definite function and there exist positive constants m_{i_1} and m_{i_2} such that $m_{i_1} I \le M_i(q_i) \le m_{i_2} I$, with $i = m, s$.*

Property 1.2 *The matrix $\dot{M}(q) - 2C(q, \dot{q})$ is skew-symmetric, i.e., $\dot{M}(q) = C(q, \dot{q}) + C^T(q, \dot{q})$.*

If matrix A satisfies $A = -A^T$, it is called skew-symmetric. Property 1.2 means that $x^T\left(\dot{M}(q) - 2C\left(q, \dot{q}\right)\right)x = 0$ for an arbitrary vector x.

Property 1.3 *There exist positive scalars β_i such that $U_i(q_i) \ge \beta_i$, where $U_i(q_i)$ is the potential energy of the manipulator satisfying $G_i(q_i) = \frac{\partial U_i(q_i)}{\partial q_i}$.*

Property 1.4 *For all $q_i, x, y \in R^n$, there exists a positive scalar a_i such that $\|C_i(q_i, x)y\| \le a_i \|x\| \|y\|$.*

Property 1.5 *If $\ddot{q}_i$ and $\dot{q}_i$ are bounded, the time derivative of $C_i(q_i, \dot{q}_i)$ is bounded.*

Property 1.6 *The gravity vector $G_i(q_i)$ is bounded as it consists of sinusoidal function q_i. There exists a positive scalar μ_{g_i} such that $\|G_i(q_i)\| \le \mu_{gi}$.*

Property 1.7 *The Lagrange dynamics are linearly parameterizable. Thus* $M_i(q_i)\ddot{q}_i + C_i(q_i, \dot{q}_i)\dot{q}_i + G_i(q_i) = Y_i(q_i, \dot{q}_i, \ddot{q}_i)\theta_i$, *where* $Y_i(q_i, \dot{q}_i, \ddot{q}_i) \in R^{n \times p}$ *are matrices of known functions and* $\theta_i \in R^p$ *are constant vectors of manipulator physical parameters (link masses, moments of inertia, etc.).*

These important properties of teleoperation system will be used in the following chapters frequently. To avoid duplication, these properties will not be presented in the following chapters, but will be applied directly.

Based on the above teleoperation model, numerical control schemes are provided. Let us have a brief rundown of recent control methodologies used for teleoperation systems with model uncertainty and unknown time-varying delay.

1.2.2 Control Methods for Teleoperation System

As an important research field of robotics, the research on the control methods of teleoperation system has achieved considerable results in recent years. Numerical effective control algorithms have also been proposed. In general cases, these algorithms are presented based on corresponding control structures. For a teleoperation system, four kinds of teleoperation system control structure frequently used are shown as follows:

(i) Position error structure [19, 20]: the controllers for the master and the slave are designed based on each other's position errors, the objective of this structure is to make the best of to minimize the position errors.
(ii) Directed force feedback structure [21]: the slave controller is designed with the position error between the master and the slave, yet the master controller is designed by employing the force error between the master and the slave.
(iii) Shared compliant control structure [22]: the master controller is designed with the force error between the master and the slave, yet the controller designed for the slave composed of position error and the measured force between the slave and the environment.
(iv) Four-channel control structure [23, 24]: the force and position measured by the sensors are exchanged between the master side and the slave side, respectively.

Comparing above control structures, it can be observed that the position-error structure does not need a force sensor, thus it is the simplest. Therefore, based on position-error structure a series of different control approaches are proposed. Moreover, these control approaches can be roughly divided into two kinds: linear control methods and nonlinear control methods.

Next some common linear control methods and nonlinear control methods are presented as follows:

(i) Linear control schemes

Although the teleoperation system itself is a complex nonlinear system, several linear control strategies have been proposed by researchers by simplifying the system model

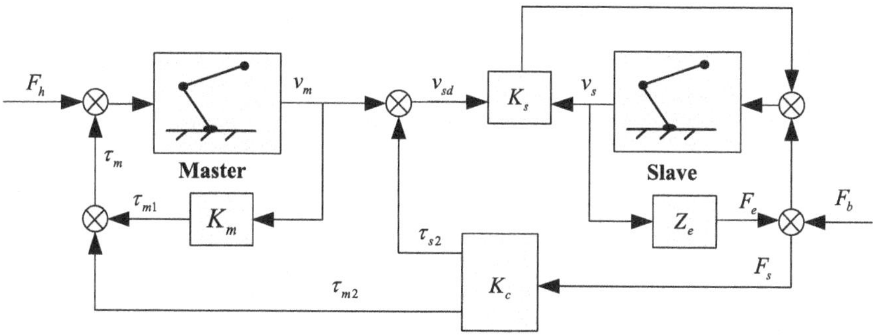

Fig. 1.3 H_∞ control for a teleoperation system

by using advanced linearization technology. Linear control schemes for teleoperation system contain Linear quadratic optimal control strategy, H_∞ control strategy, damping injected control strategy and so on.

(a) Linear quadratic optimal control strategy

For a networked teleoperation system, when the communication time delays between master and slave satisfy some limitations, linear quadratic optimal control strategy can be used by regarding the teleoperation system as a time-delay feedback system with state and output [25]. Moreover, in case that the communication time of the system is assumed to be random independent, backstepping technologies and like-Lyapunov methods can be used to design a linear quadratic optimal control algorithm from a random control point of view [26]. Consider the constant time delays, [27] proposed a new linear quadratic gaussian (LQG) control scheme to guarantee the stability of the closed-loop teleoperation system no matter if it's a free movement or on a collision with something. Consider the contradiction between stability and transparency, a new LQG algorithm was designed for multi-DOF teleoperation system [28], then both stability and transparency can be achieved. Even though the Optimal control of multiple variables can be achieved by applying linear quadratic optimal control strategies, the optimal index function should be reasonably designed. In reality, the selection of optimal index function is very difficult. What's worse, sufficient understanding, harsh limitations and assumptions of the main communication delay is needed when designed LQG control method. This inevitably leads to that the applications of linear quadratic optimal control methods have great limitations in reality.

(b) H_∞ control scheme

H_∞ control method can reduce the influence of bounded disturbance on the system to the degree of expectation regardless of the nature of the disturbance [29, 30]. The control block diagram of the teleoperation system with H_∞ control scheme is shown in Fig. 1.3.

In the above figure, F_h and F_e are operating forces at local side and remote side, respectively. F_b denotes the external disturbance at remote side. K_m and K_s represent

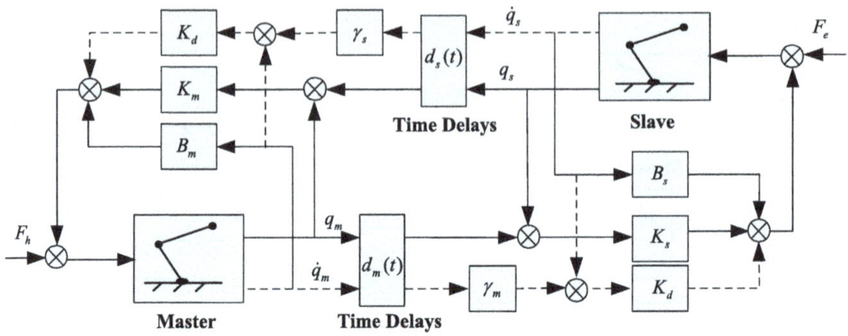

Fig. 1.4 General control scheme for the P+d (dotted lines disabled) and PD+d (dotted lines enabled) controllers

the controllers for master and slave, respectively. The objective of the H_∞ control is to ensure the stability of master-slave system and minimize the velocity error $v_m - v_s$ by designing K_c. In other words, the norm of the system transfer function matrix of the system input signal to the controlled output signal is to be minimized by choosing K_c, then the robustness of the closed-loop teleoperation system can be obtained.

With numerical simulation and experimental results, it is found that for a teleoperation system with small and bounded communication time delays, not only can the stability and the influence of the time delays be minimized, but also some performances can be satisfied. Moreover, certain robustness to uncertainties and other disturbances cam be achieved. However, it is worthwhile to notice that for the case of time delays with uncertain upper bounds, H_∞ control can not fulfill this control task. Worse still, conservative control performance will be obtained with large time delays.

(c) Damping injection scheme

The damping inserted control scheme can also be rewritten as P+d (Proportion plus damping) or PD+d (Proportion Differential plus damping) control. The term d represent the damping term, which is used to consume the redundant energy of the system. Then the stability of the master-slave system can be ensured. In earlier research, researchers have found that P+d control can stabilize the teleoperation system under small time delays by performing a number of experiments [31, 32]. Yet strict theoretical proof was not given. The first strict and detailed proof with PD+d control was given by Lee in 2006 [33]. Furthermore, the proof for P+d control was also provided by Nuño in 2008 [34], and delay-dependent stability conditions were presented. With delay-dependent stability conditions not only are the conservativeness reduced, but also theoretical guidance for the selection of controller parameters are provided. Furthermore, in [35] the delay-dependent stability criteria of teleoperation systems with asymmetric time-varying delays by using the linear matrix inequality (LMI) was shown. Figs. 1.4 shows the schematics of the P+d and PD+d controller schemes.

General PD+d controller with time-varying delays is given as

$$\begin{aligned}\tau_m &= K_d(\dot{q}_m - \gamma_s \dot{q}_s(t - d_s(t))) + K_m e_m + B_m \dot{q}_m \\ \tau_s &= -K_d(\dot{q}_s - \gamma_m \dot{q}_m(t - d_m(t))) - K_s e_s - B_s \dot{q}_s\end{aligned}$$

where, K_d, K_m, K_s, B_m and B_s are positive constants, $\gamma_i^2(t) = 1 - \dot{d}_i(t)$, $i = m, s$, and e_m, e_s are defined as $e_m = q_m - q_s(t - d_s(t))$ and $e_s = q_s - q_m(t - d_m(t))$.

The general P+d controller with time–varying delays is presented as

$$\begin{aligned}\tau_m &= K_m e_m + B_m \dot{q}_m \\ \tau_s &= -K_s e_s - B_s \dot{q}_s\end{aligned}$$

Because of simple structure, damping injection control has been studied and applied widely in recent years [36, 37]. However, as remote operation tasks become more and more complex, the demands on the system performance increases. Simple linear control methods expose their limitations. Moreover, by considering system uncertainties and external disturbances, it is clear that linear control algorithms are difficult to provide satisfactory control performances.

(ii) Nonlinear control schemes

Due to complex nonlinear characteristics of teleoperation system, linear control methods have many disadvantages in practical applications. Therefore, researchers further study the design of the nonlinear control strategy of teleoperation system. The nonlinear control methods include sliding mode control, adaptive control, intelligent control and so on.

(a) Sliding mode control scheme

Due to the properties that robustness against system parametric uncertainties and strong nonlinearities, the sliding mode control has been applied to teleoperation system successfully. Moreover, it has been noticed that sliding mode control also can be used to deal with time delays. The main idea of such a control method is to develop a sliding mode surface $s(t)$ composed of the position error and velocity error. After that, sliding mode control scheme will be designed to drive the trajectories to reach the desired sliding mode surface in finite time. the teleoperation system, with defined position errors $e_m = q_m - q_s(t - d_s(t))$, $e_s = q_s - q_m(t - d_m(t))$ and velocity errors $\dot{e}_m = \dot{q}_m - \dot{q}_s(t - d_s(t))$, $e_s = \dot{q}_s - \dot{q}_m(t - d_m(t))$, the corresponding sliding mode surfaces for master and slave are designed as $s_m = \dot{e}_m + \lambda_m e_m$ and $s_s = \dot{e}_s + \lambda_s e_s$. Then sliding mode control scheme should be designed to ensure s_m and s_s converge to zero in finite-time. In Fig. 1.5, k_{m1}, k_{m2}, k_{s1}, k_{s2} are controller parameters and $sign(\cdot)$ denotes symbolic function.

In fact, related studies have been reported in this area. Sliding mode control was applied to 1-DOF teleoperation system without time delays in [38]. Then the result was extended for teleoperation system with time-varying delays in [39, 40]. In these literatures, modified sliding mode controllers with nonlinear gains can be

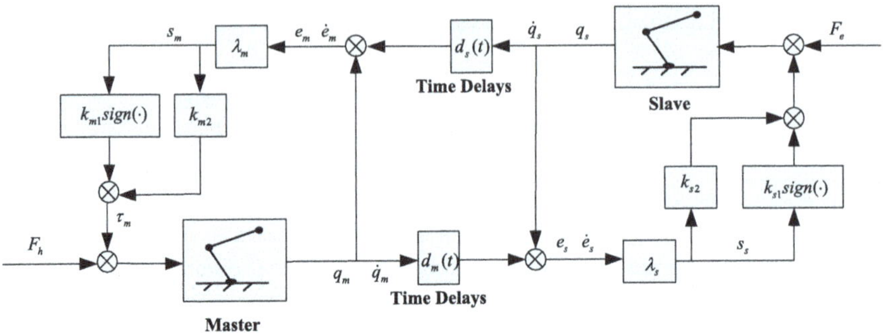

Fig. 1.5 General control scheme for general slding mode controller

chosen independently of the time-delay variations were provided. Moreover, some other related works have also been reported, for example, discrete-time sliding mode control, high-order sliding mode impedance control, to name a few [41, 42]. In recent year, terminal sliding mode based finite time controllers and finite time observers were proposed for nonlinear teleoperation system to provide faster convergence rate and higher convergence precision [43, 44].

(b) Adaptive control scheme

Due to existence of various system uncertainties, adaptive control scheme has been widely deployed for teleoperation system. The application of adaptive control depends on two preconditions: system model can linearize with the adaptive parameter θ and the human operator defined by θ (the estimation of θ) is passive. In Fig. 1.6, the general adaptive control scheme is shown. In [45] and [46], strict and detailed stability analysis has been reported for nonlinear teleoperation system under constant time delays by deploying an adaptive controller. This result has been extended to the case with time-varying delays. New adaptive control algorithms were designed for teleoperation system with symmetrical time-varying delays and asymmetrical time-varying delays, respectively in [47, 48]. Recently, finite-time adaptive control scheme was also proposed in [49]. In Fig. 1.6, general adaptive control scheme is presented. It is worthwhile mentioning that adaptive control relies the precision of the system model, which to some extent limits its application. Worse still, adaptive control can only handle fixed or slow time-varying system parameters uncertainties. This restricts the sorts of systems amenable to be controlled. For uncertain structure, complex parameters and nonlinear time-varying systems, adaptive control is difficult to provide satisfactory control performance.

(d) Intelligent control scheme

Recently, consider hard system nonlinearities, system uncertainties and external disturbances, a serial intelligent control schemes were proposed by applying Neural Networks (NNs) in [50, 51] and Fuzzy Logic system (FLs) in [52, 53]. On one hand, NNs has massively parallel, distributed storage and processing, self-organizing, adaptive and self-learning abilities. Thus it is especially suitable for processing a number

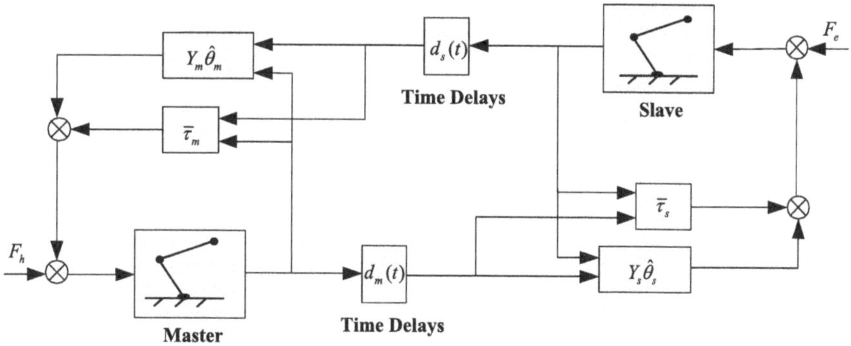

Fig. 1.6 General adaptive control scheme

of factors and conditions which are needed to be considered at the same time, inaccurate and fuzzy information processing problems. On other hand, FLS is the basic idea of fuzzy rules of using computer to realize human's thinking. Strong robustness of fuzzy control system makes the effect of disturbance and parameter variation on the control effect is weakened greatly. Therefore, it is especially suitable for nonlinear, time-varying and pure time-delay system control. The above advantages of NNs and FLs have led to the research of many scholars and is being gradually recognized and applied in the field of engineering. The intelligent control schemes are also proposed for the teleoperation system. For example, neural network wave-variable control method [54], new neural network passivity-based control strategy was designed in [55] to improve the performances of position tracking and force tracking. Consider single-master-multiple-slave teleoperation system composed of mobile robots, [56] designed a new neural network control scheme to ensure stability in the presence of time-varying delays and dead-zone input nonlinearity. [57] proposed a new adaptive neural networks control scheme to guarantee the stability of the teleoperation system with stochastic time delays, just to name a few. Similar with neural networks, there are also new numerical control algorithms based on FLs that have been designed and applied to teleoperation system, see references [58–63]. Moreover, in some literatures, NNs and FLs are combined to further improve the control performances [64–66]. In Fig. 1.7, general intelligent control scheme is shown.

More recently, compared with the single-master-single slave teleoperation system, the teleoperation with multiple robots with above control schemes are drawing more attention now [18, 67, 68].

1.3 Problems Studied in this Book

In this book, four major issues for nonlinear teleoperation systems will be studied. Since with the introduction of computer communication networks into teleoperation systems, time delays are unavoidable when data are exchanged between master and

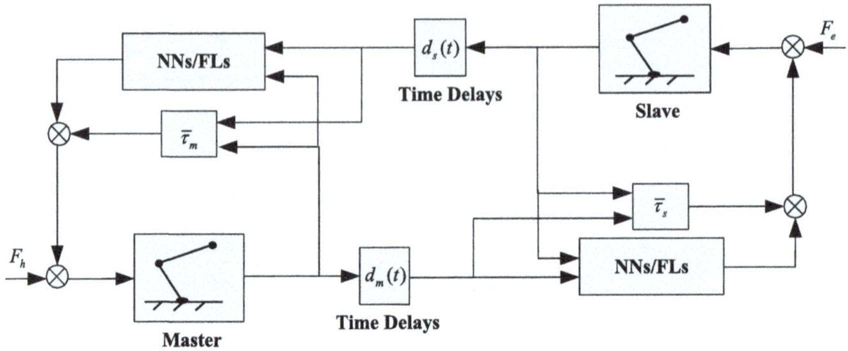

Fig. 1.7 General intelligent control scheme

slave sites. And it has been proved by numerical researchers that time delays can destabilize the whole system if they are not well addressed [69–71]. Therefore, consider communication time delays, stability analyses of closed-loop teleoperation by using Lyapunov Krasovskii functional method are shown in Part I. Then stability criteria for master-slave teleoperation system are presented as specific linear-matrix-inequality (LMI) conditions. For a teleoperation system, besides system stability, the control performances on transparency, convergence rapidity, system robustness and convergence accuracy also should be considered. Based on these issues, Part II further concerns with high-performance controller design problems for nonlinear teleoperation system. Four different controlling strategies: delayed slave torque feedback control (providing enhanced transparency performance), P+d like control (providing exponential convergence), finite time control (providing finite-time convergence) and prescribed performance control (providing prescribed steady-state and transient-state performances) are proposed. Furthermore, we consider the case that the velocity information is not available. High-gain velocity observer and finite time velocity observer are designed for teleoperation system in Part III, respectively. Finally, we consider practical applications. Like many applications of control systems, actuator output of teleoperation system has a limited amplitude, i.e., is subject to actuator saturation. Controllers that ignore actuator saturation may cause undesirable responses and even closed-loop system instability. Therefore, control schemes that take actuator saturation into account at the design outset are designed in Part IV.

1.3.1 Stability Analysis for Teleoperation System

For a teleoperation system, the communication channel connecting master and slave often involve large distances or impose limited data transfer. Such situations can result in substantial delays between the time a command is sent by a operator and the time a command is executed by the remote manipulator, idem for the reflected interaction. These time delays affect the overall stability of the teleoperator [72–74]. Consider inevitable communication time delays, many methods have been presented to address the stability problems associated with time delays, for which passivity based methodology is a prevalent one. Based on passivity theory and scattering approach, stability-analysis and controller-design problems were extensively studied in [75–77]. Lee and Huang [78] proposed a passive set-position-modulation method and applied the same to Internet teleoperation with varying delay, packet loss and haptics with slow and variable-rate data update. For cases of nonpassive environments, input-to-state-stable (ISS) properties are effective for the torque design in teleoperation system. By applying the ISS idea to a teleoperation system, Polushin et al. [79, 80] proved boundedness of position and velocity synchronization errors between master and slave sites. Recently, Lee and Spong [33] considered the position-coordination control problem for teleoperation systems. With the L_∞-stable maps required from velocity to force, a proportional-derivative (PD) controller was proposed. Nuño et al. [34] relaxed the condition a little bit and designed a P controller without delayed-derivative action. It is proved in [34] that the stability of closed loop can be achieved with controller-design parameters satisfying specific conditions. While these approaches are very inspiring, time delays considered in [33] and [34] are assumed to be constant, and the time delay is required to be symmetric.

Due to the complexity of the communication network, however, the delays of data packets are not only time-varying but also asymmetric for forward and backward paths. The reason is as follows. First, network bandwidth available for a teleoperation application is essentially dynamic and time-varying; thus, the delay is time-varying for data transmission in the network. Second, the data packets in the forward and backward channels may go through different network paths, thereby leading the forward and backward delays to not be equal. Consider the asymmetric and time-varying communication delays, the stability analyses with PD+d (proportional-derivative plus damping) controller and P+d controller for a nonlinear teleoperation system are presented in Chaps. 2 and 3. Moreover, in practice, a time-varying interval delay is often encountered, that is, the range of delay may vary in an interval. Therefore, it is significant to investigate the stability of the system with interval time-varying delays. Thus, the stability analysis for teleoperation with interval time-varying delays is provided in Chap. 4. Finally, owing to limited transmission capacity of the network and some devices in the closed-loop system, data transmitted in practical networked teleoperation system should be quantized before they are sent between local side and remote side. Therefore to achieve better performance of the considered systems, the effect of data quantization on the stability of teleoperation system is considered in Chap. 5.

1.3.2 Control Scheme Design for Teleoperation System

In the aforementioned part, we discussed the stability analysis of teleoperation system. As we known from a control theoretic point of view the main goals of a teleoperation system are twofold: stability and telepresence. For enhancement of transparency performances, slave force feedback is employed to design the controller for a master in Chap. 6. Then the human operator can feel the remote external environment. Furthermore, we have noticed that in above chapters, asymptotic convergence and stability of closed-loop systems are achieved. In practice, exponential convergence will ensure high convergence speed compared with asymptotic convergence. Thus, a new control algorithm by adding enhanced damping terms in Chap. 7 to guarantee the synchronization errors between master and slave converge to zero exponentially. Moreover, it is noticed that for some demanded teleoperation applications, it is undesirable that synchronization errors between master and slave converge to zero when time is infinite for asymptotic convergence or exponential convergence. It is more expected that the slave follows the master quickly and synchronization error tends to zero in finite time [81–83]. It is obvious that this objective can not be achieved with the existing teleoperation design schemes. Considering this problem, finite time control problems are investigated for teleoperation system by applying terminal sliding modes in Chap. 8.

Furthermore, even though many effective control algorithms have been designed for teleoperation system, only steady-state performance is considered. Few control schemes consider transient-state performance for nonlinear teleoperation system. It has been early recognized that transient-state control performance deserves further research [84–86]. Therefore, in Chap. 9 error constrained control is employed for teleoperation system. With the new controller, not only transient-state and steady-state performances can be guaranteed, but also finite time convergence can be achieved.

1.3.3 Velocity Observer Design for Teleoperation System

Like most of mechanical systems, the measurement of joint velocity in teleoperation system is very difficult and increases the hardware setup requirement. To guarantee the stability of closed-loop teleoperation system in the case that the velocity information is unavailable. Two different velocity observers are designed in Part III of this book.

In Chap. 10, a simple but effective high-gain velocity observer is designed for nonlinear teleoperation system. High-gain velocity observer has been employed in several nonlinear systems for estimating velocity signals [87, 88]. Furthermore, consider the velocity estimation speed, terminal sliding mode (TSM) based finite time velocity observer [89, 90] is designed via neural networks in Chap. 11. Compared with existing velocity observers [91, 92], finite-time velocity observer will provide faster convergence and higher precision velocity estimation. Moreover, strong robustness with respect to uncertain dynamics can also be obtained by using TSM.

1.3.4 Constrained Input for Teleoperation System

It well known that actuator saturation is inevitable in practical feedback control systems. The appearance of actuator saturation may seriously degrade system performance, and even induces instability in closed-loop system. On that account, input restraint characteristics earned widespread respect and achieved inspiring research results [93–95]. In this book, it mainly aims at the study and analysis of systems with input saturation. In Chap. 12, we consider that when the system state is far from the equilibrium point, the actuator with bounded gain cannot render the system state to the equilibrium point. How to estimate the DOA is still an open problem for nonlinear teleoperation system. Physical restrictions exist in robot motors, thus saturation problems should be considered in the torque design of master-slave robots in teleoperation. Moreover, consider that the communication delays often vary in an interval in practice, i.e., the lower bound determined by the physical distance and the upper bound determined by the congestion level. Therefore, it is of great significance to consider the interval time-varying delay in teleoperation system. Finally, relationship is established among the control parameters, the delay bounds, the saturation levels and the initial values of system state. The optimization of the allowable bounds for the time delays and the estimation of DOA can be performed easily. In Chap. 13, some complicated situations are considered. First, a new fast terminal sliding-mode velocity observer is proposed to estimate the unknown velocity signals for the teleoperation system. Then an adaptive SP+Sd-type (saturated proportion plus saturated damping) controller is designed with estimated velocity information by assuming that the gravity term is unknown. Further, by choosing Lypunov-Krasovskii functional, it shows that the teleoperation system is stable under specific linear matrix inequality conditions.

1.4 Summary

As we know nonlinear teleoperation system have been applied in many fields, information transmission time delays are inherent in teleoperation system and system uncertainties are not avoidable, so it is important to investigate the stability analysis and control of uncertain nonlinear teleoperation systems with time delays. In this book, we do some work about several kinds of commonly encountered nonlinear teleoperation systems, such as the stability analysis of teleoperation system with time delays, high-performance control scheme design, systems with nonlinear input and so on. The results in this book will no doubt advance the study of nonlinear teleoperation control theory, enrich the content of nonlinear control theory and lay the foundation for decentralized control theory in nonlinear large scale systems.

References

1. G. Niemeyer, J.J.E. Slotine, Telemanipulation with time delays. Int. J. Robot. Res. **23**(9), 873–890 (2004)
2. A. Rovetta, R. Sala, X. Wen, A. Togno, Remote control in telerobotic surgery. IEEE Trans. Syst. Man Cybern. **26**(4), 438–444 (1996)
3. A.K. Bejczy, Sensors, controls, and man-machine interface for advanced teleoperation. Science **208**(4450), 1327–1335 (1980)
4. O. Toshiki, W. Atsushi, Input-output linearization of nonlinear systems with time delays in state variables. Int. J. Syst. Sci. **29**(6), 573–578 (1998)
5. T.B. Sheridan, Telerobotics. Automatica **25**(4), 487–507 (1989)
6. A.A. El Kalam, A. Ferreira, F. Kratz, Bilateral teleoperation system using qos and secure communication networks for telemedicine applications. IEEE Syst. J. **10**(2), 709–720 (2016)
7. P.M. Kebria, H. Abdi et al., Control methods for internet-based teleoperation systems: a review. IEEE Trans. Hum. Mach. Syst. **49**(1), 32–46 (2019)
8. P.F. Hokayem, M.W. Spong, Bilateral teleoperation: an history survery. Automatica **42**(12), 2035–2057 (2006)
9. R. Kong, X. Dong, X. Liu, Position and force control of teleoperation system based on phantom omni robots. Int. J. Mech. Eng. Robot. Res. **5**(1), 57–61 (2016)
10. S. Soyguder, T. Abut, Haptic industrial robot control with variable time delayed bilateral teleoperation. Ind. Robot: Int. J. **43**(4), 390–402 (2016)
11. D. Sun, F. Naghdy, H. Du, Wave-variable-based passivity control of four-channel nonlinear bilateral teleoperation system under time delays. IEEE/ASME Trans. Mechatron. **21**(1), 238–253 (2016)
12. P. Pitakwatchara, Wave correction scheme for task space control of time-varying delayed teleoperation systems. IEEE Trans. Control Syst. Technol. **26**(6), 2223–2231 (2018)
13. X.J. Jing, Y.C. Wang, D.L. Tan, Control of time-delayed tele-robotic systems: review and analysis. ACTA Autom. Sin. **30**(2), 214–223 (2004)
14. D. Heck, A. Saccon, R. Beerens, H. Nijmeijer, Direct force-reflecting two-layer approach for passive bilateral tteleoperation with time delays. IEEE Trans. Robot. **34**(1), 194–206 (2018)
15. R. Kelly, V. Santibáñez, A. Loria. *Control of Robot Manipulators in Joint Space*. (Springer, 2005)
16. M.W. Spong, S. Hutchinson, M. Vidyasagar, *Robot Modeling and Control* (Wiley, 2005)
17. D.H. Zhai, Y.Q. Xia, A novel switching-based control framework for improved task performance in teleoperation system with asymmetric time-varying delays. IEEE Trans. Cybern. **48**(2), 625–638 (2018)
18. D.H. Zhai, Y.Q. Xia, Multilateral telecoordinated control of multiple robots with uncertain kinematics. IEEE Trans. Neural Netw. Learn. Syst. **29**(7), 2808–2822 (2018)
19. I.G. Polushin, P.X. Liu, C.H. Lung et al., Position-error based schemes for bilateral teleoperation with time delay: theory and experiments. J. Dyn. Syst. Measurement Control **132**(3), 1–11 (2010)
20. S.A. Deka, D.M. Stipanovi, T. Kesavadas, Stable bilateral teleoperation with bounded control. IEEE Trans. Control Syst. Technol. **1**(1), 1–10 (2018)
21. W.R. Ferrell, Remote manipulation with transmission delay. IEEE Trans. Human Factors Electron. **1**, 24–32 (1965)
22. W.S. Kim, Shared compliant control: a stability analysis and experiments, in *Proceedings of the IEEE International Conference on Systems, Man and Cybernetics*, pp. 620–623 (1990)
23. Y. Yokokohji, T. Yoshikawa, Bilateral control of master-slave manipulator for ideal kinesthetic coupling-formulation and experiment. IEEE Trans. Robot. Autom. **10**(5), 605–619 (1994)
24. Hashtrudi-Zaad, S.E. Salcudean, On the use of local force feedback for teleoperation, in*IEEE International Conference on Robotics and Automation*, pp. 1863–1869 (1990)
25. G.C. Walsh, H. Ye, Scheduling of networked control systems. IEEE Control Syst. **21**(1), 57–65 (2001)

26. A. Ray, Output feedback control under randomly varying distributed delays. J. Guidance Control Dyn. **17**(4), 57–65 (1994)
27. S. Sirouspour, A. Shahdi, Bilateral teleoperation under communication time delay using an lqg controller, in *Proceeding of IEEE Conference on Control Applications*, pp. 1257–1262 (2005)
28. P.G. Griffiths, A.M. Okamura, Defining performance tradeoffs for multi-degree-of-freedom bilateral teleoperators with lqg control, in *Proceeding of 49th IEEE Conference on Decision and Control (CDC)*, pp. 3542–3547 (2010)
29. J. Yan, S.E. Salcudean, Teleoperation controller design using h-infinity optimization with application to motion-scaling. IEEE Trans. Control Syst. **4**(3), 244–258 (1996)
30. H.P. Du, H-infinity state-feedback control of bilateral teleoperation systems with asymmetric time-varying delays. IET Control Theory Appl. **7**(4), 594–605 (2013)
31. S. Lee, H.S. Lee, Modeling, design, and evaluation of advanced teleoperator control systems with short time-delay. IEEE Trans. Robot. Autom. **9**(5), 607–623 (1993)
32. K. Hastrudi-Zaad, S.E. Salcudean, Analysis of control architectures for teleoperation systems with impedance/admittance master and slave manipulators. Int. J. Robot. Res. **20**(6), 419–445 (2001)
33. D. Lee, M.W. Spong, Passive bilateral teleoperation with constant time delay. IEEE Trans. Robot. Autom. **22**(2), 269–281 (2006)
34. E. Nuño, R. Ortega, N. Barabanov, and L. Basañez, A globally stable pd controller for bilateral teleoperators. IEEE Trans. Robot. **24**(3), 753–758 (2008)
35. C.C. Hua, X.P. Liu, Delay-dependent stability criteria of teleoperation systems with asymmetric time-varying delays. IEEE Trans. Robot. **26**(5), 925–932 (2010)
36. X. Yang, C.C. Hua, J. Yan et al., Synchronization analysis for nonlinear bilateral teleoperator with interval time-varying delay. Int. J. Robust Nonlinear Control **25**(3), 2142–2161 (2014)
37. X. Yang, C.C. Hua, J. Yan et al., New stability criteria for networked teleoperation system. Inf. Sci. **233**(2), 244–254 (2013)
38. P. Buttolo, P. Braathen, and B. Hannaford, Sliding control of force reflecting teleoperation: preliminary studies. Presence Teleoperators Virtual Environ. **3**(2), 158–172 (1994)
39. R. Moreau, M.T. Pham, M. Tavakoli et al., Sliding-mode bilateral teleoperation control design for mastercslave pneumatic servo systems. Control Eng. Pract. **20**(6), 584–597 (2012)
40. X. Liu, X. Dong, 4-channel sliding mode control of teleoperation systems with disturbances. Asian J. Control **17**(4), 1267–1273 (2015)
41. S. Ganjefar, M.H. Sarajchi, M.T.H. Beheshti, Adaptive sliding mode controller design for nonlinear teleoperation systems using singular perturbation method. Nonlinear Dyn. **81**(3), 1–18 (2015)
42. S. Ganjefar, M.H. Sarajchi, M. Hosseini, Teleoperation systems design using singular perturbation method and sliding mode controllers. J. Dyn. Syst. Measurement Control **136**(5), 562–576 (2014)
43. Y.N. Yang, C.C. Hua, H.F. Ding et al., Finite-time coordination control for networked bilateral teleoperation. Robotica **33**(2), 451–462 (2015)
44. Y.N. Yang, C.C. Hua, X.P. Guan, Finite-time synchronization control for bilateral teleoperation under communication delays. Robot. Comput. Integr. Manuf. **31**, 61–69 (2015)
45. E. Nuño, R. Ortega, L. Basañez, An adaptive controller for nonlinear bilateral teleoperators. Automatica **46**(1), 155–159 (2010)
46. E. Nuño, R. Ortega, L. Basañez, Erratum to an adaptive controller for nonlinear bilateral teleoperators [automatica 46 (2010) 155-159]. Automatica **47**(5), 1093–1094 (2011)
47. F. Hashemzadeh, I. Hassanzadeh, M. Tavakoli, G. Alizadeh, Adaptive control of nonlinear teleoperation systems with varying asymmetric time delays, in *In Proceeding of IEEE/RSJ International Conference on Intelligent Robots and Systems*, pp. 3023–3028 (2012)
48. S. Islam, X.P. Liu, A.E.I. Saddik, Nonlinear adaptive control for teleoperation systems with symmetrical and unsymmetrical time-varying delay. Int. J. Syst. Sci. **46**(16), 2928–2938 (2015)
49. D.H. Zhai, Y.Q. Xia, Adaptive finite-time control for nonlinear teleoperation systems with asymmetric time-varying delays. Int. J. Robust Nonlinear Control **26**(12), 2586–2607 (2015)
50. S. Haykin. *Neural Networks: A comprehensive foundation*. (Prentice Hall PTR, 1998)

51. F.L. Lewis, A. Yesildirak, S. Jagannathan, *Neural Network Control of Robot Manipulators and Nonlinear Systems* (Taylor & Francis, 1998)
52. G.J. Klir, B. Yuan, *Fuzzy sets and fuzzy logic: theory and applications* (Prentice-Hall, 1994)
53. C.C. Lee, Fuzzy logic in control systems: fuzzy logic controller. IEEE Trans. Syst. Man Cybern. **20**(2), 404–418 (1990)
54. S.G. Yoo, K.T. Chong, Adaptive wave variables for bilateral teleoperation using neural networks. Neural Comput. Appl. **25**(6), 1249–1262 (2014)
55. A. Forouzantabar, H.A. Talebi, A.K. Sedigh, Adaptive neural network control of bilateral teleoperation with constant time delay. Nonlinear Dyn. **67**(2), 1123–1134 (2012)
56. Z.J. Li, C.Y. Su, Neural-adaptive control of single-master-multiple-slaves teleoperation for coordinated multiple mobile manipulators with time-varying communication delays and input uncertainties. IEEE Trans. Neural Netw. Learn. Syst. **24**(9), 1400–1413 (2013)
57. Z.J. Li, Y.Q. Xia et al., Neural network-based control of networked trilateral teleoperation with geometrically unknown constraints. IEEE Trans. Cybern. **46**(5), 1051–1054 (2016)
58. X. Yang, C.C. Hua, J. Yan et al., A new master-slave torque design for teleoperation system by t-s fuzzy approach. IEEE Trans. Control Syst. Technol. **23**(4), 1611–1619 (2014)
59. Z.J. Li, L. Ding, H.B. Gao et al., Trilateral teleoperation of adaptive fuzzy force/motion control for nonlinear teleoperators with communication random delays. IEEE Trans. Fuzzy Syst. **21**(4), 610–624 (2013)
60. Z.J. Li, Y.Q. Xia, F.C. Sun, Adaptive fuzzy control for multilateral cooperative teleoperation of multiple robotic manipulators under random network-induced delays. IEEE Trans. Fuzzy Syst. **22**(2), 437–450 (2014)
61. D.H. Zhai, Y.Q. Xia, Adaptive fuzzy control of multilateral asymmetric teleoperation for coordinated multiple mobile manipulators. IEEE Trans. Fuzzy Syst. **24**(1), 57–70 (2015)
62. Z.Y. Lu, P.F. Huang, Z.X. Liu, Predictive approach for sensorless bimanual teleoperation under random time delays with adaptive fuzzy control. IEEE Trans. Ind. Electron. **65**(3), 2439–2448 (2018)
63. D. Sun, Q.F. Liao, H.L. Ren, Type-2 fuzzy modeling and control for bilateral teleoperation system with dynamic uncertainties and time-varying delays. IEEE Trans. Ind. Electron. **65**(1), 447–459 (2018)
64. S. Manifar, M.A. Shoorhedeli, Application of ga based neuro-fuzzy automatic generation for teleoperation systems, in *Conference In Proceeding of International Conference on Digital Content, Multimedia Technology and ITS Applications*, pp. 466–479 (2010)
65. R. Mellah and R. Toumi, Compensatory neuro-fuzzy control of bilateral teleoperation system, in *Proceeding of 20th International Conference on Methods and MODELS in Automation and Robotics*, pp. 24–27 (2015)
66. H. Shao, K. Nonami, T. Wojtaea et al., Neuro-fuzzy position control of demining tele-operation system based on rnn modeling. Robot. Comput. Integr. Manuf. **22**(1), 25–32 (2006)
67. Y.N. Yang, C.C. Hua, X.P. Guan, Multi-manipulators coordination for bilateral teleoperation system using fixed-time control approach. Int. J. Robust Nonlinear Control **28**(18), 5667–5687 (2018)
68. M. Shahbazi, S.F. Atashzar, R.V. Patel, A systematic review of multilateral teleoperation systems. IEEE Trans. Haptics **11**(3), 338–356 (2018)
69. R.J. Anderson, M.W. Spong, Bilateral control of teleoperators with time delay. IEEE Trans. Autom. Control **34**(5), 494–501 (1989)
70. W.R. Ferrell, Delayed force feedback. Hum. Factors **8**(5), 449–455 (1966)
71. N. Chopra, M. W. Spong, S. Hirche, M. Buss. Bilateral teleoperation over the internet: the time varying delay problem, in *In Proceeding of American Control Conference*, pp. 155–160 (2003)
72. Y. He, G. Liu, D. Rees, M. Wu, Improved delay-dependent stability criteria for systems with nonlinear perturbations. Eur. J. Control **13**(4), 356–365 (2007)
73. C.C. Hua, G. Feng, X.P. Guan, Robust stabilization of a class of nonlinear time delay systems via backstepping method. Eur. J. Control **44**(2), 567–573 (2008)
74. J. Ramos, S. Kim, Humanoid dynamic synchronization through whole-body bilateral feedback teleoperation. IEEE Trans. Robot. **34**(4), 953–965 (2018)

75. P. Arcara, C. Melchiorri, Control schemes for teleoperation with time delay: a comparative study. Robot. Auton. Syst. **38**(1), 49–64 (2002)
76. K. Natori, T. Tsuji, K. Ohnishi et al., Time-delay compensation by communication disturbance observer for bilateral teleoperation under time-varying delay. IEEE Trans. Ind. Electron. **57**(3), 1050–1062 (2010)
77. N.A. Tung, N.T. Binh, T.H. Anh et al., Synchronization control of bilateral teleoperation systems by using wave variable method under varying time delay, in *2017 International Conference on System Science and Engineering (ICSSE)*, pp. 21–23 (2017)
78. D. Lee, K. Huang, Passive-set-position-modulation framework for interactive robotic systems. IEEE Trans. Robot. **26**(2), 354–369 (2010)
79. I.G. Polushin, P.X. Liu, C.H. Lung, A control scheme for stable force-reflecting teleoperation over ip networks. IEEE Trans. Syst. Man Cybern. B Cybern. **36**(4), 930–939 (2006)
80. I.G. Polushin, P.X. Liu, C.H. Lung, A force reflection algorithm for improved transparency in bilateral teleoperation with communication delay. IEEE/ASME Trans. Mechatron. **12**(3), 361–374 (2007)
81. S.H. Yu, X.H. Yu, B. Shirinzadeh, Z.H. Man, Continuous finite-time control for robotic manipulators with terminal sliding mode. Automatica **41**(11), 1957–1964 (2005)
82. F. Amato, M. Ariola, P. Dorato, Finite-time control of linear systems subject to parametric uncertainties and disturbances. Automatica **37**(9), 1459–1463 (2001)
83. Y.G. Hong, Y.S. Xu, J. Huang, Finite-time control for robot manipulators. Syst. Control Lett. **46**(4), 243–253 (2002)
84. A. Ilchmann, E.P. Ryan, S. Trenn, Tracking control: performance funnels and prescribed transient behaviour. Syst. Control Lett. **54**(7), 655–670 (2005)
85. C.P. Bechlioulis, G.A. Rovithakis, Adaptive control with guaranteed transient and steady state tracking error bounds for strict feedback systems. Automatica **45**(2), 532–538 (2009)
86. C.P. Bechlioulis, G.A. Rovithakis, Robust adaptive control of feedback linearizable mimo nonlinear systems with prescribed performance. IEEE Trans. Autom. Control **53**(9), 2090–2099 (2008)
87. A.N. Atassi, H.K. Khalil, Separation results for the stabilization of nonlinear systems using different high-gain observer designs. Syst. Control Lett. **39**(3), 183–191 (2000)
88. K.W. Lee, H.K. Khalil, Adaptive output feedback control of robot manipulators using high-gain observer. Int. J. Control **67**(6), 869–886 (1997)
89. J. Davila, L. Fridman, A. Levant, Second-order sliding-mode observer for mechanical systems. IEEE Trans. Autom. Control **50**(11), 1785–1789 (2005)
90. Y. Feng, J.F. Zheng, X.H. Yu, N.V. Truong, Hybrid terminal sliding-mode observer design method for a permanent-magnet synchronous motor control system. IEEE Trans. Ind. Electron. **56**(9), 3424–3431 (2009)
91. A. Astolfi, R. Ortega, A. Venkatraman, A globally exponentially convergent immersion and invariance speed observer for mechanical systems with non-holonomic constraints. Automatica **46**(1), 182–189 (2010)
92. A. Astolfi, R. Ortega, A. Venkatraman, A globally exponentially convergent immersion and invariance speed observer for n degrees of freedom mechanical systems, in *Proceedings of the 48h IEEE Conference on Decision and Control (CDC)*, pp. 6508–6513 (2009)
93. T.S. Hu, Z.L. Lin, *Control Systems with Actuator Saturation: Analysis and Design* (Springer, 2001)
94. T.S. Hu, Z.L. Lin, B.M. Chen, An analysis and design method for linear systems subject to actuator saturation and disturbance. Syst. Control Lett. **38**(2), 351–359 (2002)
95. V. Kapila, K. M. Grigoriadis, *Acturator Saturation Control* (Marcel Dekker, Inc, 2002)

Part I
Stability Analysis of Networked Teleoperation System

Chapter 2
Stability Analysis of Teleoperation Systems with Asymmetric Time-Varying Delays

Abstract This chapter addresses the stability analysis problem for teleoperation systems with time delays. The communication delays are assumed to be both time-varying and asymmetric, which is the case for network-based teleoperation systems. The stability analysis is performed for the controller composed of delayed position error and velocity signal. By choosing Lyapunov Krasovskii functional, we show that the master-slave teleoperation systems are stable under specific LMI conditions. With the given controller design parameters, the proposed stability criteria can be used to compute the allowable maximal transmission delay. Finally, both simulations and experiments are performed to show the effectiveness of the proposed method.

2.1 Introduction

As mentioned in Chap. 1, with the introduction of communication networks into teleoperation systems, time delays are unavoidable while data are exchanged between the master and slave sites [1–3]. And it is well known that time delays can destabilize the whole system if they are not well addressed [4–7]. Many methods have been presented for addressing the stability problems associated with time delays, for which the passivity-based methodology is a prevalent one. Based on the passivity theory and scattering approach, the stability analysis and controller design problems were extensively studied in [8–10]. Moreover, a passive set-position modulation method was proposed in [11], and applied to Internet teleoperation with varying-delay and packet-loss and haptics with slow and variable-rate data update.

For the cases of non-passive environments, the ISS (Input to State Stable) property is effective for the torques design of teleoperation systems. By applying the ISS idea to a teleoperation system, the boundedness of the position and velocity synchronization errors between master and slave sites was proved in [12, 13]. Recently, the position coordination control problem for teleoperation systems was considered in [14]. With the $\mathcal{L}_\infty$-stable maps required from velocity to force, a PD (Proportional Derivative) controller was proposed. The condition was relaxed a little bit and a P controller

C. Hua et al., *Analysis and Design for Networked Teleoperation System*,
https://doi.org/10.1007/978-981-13-7936-9_2

without the delayed derivative action was designed in [15]. It was proved in [15] that the stability of the closed-loop can be achieved with the controller design parameters satisfying specific conditions. While these approaches are very inspiring, the time delays considered in [14] and [15] are assumed to be constant, and the time delay is required to be symmetric.

Due to the complexity of the communication network, however, the delays of data packets are not only time-varying, but also asymmetric for the forward and backward paths. The reason is as follows. First, the network bandwidth available for a teleoperation application is essentially dynamic and time-varying, thus the delay is time-varying for data transmission in the network. Second, the data packets in the forward and backward channels may go through different network paths, leading that the forward and backward delays are not equal. To the best of our knowledge, there have been few results reported in the literature to address the stability problem associated with asymmetric and time-varying communication delays for teleoperation systems. In [16], the stability analysis was performed for the teleoperation system with a PD control scheme for the asymmetric and time-varying delay case.

This chapter addresses specifically the stability analysis problem for a class of teleoperation systems, in particular for network-based teleoperation systems. The forward and backward transmission time delays are assumed to be time-varying and asymmetric. By employing Lyapunov Krasovskii functional, we establish the relationships among the controller design parameters and the upper bounds of the time-varying delays. The stability criteria are given in the form of linear matrix inequalities (LMIs), which can be used to compute the allowed maximum values of forward and backward delays. Finally, simulations and experiments are both performed to show the effectiveness of the proposed main results.

2.2 Preliminaries and Problem Formulation

In this section we consider a master-slave bilateral teleoperation system as follows

$$\begin{cases} M_m(q_m)\ddot{q}_m + C_m(q_m, \dot{q}_m)\dot{q}_m + G_m(q_m) = \tau_m + F_h \\ M_s(q_s)\ddot{q}_s + C_s(q_s, \dot{q}_s)\dot{q}_s + G_s(q_s) = \tau_s - F_e \end{cases} \tag{2.1}$$

The well-known properties for the robot systems (2.1) have been shown in Chap. 1.

In this chapter, we will consider the stability problem of system (2.1) with the controller: Proportional position errors plus damping injection (see Fig. 2.1)

$$\begin{cases} \tau_m = -k_m(q_m - q_s(t - d_s(t))) - \alpha_m\dot{q}_m, \\ \tau_s = k_s(q_m(t - d_m(t)) - q_s) - \alpha_s\dot{q}_s. \end{cases} \tag{2.2}$$

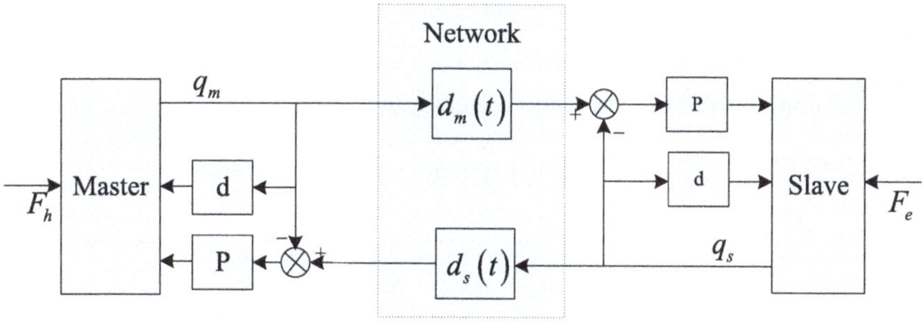

Fig. 2.1 Master and slave with position error plus damping

where $d_m(t)$ and $d_s(t)$ are forward time delay (from the master site to slave site) and backward time delay (from the slave site to master site), respectively; k_m and k_s are proportional coefficients; α_m and α_s are damping coefficients.

Assumption 2.1 The human operator and the environment are passive, that is

$$\int_0^t -\dot{q}_m^T(\sigma)F_h(\sigma)d\sigma \geq 0, \int_0^t \dot{q}_s^T(\sigma)F_e(\sigma)d\sigma \geq 0.$$

Assumption 2.2 The time delays $d_m(t)$ and $d_s(t)$ are bounded, that is, there exist positive scalars $\overline{d}_m$ and $\overline{d}_s$ such that $d_m(t) \leq \overline{d}_m$ and $d_s(t) \leq \overline{d}_s$.

Remark 2.1 Without the consideration of packet loss, the time delays of practical systems are often bounded, thus it is reasonable to assume there are upper bounds for the time delays. In [15] the stability problem was considered for system (2.1) with controller (2.2). The constant and symmetric time delays were considered and the stability criteria are developed. This chapter focuses on the time-varying and asymmetric delay case, and aims to develop the corresponding criteria for system (2.1) with (2.2). The proposed stability conditions will be given in the form of LMIs, which can be easily solved by using Matlab.

Before proposing the main results, we need the following lemma:

Lemma 2.1 *For a positive definite matrix Υ, the following inequality holds*

$$\begin{aligned} &-2a^T(t)\int_{t-d(t)}^t b(\xi)d\xi - \int_{t-d(t)}^t b^T(\xi)\Upsilon b_m(\xi)d\xi \\ &\leq \overline{d}a^T(t)\Upsilon^{-1}a(t), \end{aligned}$$

where $a(\cdot)$ and $b(\cdot)$ are vector functions, $d(t)$ is a time-varying scalar with $0 \leq d(t) \leq \overline{d}$.

Proof Using the fact $-2a^T(t)b(\xi) - b^T(\xi)\Upsilon b(\xi) \leq a^T(t)\Upsilon^{-1}a(t)$, we can obtain the above inequality easily.

2.3 Stability Analysis

In this section, the stability problems are analyzed for system (2.1) with controller (2.2).

With (2.1) and (2.2), the controlled system is rewritten as follows

$$\begin{cases} M_m(q_m)\ddot{q}_m + C_m(q_m, \dot{q}_m)\dot{q}_m + G_m(q_m) \\ = F_h - k_m(q_m - q_s(t - d_s(t))) - \alpha_m \dot{q}_m \\ M_s(q_s)\ddot{q}_s + C_s(q_s, \dot{q}_s)\dot{q}_s + G_s(q_s) \\ = k_s(q_m(t - d_m(t)) - q_s) - \alpha_s \dot{q}_s - F_e. \end{cases} \tag{2.3}$$

For (2.3), we choose the following Lyapunov functional $V = V_1 + V_2 + V_3$ with

$$\begin{aligned} V_1 = {} & \dot{q}_m^T M_m(q_m)\dot{q}_m + \frac{k_m}{k_s}\dot{q}_s^T M_s(q_s)\dot{q}_s \\ & + 2(U_m(q_m) - \beta_m) + \frac{2k_m}{k_s}(U_s(q_s) - \beta_s) \\ & - 2\int_0^t \left(\dot{q}_m^T(\sigma) F_h(\sigma) - \frac{k_m}{k_s}\dot{q}_s^T(\sigma) F_e(\sigma)\right) d\sigma \\ V_2 = {} & k_m(q_m - q_s)^T(q_m - q_s), \\ V_3 = {} & \int_{-\bar{d}_m}^0 \int_{t+\theta}^t \dot{q}_m(\xi)^T Z \dot{q}_m(\xi)\, d\xi d\theta \\ & + \int_{-\bar{d}_s}^0 \int_{t+\theta}^t \dot{q}_s(\xi)^T S \dot{q}_s(\xi)\, d\xi d\theta, \end{aligned} \tag{2.4}$$

where Z and S are positive definite matrices.

By the Property 1.2 and 1.3, the time derivative of V_1 is

$$\begin{aligned} \dot{V}_1 = & - 2k_m\dot{q}_m^T(q_m - q_s(t - d_s(t))) \\ & - 2\alpha_m\dot{q}_m^T\dot{q}_m - \frac{2k_m\alpha_s}{k_s}\dot{q}_s^T\dot{q}_s \\ & + 2k_m\dot{q}_s^T(q_m(t - d_m(t)) - q_s) \\ = & - 2\alpha_m\dot{q}_m^T\dot{q}_m - \frac{2k_m\alpha_s}{k_s}\dot{q}_s^T\dot{q}_s \\ & - 2k_m\dot{q}_m^T(q_m - q_s) - 2k_m\dot{q}_m^T(q_s - q_s(t - d_s(t))) \\ & - 2k_m\dot{q}_s^T(q_s - q_m) - 2k_m\dot{q}_s^T(q_m - q_m(t - d_m(t))) \\ = & - 2\alpha_m\dot{q}_m^T\dot{q}_m - 2k_m(q_m - q_s)^T(\dot{q}_m - \dot{q}_s) \\ & - \frac{2k_m\alpha_s}{k_s}\dot{q}_s^T\dot{q}_s - 2k_m\dot{q}_m^T\int_{t-d_s(t)}^t \dot{q}_s(\xi)\, d\xi \end{aligned}$$

$$- 2k_m\dot{q}_s^T\int_{t-d_m(t)}^{t}\dot{q}_m(\xi)\,d\xi. \tag{2.5}$$

The time derivatives of V_2 and V_3 are

$$\dot{V}_2 = 2k_m(\dot{q}_m - \dot{q}_s)^T(q_m - q_s), \tag{2.6}$$

and

$$\begin{aligned}\dot{V}_3 =&\overline{d}_m\dot{q}_m^TZ\dot{q}_m - \int_{t-\overline{d}_m}^{t}\dot{q}_m^T(\xi)Z\dot{q}_m(\xi)\,d\xi\\&+\overline{d}_s\dot{q}_s^TS\dot{q}_s - \int_{t-\overline{d}_s}^{t}\dot{q}_s^T(\xi)S\dot{q}_s(\xi)\,d\xi\\\le&\overline{d}_m\dot{q}_m^TZ\dot{q}_m - \int_{t-d_m(t)}^{t}\dot{q}_m^T(\xi)Z\dot{q}_m(\xi)\,d\xi\\&+\overline{d}_s\dot{q}_s^TS\dot{q}_s - \int_{t-d_s(t)}^{t}\dot{q}_s^T(\xi)S\dot{q}_s(\xi)\,d\xi.\end{aligned} \tag{2.7}$$

Using Lemma 2.1, we have the following inequalities

$$\begin{aligned}&- 2k_m\dot{q}_m^T\int_{t-d_s(t)}^{t}\dot{q}_s(\xi)\,d\xi - \int_{t-d_s(t)}^{t}\dot{q}_s^T(\xi)S\dot{q}_s(\xi)\,d\xi\\&\le \overline{d}_sk_m^2\dot{q}_m^T(t)S^{-1}\dot{q}_m(t),\end{aligned} \tag{2.8}$$

and

$$\begin{aligned}&- 2k_m\dot{q}_s^T\int_{t-d_m(t)}^{t}\dot{q}_m(\xi)\,d\xi - \int_{t-d_m(t)}^{t}\dot{q}_m^T(\xi)Z\dot{q}_m(\xi)\,d\xi\\&\le \overline{d}_mk_m^2\dot{q}_s^T(t)Z^{-1}\dot{q}_s(t).\end{aligned} \tag{2.9}$$

With (2.5–2.9), it gives

$$\begin{aligned}\dot{V} =&\dot{V}_1 + \dot{V}_2 + \dot{V}_3 \le \dot{q}_m^T\left(-2\alpha_mI + \overline{d}_mZ + \overline{d}_sk_m^2S^{-1}\right)\dot{q}_m\\&+\dot{q}_s^T\left(-\frac{2k_m\alpha_s}{k_s}I + \overline{d}_sS + \overline{d}_mk_m^2Z^{-1}\right)\dot{q}_s.\end{aligned} \tag{2.10}$$

With Eq. (2.10), we have the following main result:

Theorem 2.1 *If there exist positive definite matrices Z and S such that the following LMI holds*

$$\begin{bmatrix} -2\alpha_m I + \overline{d}_m Z & \overline{d}_s k_m & 0 & 0 \\ * & -\overline{d}_s S & 0 & 0 \\ * & * & -\frac{2k_m\alpha_s}{k_s} I + \overline{d}_s S & \overline{d}_m k_m \\ * & * & * & -\overline{d}_m Z \end{bmatrix} < 0, \tag{2.11}$$

where $*$ represents the transpose of corresponding matrix, then the velocities $\dot{q}_m, \dot{q}_s$ and position error $q_m - q_s$ are bounded.

Proof The LMI inequality (2.11) is equivalent to the following two inequalities

$$-2\alpha_m I + \overline{d}_m Z + \overline{d}_s k_m^2 S^{-1} < 0,$$
$$-2\frac{k_m \alpha_s}{k_s} I + \overline{d}_s S + \overline{d}_m k_m^2 Z^{-1} < 0.$$

Based on Lyapunov stability theory, $\dot{q}_i$ and $q_m - q_s$ are bounded. The proof is completed.

Theorem 2.1 shows the boundedness of the velocities and position error. If the human input and environment force of system (2.1) are zero, the asymptotical stability of the closed-loop system can be also guaranteed. We have the following theorem:

Theorem 2.2 *If there exist positive definite matrices Z and S such that LMI (2.11) holds, and human input F_h and environment force F_e are zero, then the following controller*

$$\begin{cases} \tau_m = -k_m \left(q_m - q_s \left(t - d_s \left(t\right)\right)\right) - \alpha_m \dot{q}_m + G_m \left(q_m\right), \\ \tau_s = k_s \left(q_m \left(t - d_m \left(t\right)\right) - q_s\right) - \alpha_s \dot{q}_s + G_s \left(q_s\right). \end{cases} \tag{2.12}$$

renders velocities $\dot{q}_m, \dot{q}_s$ and position error $q_m - q_s$ converge to zero asymptotically.

Proof Choosing Lyapunov functional (2.4) with the removed $2(U_m(q_m) - \beta_m) + \frac{2k_m}{k_s}(U_s(q_s) - \beta_s)$ in V_1, we can see that the time derivative of V along system (2.1) with controller (2.12) yields inequality (2.10). With (2.10) and (2.11), we have $\dot{q}_m, \dot{q}_s, q_m - q_s \in \mathcal{L}_\infty$ and $\dot{q}_m, \dot{q}_s \in \mathcal{L}_2$. By considering $q_m - q_s\left(t - d_s\left(t\right)\right) = q_m - q_s + \int_{t-d_s(t)}^{t} \dot{q}_s\left(\xi\right) d\xi$, we know that $q_m - q_s\left(t - d_s\left(t\right)\right)$ is bounded. Similarly, $q_m\left(t - d_m\left(t\right)\right) - q_s$ is also bounded. To prove the asymptotical stability, we need to show the boundedness of $\ddot{q}_m$ and $\ddot{q}_s$. With (2.1) and (2.12), one has

$$\begin{aligned} \ddot{q}_m = M_m^{-1}\left(q_m\right)(&-C_m\left(q_m, \dot{q}_m\right)\dot{q}_m - \alpha_m \dot{q}_m \\ &-k_m\left(q_m - q_s\left(t - d_s\left(t\right)\right)\right)) \end{aligned} \tag{2.13}$$

With the boundedness of $\dot{q}_m$ and $q_m - q_s\left(t - d_s\left(t\right)\right)$, $\ddot{q}_m$ is bounded. Based on Barbalat's lemma, we know that $\dot{q}_m \to 0$ as $t \to \infty$, and similarly $\dot{q}_s \to 0$ as $t \to \infty$. Now Let us prove that the tracking error $q_m - q_s$ converges to zero asymptotically.

With $\dot{q}_s$ converging to zero asymptotically, we need to show $q_m - q_s(t - d_s(t))$ converges to zero asymptotically. Based on the first equation of (2.13), we need to prove $\ddot{q}_m \to 0$. Now, we need to show that the time derivative of $\ddot{q}_m$ is bounded. By differentiation of the above Eq. (2.13) on both sides, we have

$$\begin{aligned}\dddot{q}_m =& \frac{dM_m^{-1}(q_m)}{dt}(-C_m(q_m, \dot{q}_m)\dot{q}_m - \alpha_m \dot{q}_m - k_m(q_m - q_s(t - d_s(t)))) \\ &+ M_m^{-1}(q_m)(-C_m(q_m, \dot{q}_m)\ddot{q}_m - \dot{C}_m(q_m, \dot{q}_m)\dot{q}_m - \alpha_m \ddot{q}_m \\ &- k_m(\dot{q}_m - (1 - \dot{d}_s)\dot{q}_s(t - d_s(t))))\end{aligned} \tag{2.14}$$

By using Properties 1.1, 1.2, 1.4 and 1.5, we have $\dddot{q}_m \in \mathcal{L}_\infty$ with the boundedness of $\ddot{q}_m$, $\dot{q}_m$ and $\dot{q}_s$. Then, $\ddot{q}_m \to 0$ as $t \to \infty$ based on Barbalat's lemma and further $q_m - q_s \to 0$ as $t \to \infty$ The proof is completed.

Remark 2.2 For the teleoperation system, the environment force and the human input are very complex. In this section, we assume that the forces satisfy the passive condition (Assumption 2.1). With this condition, we can prove the boundedness of $\dot{q}_m$, $\dot{q}_s$ and $q_m - q_s$. Using the controller (2.12), one knows that the position error converges to zero when F_h and F_e converge to zero based on Eq. (2.13). It is well known that the transparency performance is very important for a teleoperation design. For the stability case ($\ddot{q}_m$, $\ddot{q}_s$, $\dot{q}_m$ and $\dot{q}_s$ converges to zero), we have $F_h = k_m(q_m - q_s(t - d_s(t)))$ and $F_e = k_s(q_m(t - d_m(t)) - q_s)$. So, we can see that F_h reflects the environment force torque F_e. For a special case that F_e is a scalar and $\dot{q}_s = 0$ (a wall stops a slave), the designed controller (2.12) could not render the tracking error $q_m - q_s$ converge to zero asymptotically, but the boundedness of the error can still be guaranteed.

Remark 2.3 From (2.11), one sees that the inequalities contain the bound values of the forward and backward time delays. For given one of the delay bound parameters, we can obtain the allowable maximum value of the other delay by using the LMI optimization tool box of Matlab. For example, if $\overline{d}_m$ is known, we can obtain the allowable maximum value $\overline{d}_s$. In addition, the LMI doesn't require inertial parameters of master/slave robots, which is a nice property for practical implementation.

Remark 2.4 It was assumed in [14, 15] that the forward and backward time delays are symmetric and constant. The corresponding stability conditions were developed for system (2.1). Obviously, with $\mu = 0$ and $\overline{d}_m = \overline{d}_s$ the LMI condition is identical to the stability condition in [14, 15], thus the result presented in this paper actually includes the symmetric and constant time delay as a special case.

Remark 2.5 In this chapter, we employ a Lyapunov Krasovskii functional to derive the stability condition. Although we investigate the case that the control parameters k_m,k_s, α_m and α_s are scalars, the method is also applicable to the case that they are matrices. This method can be used for the stability analysis for other controllers by constructing the similar Lyapunov Krasovskii functional.

2.4 Simulations and Experiments

In this section, simulation results are first presented to verify the effectiveness of the main result. Then, the experiments performed on a teleoperation system consisting of two Phantom Premium 1.5A robot manipulators are shown.

2.4.1 Simulation on a Teleoperation System of 2-DOF Planar Manipulators

The simulations are performed on 2-DOF manipulators:

$$\begin{gathered} M_m(q_m)\ddot{q}_m + C_m(q_m, \dot{q}_m)\dot{q}_m + G_m(q_m) = \tau_m + F_h \\ M_s(q_s)\ddot{q}_s + C_s(q_s, \dot{q}_s)\dot{q}_s + G_s(q_s) = \tau_s - F_e \end{gathered}.$$

where

$$\begin{aligned} M_m(q) = M_s(q) &= \begin{bmatrix} M_{11} & M_{12} \\ * & M_{22} \end{bmatrix}, \\ C_m\left(q, \dot{q}\right) = C_s\left(q, \dot{q}\right) &= \begin{bmatrix} C_{11} & C_{12} \\ C_{21} & C_{22} \end{bmatrix}, \\ G_m(q_m) = G_s(q_s) &= \begin{bmatrix} g_1 \\ g_2 \end{bmatrix}, \end{aligned}$$

in which

$$\begin{aligned} M_{11} &= (2L_1 \cos q_2 + L_2) L_2 m_2 + L_1^2 (m_1 + m_2), \\ M_{12} &= L_2^2 m_2 + L_1 L_2 m_2 \cos q_2, M_{22} = L_2^2 m_2, \\ C_{11} &= -L_1 L_2 m_2 \sin(q_2)\dot{q}_2, C_{21} = L_1 L_2 m_2 \sin(q_2), \\ C_{12} &= -L_1 L_2 m_2 \sin(q_2)\left(\dot{q}_1 + \dot{q}_2\right), C_{22} = 0, \\ g_1 &= g\left(m_2 L_2 \sin(q_1 + q_2) + (m_1 + m_2) L_1 \sin(q_1)\right), \\ g_2 &= g m_2 L_2 \sin(q_1 + q_2). \end{aligned}$$

For simulation, we choose the parameters $m_1 = 10\,\mathrm{kg}$, $m_2 = 5\,\mathrm{kg}$, $L_1 = 0.7\,\mathrm{m}$, $L_2 = 0.5\,\mathrm{m}$, $g = 9.81\,\mathrm{m/s^2}$. We apply a human force F to the master site on the Y-direction, which is shown in Fig. 2.2. The human input is zero at 0 s, and then increases to 30 N at 10 s. The force is kept constant from 10 s to 40 s, and it decreases to zero from 40 s to 50 s. In the slave site, there is a wall at $y = 0.5$ m. When the slave joint reaches the wall and moves further, the feedback force is $10000 * (y - 0.5)\,N$. It is clear that the wall is very stiff.

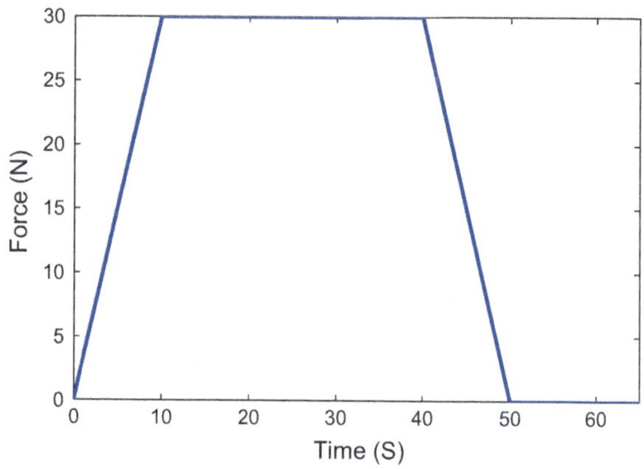

Fig. 2.2 The human force input F

With the human force input, we know the torque $F_h = -J_m^T * \left[0\ 1\right]^T * F$, where J_m is the Jacobian matrix of the master manipulator. The torque F_e is $F_e = -J_s^T * \left[0\ 1\right]^T * 10000N * (y - 0.5)$. The simulation results are used to verify the follows: (i) When we move the master robot, does the slave follow the master? (ii) When the slave contacts the wall, does the feedback torque F_e from the wall reflect the human input F_h? (iii) When the human input force changes to zero, does the error between the master and slave positions disappear? For (i) and (iii), our aim is to show the stability of the closed-loop system. For (ii), it is used to show the slave torque changing with the changes of the master torque. If we see F_h as the environment force, the change of F_e can be used to show the transparency performance. To show the effectiveness of the proposed delay-dependent controller, all the above questions will be checked with the computed maximum allowable time delays.

First, we design the controller (2.12) with the parameters $\alpha_m = \alpha_s = k_m = k_s = 100$. For the symmetric time delay case $\overline{d}_m = \overline{d}_s$, we solve LMI (2.11) and obtain the allowable maximum time delays $\overline{d}_m = \overline{d}_s = 0.9999$ s. This result is identical to that of [15] for the symmetric time delays. When $\overline{d}_m = 0.5$ s, we obtain the allowable maximum value of backward delay $\overline{d}_s = 1.4999$ s by solving LMI (2.11). In simulation, we choose $q_m(0) = \dot{q}_m(0) = q_s(0) = \dot{q}_s(0) = \left[0\ 0\right]^T$, and the time delay $d_m(t) = 0.3 + 0.2\sin(t), d_s(t) = 0.9999 + 0.5\sin(t)$. The simulation results are shown in Figs. 2.3, 2.4, 2.5 and 2.6. Figure 2.3 shows the Y-position of the master and the slave joints, and Fig. 2.4 shows the human torque F_h and environment force F_e (the wall feedback torque), and the error between the two torques is shown in Fig. 2.5. Figure 2.6 shows the positions of the q_m and q_s. When the human force increases, the Y-positions of master and slave joints move forward, which can be seen from Fig. 2.3. The torque F_h increases with the augment of F, while $F_e = 0$ at the beginning for no touch of the obstacle. In 9.5 s, the slave joint reaches the wall

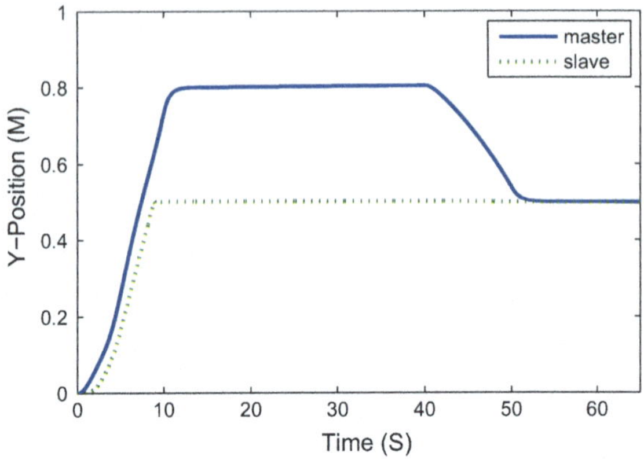

Fig. 2.3 The Y-position of master and slave with control (2.12)

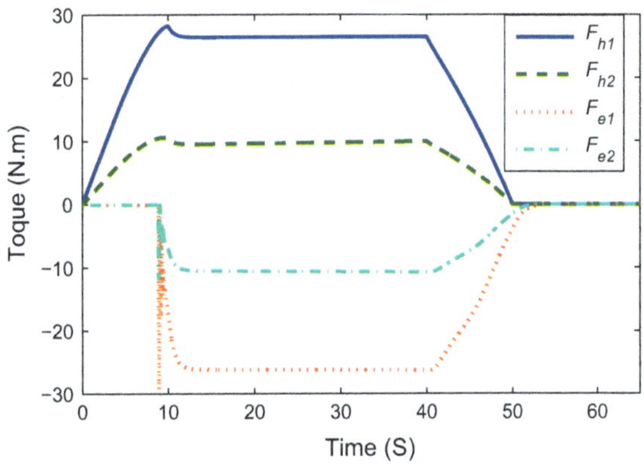

Fig. 2.4 The torque of F_h and F_e with control (2.12)

at $y = 0.5\,\text{m}$ and could not move any more, while the master joint moves further because of the human input force, and finally it reaches $y \approx 0.8\,\text{m}$ in $10s$, see Fig. 2.3. In Fig. 2.4, we can see that a large environmental torque pulse appears in about 9.5 s for the touch of the wall. From 9.5 s to 10 s, we can see that environment torque F_e increases quickly, and $F_e = -F_h$ at 10 s. With the force F keeps 30 N from 10 s to 40 s, we find that torques F_h and F_e also keep the former values, and the master joint and slave joint do not move any more on the Y-direction during this time interval. With the decrease of F from 40 s, F_h and F_e also reduces, and the Y-position of master moves backward, which can be seen from Figs. 2.3 and 2.4. When $F = 0$ at $t = 50\,\text{s}$, $F_h = 0$ and $F_e = 0$, and the master reaches $y = 0.5\,\text{m}$, and the positions of the two systems synchronize. If the controller is with the gravity compensator and there is no negative force to push the master back, the positions of the two joints will not move any more.

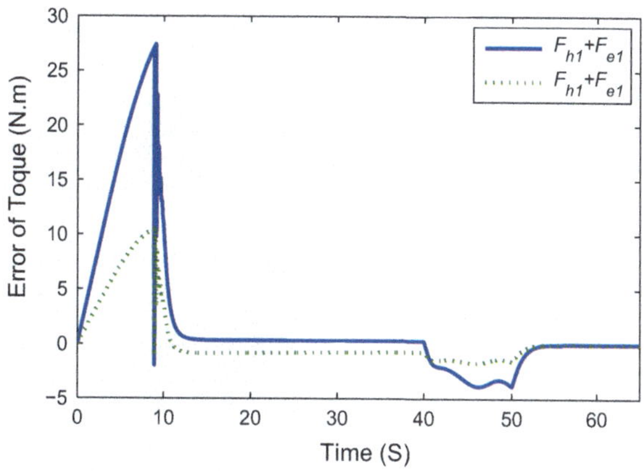

Fig. 2.5 The torque error between master and slave with control (2.12)

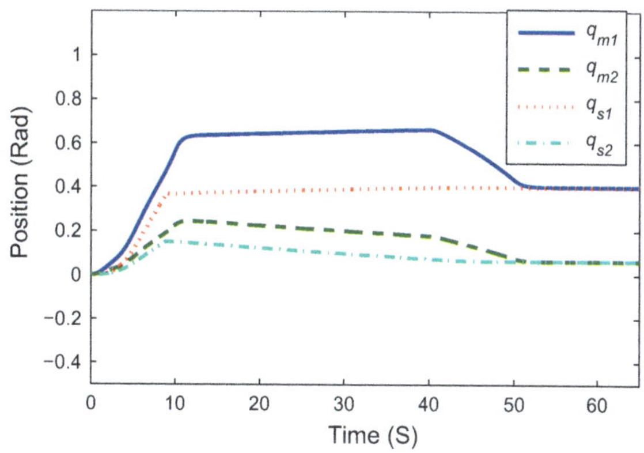

Fig. 2.6 The position of q_m and q_s with control (2.12)

2.4.2 *Experiment on a Teleoperated Pair of 3-DOF PHANToM Manipulator*

The teleoperation system for the experiments consists of two Phantom Premium 1.5A robots (SensAble Technologies Inc.) which are of 3-DOF, see Fig. 2.7. The two robotic manipulators are controlled by two computers which are connected via the Internet network. The network environment can be set in the network simulator block. For the master side (right in Fig. 2.7), the manipulator is equipped with standard passive stylus and thimble gimbal. For the slave site (left in Fig. 2.7), a marker is attached to the last link of the manipulator. We put a wood box at -0.06 m on the X-direction of the slave manipulator, which represents a stiff wall. We apply the human input to the master manipulator and push the last joint to move on the X-direction. The slave joint follows the movement of the master joint, and then the slave joint touches the stiff box and could not move any more. We move the master joint further and can feel the feedback force, then move the master joint to the original position. Our objective here is to verify the effectiveness of the proposed controller (2.12) and

Fig. 2.7 Teleoperation system

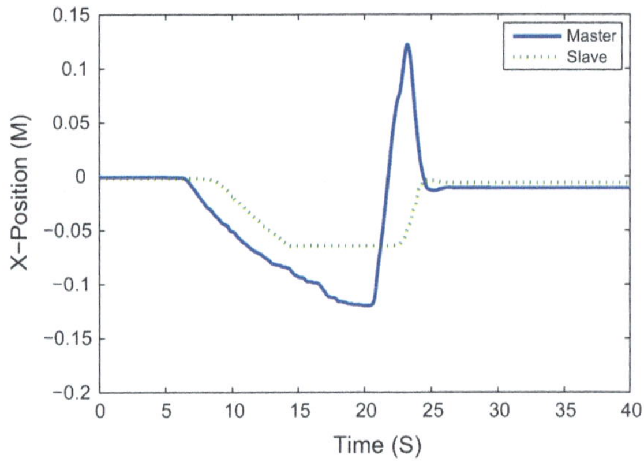

Fig. 2.8 The X-position of master and slave with control (2.12)

check whether the slave joint follows the movement of the master joint and whether they can return eventually to the original position.

With the use of the proposed controller (2.12) and same parameters above, the movement of the master and slave joints on the X-direction is shown in Fig. 2.8. In Fig. 2.9, the positions of q_m and q_s are shown. It depicts that the slave joint reaches -0.06 m and could not move any more. At the master site, we can feel the feedback force. With the master joint back to the origin, the slave manipulator follows the movement and reaches the origin. Thus the experimental results further show the effectiveness and correctness of the proposed main results.

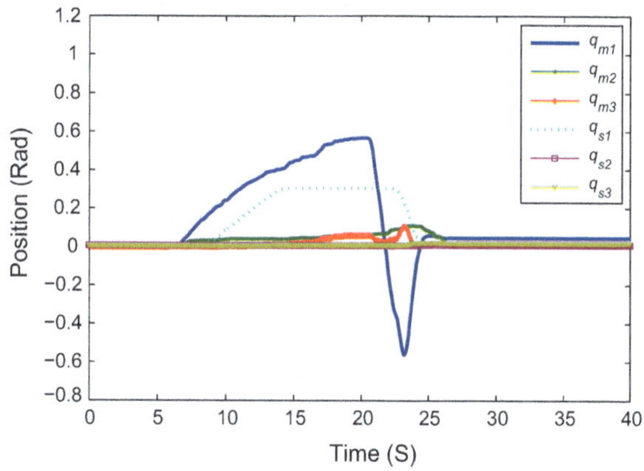

Fig. 2.9 The position of master and slave with control (2.12)

2.5 Summary

This chapter addresses the stability analysis problem for teleoperation systems with asymmetric and time-varying delays. We consider the class of controllerS as proportional position errors plus damping injection controller. We use a Lyapunov Krasovskii functional to derive the delay-dependent stability criteria, which are given in the form of LMIs. The maximum allowable time delay values can be obtained by solving the LMIs. The LMI-based method proposed in this chapter presents a more flexible tool for the control design of teleoperation systems. The simulations and experiments performed verify well the theoretical results and demonstrate their effectiveness.

References

1. R.J. Anderson, M.W. Spong, Bilateral control of teleoperators with time delay. IEEE Trans. Autom. Control **34**(5), 494–501 (1989)
2. W.R. Ferrell, Delayed force feedback. Hum. Factors **8**(5), 449–455 (1966)
3. N. Chopra, M.W. Spong, S. Hirche, M. Buss, Bilateral teleoperation over the internet: the time varying delay problem, in *Proceeding of American Control Conference*, pp. 155–160 (2003)
4. K. Gu, V.L. Kharitonov, J. Chen, *Stability of Time Delay Systems* (2003)
5. Y. He, G. Liu, D. Rees, New delay-dependent stability criteria for neural networks with time-varying delay. IEEE Trans. Neural Netw. Learn. Syst. **18**(1), 310–314 (2007)
6. Y. He, G. Liu, D. Rees, M. Wu, Improved delay-dependent stability criteria for systems with nonlinear perturbations. Eur. J. Control **13**(4), 356–365 (2007)
7. C.C. Hua, G. Feng, X.P. Guan, Robust stabilization of a class of nonlinear time delay systems via backstepping method. Eur. J. Control **44**(2), 567–573 (2008)
8. P. Arcara, C. Melchiorri, Control schemes for teleoperation with time delay: A comparative study. Robot. Auton. Syst. **38**(1), 49–64 (2002)
9. G. Niemeyer, J.E. Slotine, Stable adaptive teleoperation. IEEE J. Ocean. Eng. **16**(1), 152–162 (1991)

10. N.A. Tung, N.T. Binh, T.H. Anh et al., Synchronization control of bilateral teleoperation systems by using wave variable method under varying time delay, in *2017 International Conference on System Science and Engineering (ICSSE)*, pp. 21–23 (2017)
11. D. Lee, K. Huang, Passive-set-position-modulation framework for interactive robotic systems. IEEE Trans. Robot. **26**(2), 354–369 (2010)
12. I.G. Polushin, P.X. Liu, C.H. Lung, A control scheme for stable force-reflecting teleoperation over ip networks. IEEE Trans. Syst. Man Cybern. B Cybern. **36**(4), 930–939 (2006)
13. I.G. Polushin, P.X. Liu, C.H. Lung, A force reflection algorithm for improved transparency in bilateral teleoperation with communication delay. IEEE/ASME Trans. Mechatron. **12**(3), 361–374 (2007)
14. D. Lee, M.W. Spong, Passive bilateral teleoperation with constant time delay. IEEE Trans. Robot. Autom. **22**(2), 269–281 (2006)
15. E. Nuño, R. Ortega, N. Barabanov, L. Basañez, A globally stable pd controller for bilateral teleoperators. IEEE Trans. Robot. **24**(3), 753–758 (2008)
16. E. Nuño, L. Basañez, R. Ortega, M.W. Spong, Position tracking for non-linear teleoperators with variable time delay. Int. J. Robot. Res. **28**(7), 895–910 (2009)

Chapter 3
Convergence Analysis of Teleoperation Systems with Nonpassive Forces

Abstract This chapter addresses the stability problem of a class of teleoperation systems with nonpassive forces. The communication delays are also assumed to be both time-varying and unsymmetric. We consider the usual case that the master and the slave manipulators are coupled using a PD control strategy. By using a Lyapunov Krasovskii functional, we show that the master-slave teleoperation system is asymptotically stable under specific LMI conditions. With the given PD parameters, the values of allowable maximum time delays can be obtained. Finally, simulations are performed to show the effectiveness of the proposed method.

3.1 Introduction

As mentioned in Chap. 2 and the Introduction parttime delays in the communication channel can destabilize the whole teleoperation system if they are not well compensated for [1–4]. It is even worse for the network-based (such as Internet-based) teleoperation since in this case the delays are time-varying, irregular and unsymmetric [1, 2, 4]. Many methods have been proposed for the controller design of network-based teleoperation systems with time delays [5], among which the approaches based on the passivity theory are prevalent and have been extensively studied [4–6]. The passivity-based schemes, however, require that both the human operator and the environment are passive, which may not be satisfied for many teleoperation applications. The ISS (Input to State Stable) theory does not impose the assumption of passivity on any part of the system, so it is a very suitable framework to deal with those non-passive cases. By applying the ISS (Input to State Stable) idea to teleoperation systems, [7–9] proved the boundedness of the position and velocity synchronization errors between master and slave sites with non-passive human input and environment forces. However, there is little work addressing the asymptotical stability behavior of non-passive teleoperation systems. For teleoperation systems with a constant human operator input and symmetric time-invariant delays, the asymptotical stability was analyzed in [10]. Due to the complexity of the communication network, however, the

C. Hua et al., *Analysis and Design for Networked Teleoperation System*,
https://doi.org/10.1007/978-981-13-7936-9_3

delays of data packets are not only time-varying, but also unsymmetric. The reason exists in the fact that the data packets in the forward and backward channels may go through different network paths, leading that the forward and backward delays are not equal. Therefore, it is very important to investigate the asymmetry of time delays and its impact on the stability of network-based teleoperation systems. To the best of our knowledge, there have been very few results reported in the literature in this regard.

This chapter deals with the stability analysis problem for a class of teleoperation systems. The main contents of this chapters are summarized as: (i) nonpassive forces which ave rarely been investigated in the literature are considered; (ii) compared with existing work, we assume that the forward and the backward transmission time delays are both time-varying and unsymmetric, which is closer to real network communications. We focus on the case that the master and the slave manipulators are coupled using a PD control strategy, which is widely deployed in practical applications. By employing Lyapunov Krasovskii functional, we build the relationships among the PD controller design parameters and the upper bounds of the time-varying delays. The stability condition is given in the form of linear matrix inequality (LMI), which can be used to compute the allowed maximum values of forward and backward delays. Finally, simulations are performed to show the effectiveness of the developed main results.

3.2 Preliminaries and Problem Formulation

In this chapter, a master-slave system given by the following model is considered

$$\begin{cases} M_m(q_m)\ddot{q}_m + C_m\left(q_m, \dot{q}_m\right)\dot{q}_m + G_m(q_m) = \tau_m + F_h \\ M_s(q_s)\ddot{q}_s + C_s\left(q_s, \dot{q}_s\right)\dot{q}_s + G_s(q_s) = \tau_s - F_e \end{cases} \tag{3.1}$$

The properties for the system (3.1) also have been shown in Chap. 1.

For the teleoperation system (3.1), we consider the case that the human operator force F_h and the environmental force are given by

$$F_h = k_0 - k_h q_m - \overline{k}_h \dot{q}_m,\ F_e = k_e q_s + \overline{k}_e \dot{q}_s \tag{3.2}$$

where $k_0, k_h, \overline{k}_h, k_e$ and $\overline{k}_e$ are positive scalars.

Remark 3.1 We can see that the human operator force F_h contains the constant k_0, thus the passive condition is not satisfied. This paper aims to develop the asymptotical stability conditions of system (3.1) under a PD controller.

For the master and the slave systems, we design the following PD control strategies

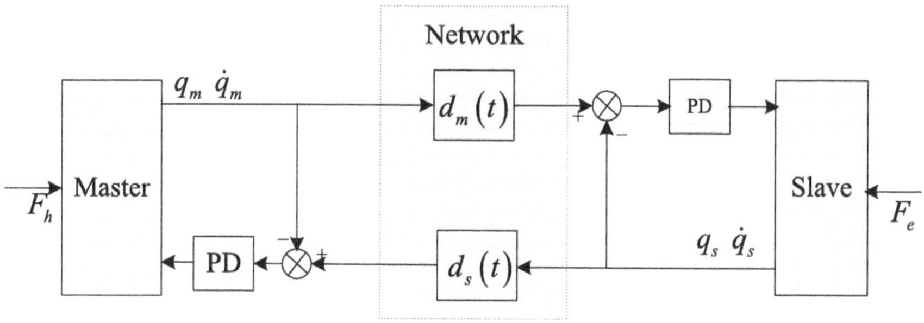

Fig. 3.1 The teleoperation system with unsymmetric time-varying delays

$$\begin{cases} \tau_m = G_m(q_m) - k_d(\dot{q}_m - \dot{q}_s(t - d_s(t))) \\ \qquad -k_p(q_m - q_s(t - d_s(t))) - \alpha_m \dot{q}_m, \\ \tau_s = G_s(q_s) + k_d(\dot{q}_m(t - d_m(t)) - \dot{q}_s) \\ \qquad +k_p(q_m(t - d_m(t)) - q_s) - \alpha_s \dot{q}_s. \end{cases} \tag{3.3}$$

where $d_m(t)$ and $d_s(t)$ are forward time delay (from the master site to slave site) and backward time delay (from the slave site to master site), respectively; α_m and α_s are positive scalars; k_p and k_d are positive proportional and differential parameters. The frame of the teleoperation system is shown Fig. 3.1.

Our objective is thus to give the asymptotic stability condition for system (3.1) with controller (3.3).

For the unsymmentric time delays, we impose the following condition:

Assumption 3.1 The time-varying delays of the system satisfy $d_m(t) \le \overline{d}_m, d_s(t) \le \overline{d}_s$, $\dot{d}_m(t) \le \mu_1 < 1$ and $\dot{d}_s(t) \le \mu_2 < 1$, where $\overline{d}_m, \overline{d}_s, \mu_1$ and μ_2 are positive scalars.

Remark 3.2 This assumption is similar to that in [7], which is very general for the time delay in the communication channel. First, the time delays of practical systems are often bounded, thus it is reasonable to assume the upper bounds on the time delays. Second, for any communication channel/communication protocol where a received signal is used by the teleoperation system (i.e. not discarded) until a fresher (more recently sent) packet arrives, time delays cannot grow faster than the time. In other words, the time derivative of the delay is less than 1. We can see that the imposed conditions are actually very loose.

With the torques (3.3), the whole system is

$$\begin{cases} M_m(q_m)\ddot{q}_m + C_m\left(q_m, \dot{q}_m\right)\dot{q}_m = k_0 - k_h q_m - \overline{k}_h \dot{q}_m - \alpha_m \dot{q}_m \\ \quad -k_d\left(\dot{q}_m - \dot{q}_s(t - d_s(t))\right) - k_p(q_m - q_s(t - d_s(t))), \\ M_s(q_s)\ddot{q}_s + C_s\left(q_s, \dot{q}_s\right)\dot{q}_s = -k_e q_s - \overline{k}_e \dot{q}_s - \alpha_s \dot{q}_s \\ \quad +k_d\left(\dot{q}_m(t - d_m(t)) - \dot{q}_s\right) + k_p(q_m(t - d_m(t)) - q_s). \end{cases} \tag{3.4}$$

From (3.4), one knows that the equilibrium points of position q_m^* and q_s^* satisfy

$$\begin{aligned} k_0 &= k_p\left(q_m^* - q_s^*\right) + k_h q_m^*, \\ 0 &= k_p\left(q_m^* - q_s^*\right) - k_e q_s^*, \end{aligned} \tag{3.5}$$

then $q_m^* = \left(k_e\left(k_p + k_e\right)^{-1} k_p + k_h\right)^{-1} k_0$ and $q_s^* = \left(k_p + k_e\right)^{-1} k_p q_m^*$.

In the following section, we will analyze the stability problem of (3.4).

3.3 Stability Analysis

With (3.5), we let $x = q_m - q_m^*$ and $y = q_s - q_s^*$, then (3.4) is rewritten as follows

$$\begin{cases} M_m(q_m)\ddot{x} + C_m\left(q_m, \dot{q}_m\right)\dot{x} = -k_h x - \overline{k}_h \dot{x} \\ -k_d\left(\dot{x} - \dot{y}(t - d_s(t))\right) - k_p(x - y(t - d_s(t))) - \alpha_m \dot{x} \\ M_s(q_s)\ddot{y} + C_s\left(q_s, \dot{q}_s\right)\dot{y} = k_d\left(\dot{x}(t - d_m(t)) - \dot{y}\right) \\ +k_p(x(t - d_m(t)) - y) - k_e y - \overline{k}_e \dot{y} - \alpha_s \dot{y} \end{cases} \tag{3.6}$$

Now we present the main result of the chapter as follows:

Theorem 3.1 *For given parameters μ_1 and μ_2, if k_h and k_e are not zero simultaneously and there exist positive matrices H_1, H_2, Z_1, Z_2 such that the following LMI holds*

$$\overline{\Theta} = \begin{bmatrix} \overline{\Theta}_{11} & 0 & 0 & k_d I & \overline{d}_s k_p I & 0 \\ * & -(1-\mu_1)H_1 & k_d I & 0 & 0 & 0 \\ * & * & \overline{\Theta}_{33} & 0 & 0 & \overline{d}_m k_p I \\ * & * & * & -(1-\mu_2)H_2 & 0 & 0 \\ * & * & * & * & -\overline{d}_s Z_2 & 0 \\ * & * & * & * & * & -\overline{d}_m Z_1 \end{bmatrix} < 0$$

where $*$ represents the transpose of the corresponding matrix, $\overline{\Theta}_{11} = -2(k_d + \alpha_m + \overline{k}_h)I + H_1 + \overline{d}_m Z_1$ and $\overline{\Theta}_{33} = -2(k_d + \alpha_s + \overline{k}_e)I + \overline{d}_s k_p^2 Z_1^{-1} + H_2 + \overline{d}_s Z_2$, then the whole system is asymptotically stable.

Proof For (3.6), with the Property 1.1 we choose the following Lyapunov functional

$$
\begin{aligned}
V_1 &= \dot{x}^T M_m(q_m)\dot{x} + \dot{y}^T M_s(q_s)\dot{y}, \\
V_2 &= k_p(x-y)^T(x-y) + k_e y^T y + k_h x^T x, \\
V_3 &= \int_{t-d_m(t)}^{t} \dot{x}(\xi)^T H_1 \dot{x}(\xi)\, d\xi \\
&\quad + \int_{t-d_s(t)}^{t} \dot{y}(\xi)^T H_2 \dot{y}(\xi)\, d\xi, \\
V_4 &= \int_{-\overline{d}_m}^{0}\int_{t+\theta}^{t} \dot{x}(\xi)^T Z_1 \dot{x}(\xi)\, d\xi d\theta \\
&\quad + \int_{-\overline{d}_s}^{0}\int_{t+\theta}^{t} \dot{y}(\xi)^T Z_2 \dot{y}(\xi)\, d\xi d\theta.
\end{aligned}
$$

By using Property 1.2, the time derivative of V_1 is

$$
\begin{aligned}
\dot{V}_1 =& -2k_d\dot{x}^T\left(\dot{x} - \dot{y}(t-d_s(t))\right) - 2\left(\overline{k}_e + \alpha_s\right)\dot{y}^T\dot{y} \\
& - 2k_p\dot{x}^T(x - y(t-d_s(t))) - 2\left(\alpha_m + \overline{k}_h\right)\dot{x}^T\dot{x} \\
& - 2k_h x^T\dot{x} + 2k_d\dot{y}^T\left(\dot{x}(t-d_m(t)) - \dot{y}\right) \\
& + 2k_p\dot{y}^T(x(t-d_m(t)) - y) - 2k_e y^T\dot{y} \\
=& -2\left(k_d + \alpha_m + \overline{k}_h\right)\dot{x}^T\dot{x} + 2k_d\dot{x}^T\dot{y}(t-d_s(t)) \\
& - 2k_h x^T\dot{x} - 2k_p\dot{x}^T(x-y) - 2k_p\dot{x}^T\int_{t-d_s(t)}^{t}\dot{y}(\xi)\, d\xi \\
& - 2\left(k_d + \overline{k}_e + \alpha_s\right)\dot{y}^T\dot{y} + 2k_d\dot{y}^T\dot{x}(t-d_m(t)) \\
& + 2k_p\dot{y}^T(x-y) - 2k_p\dot{y}^T\int_{t-d_m(t)}^{t}\dot{x}(\xi)\, d\xi - 2k_e y^T\dot{y}.
\end{aligned}
\tag{3.8}
$$

The time derivatives of V_2, V_3, V_4 are

$$
\dot{V}_2 = 2k_p(x-y)^T\left(\dot{x} - \dot{y}\right) + 2k_e y^T\dot{y} + 2k_h x^T\dot{x}, \tag{3.9}
$$

$$
\begin{aligned}
\dot{V}_3 \leq& \dot{x}(t)^T H_1\dot{x}(t) + \dot{y}(t)^T H_2\dot{y}(t) \\
& - (1-\mu_1)\dot{x}^T(t-d_m(t))H_1\dot{x}(t-d_m(t)) \\
& - (1-\mu_2)\dot{y}^T(t-d_s(t))H_2\dot{y}(t-d_s(t)),
\end{aligned}
\tag{3.10}
$$

$$\begin{aligned}\dot{V}_4 &= \overline{d}_m \dot{x}^T Z_1 \dot{x} - \int_{t-\overline{d}_m}^{t} \dot{x}^T(\xi) Z_1 \dot{x}(\xi)\, d\xi \\ &\quad + \overline{d}_2 \dot{y}^T Z_2 \dot{y} - \int_{t-\overline{d}_s}^{t} \dot{y}^T(\xi) Z_2 \dot{y}(\xi)\, d\xi \\ &\leq \overline{d}_1 \dot{x}^T Z_1 \dot{x} - \int_{t-d_m(t)}^{t} \dot{x}^T(\xi) Z_1 \dot{x}(\xi)\, d\xi \\ &\quad + \overline{d}_2 \dot{y}^T Z_2 \dot{y} - \int_{t-d_s(t)}^{t} \dot{y}^T(\xi) Z_2 \dot{y}(\xi)\, d\xi. \end{aligned} \tag{3.11}$$

With (3.8)–(3.11) and the following two inequalities

$$\begin{aligned} -2k_p \dot{x}^T \int_{t-d_s(t)}^{t} \dot{y}(\xi)\, d\xi - \int_{t-d_s(t)}^{t} \dot{y}^T(\xi) Z_2 \dot{y}(\xi)\, d\xi &\leq \overline{d}_s k_p^2 \dot{x}^T Z_2^{-1} \dot{x}, \\ -2k_p \dot{y}^T \int_{t-d_m(t)}^{t} \dot{x}(\xi)\, d\xi - \int_{t-d_m(t)}^{t} \dot{x}^T(\xi) Z_1 \dot{x}(\xi)\, d\xi &\leq \overline{d}_m k_p^2 \dot{y}^T Z_1^{-1} \dot{y}, \end{aligned}$$

one has

$$\dot{V} = \dot{V}_1 + \dot{V}_2 + \dot{V}_3 + \dot{V}_4 \leq \eta^T \Theta \eta, \tag{3.12}$$

where $\eta = \left[\dot{x}^T \; \dot{x}^T(t - d_m(t)) \; \dot{y}^T \; \dot{y}^T(t - d_s(t)) \right]^T$, and

$$\Theta = \begin{bmatrix} \Theta_{11} & 0 & 0 & k_d I \\ * & -(1-\mu_1) H_1 & k_d I & 0 \\ * & * & \Theta_{33} & 0 \\ * & * & * & -(1-\mu_2) H_2 \end{bmatrix}$$

in which $\Theta_{11} = -2\left(k_d + \alpha_m + \overline{k}_h\right) I + \overline{d}_s k_p^2 Z_2^{-1} + H_1 + \overline{d}_m Z_1$ and $\Theta_{33} = -2\left(k_d + \overline{k}_e + \alpha_s\right) I + \overline{d}_m k_p^2 Z_1^{-1} + H_2 + \overline{d}_s Z_2$. By using Schur complement, we know that $\Theta < 0$ is equivalent to $\overline{\Theta}_{(6\times 6)} < 0$. Thus (2.12) gives that $\dot{x}(t)$ and $\dot{y}(t)$ converges asymptotically to zero as $t \to \infty$ based on Lyapunov stability theory, then $x(t)$ and $y(t)$ converges to constants as $t \to \infty$. Further (3.6) gives

$$\begin{cases} k_p x = \left(k_e + k_p\right) y \\ k_p y = \left(k_p + k_h\right) x \end{cases}$$

Considering that k_h and k_e are not zero simultaneously, we get that x, y converge to zero asymptotically. The proof is completed.

Remark 3.3 From (3.7), one sees that the inequalities contain the PD parameters and the bound values of the forward and backward time delays. For given parameters μ_1, μ_2, k_p, k_d and one of delay bound parameters, we can calculate the allowable maximum value of the other delay by using the LMI optimization tool box of

Matlab. For instance, if μ_1, μ_2, k_p, k_d and $\overline{d}_m$ are known, we can obtain the allowable maximum value $\overline{d}_s$.

Remark 3.4 In [10], the convergence problem was considered for the case that the human force $F_h = k_0$ and it is assumed that the forward and backward time delays are symmetric and constant. By rigorous mathematical analysis, they give the stability conditions for system (3.1) with PD controller (3.3). Obviously, letting $\mu_1 = \mu_2 = 0$ and $d_m = d_s$, the LMI condition is identical to the stability condition in [10], so, our result includes the symmetric and constant time delay in [10] as a special case.

3.4 Simulation on a Teleoperation System of 2-DOF Planar Manipulators

In this section, the simulation results are shown to verify the effectiveness of the main result. The simulations are performed on 2-DOF manipulators:

$$M_m(q_m)\ddot{q}_m + C_m\left(q_m, \dot{q}_m\right)\dot{q}_m + G_m(q_m) = F_h + \tau_m$$
$$M_s(q_s)\ddot{q}_s + C_s\left(q_s, \dot{q}_s\right)\dot{q}_s + G_s(q_s) = \tau_s - F_e$$

where

$$M_m(q) = M_s(q) = \begin{bmatrix} M_{11} & M_{12} \\ * & M_{22} \end{bmatrix},$$
$$C_m\left(q, \dot{q}\right) = C_s\left(q, \dot{q}\right) = \begin{bmatrix} C_{11} & C_{12} \\ C_{21} & C_{22} \end{bmatrix}.$$
$$G_m(q_m) = G_s(q_s) = \begin{bmatrix} g_1 \\ g_2 \end{bmatrix},$$

in which

$$\begin{aligned} M_{11} &= (2L_1 \cos q_2 + L_2) L_2 m_2 + L_1^2 (m_1 + m_2), \\ M_{12} &= L_2^2 m_2 + L_1 L_2 m_2 \cos q_2, M_{22} = L_2^2 m_2, \\ C_{11} &= -L_1 L_2 m_2 \sin(q_2)\dot{q}_2, C_{21} = L_1 L_2 m_2 \sin(q_2), . \\ C_{12} &= -L_1 L_2 m_2 \sin(q_2)\left(\dot{q}_1 + \dot{q}_2\right), C_{22} = 0, \\ g_1 &= g(m_2 L_2 \sin(q_1 + q_2) + (m_1 + m_2) L_1 \sin(q_1)), \\ g_2 &= g m_2 L_2 \sin(q_1 + q_2). \end{aligned}$$

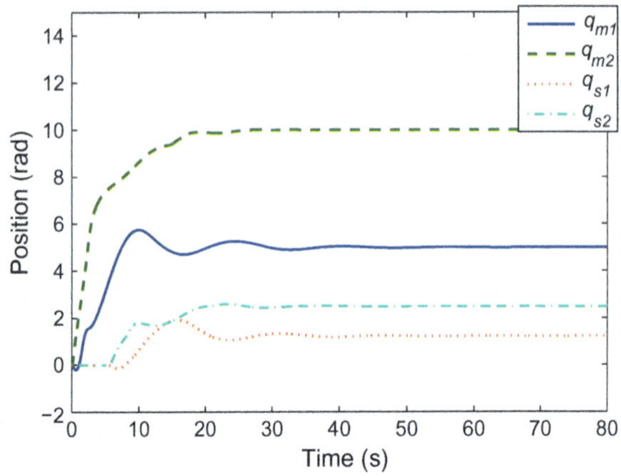

Fig. 3.2 The state response with $d_m(t) = 5.7 + 0.3\sin(t)$ and $d_s(t) = 8.4860 + 0.5\sin(t)$

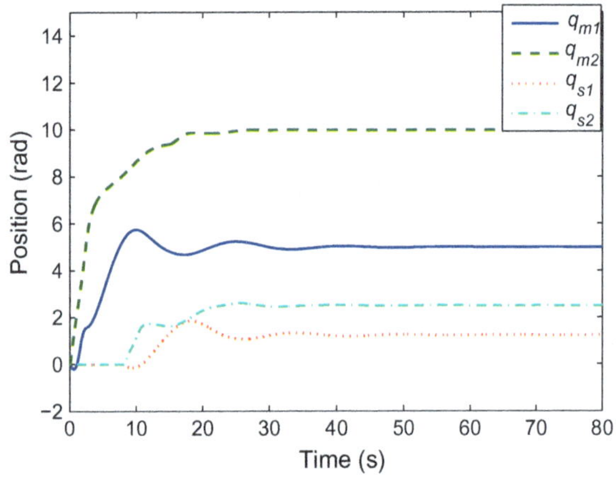

Fig. 3.3 The state response with $d_m(t) = 7.6860 + 0.3\sin(t)$ and $d_s(t) = 6.5 + 0.5\sin(t)$.

Choose $k_0 = [5\ 10]^T$, $k_d = 0.25$, $k_p = 0.25$, $k_e = 0.75$, $\overline{k}_e = 0.1$, $k_d = 1$, $\alpha_m = 1$, $\alpha_s = 0.9$, $k_h = 13/16$, $\overline{k}_h = 1$, $\overline{ke} = 1$, $\mu_1 = 0.3$, $\mu_2 = 0.5$. Based on (3.5), the equilibriums are $q_m^* = \left[5\ 10\right]^T$ and $q_s^* = \left[1.25\ 2.5\right]^T$.

With the parameters and forward delay $\overline{d}_m = 6$, solving LMI (3.7) gives the maximum allowed backward delay $\overline{d}_s = 8.9860$. Choose $\overline{d}_s = 7$, we can compute the maximum allowed backward delay value $\overline{d}_m = 7.9860$.

The parameters of the manipulators are $m_1 = 10\,\text{kg}$, $m_2 = 5\,\text{kg}$, $L_1 = 0.7\,\text{m}$, $L_2 = 0.5\,\text{m}$, $g = 9.81\,\text{m/s}^2$. The initial values of $q_m = \dot{q}_m = q_s = \dot{q}_s = \left[0\ 0\right]^T$. Fig. 3.2 shows the response of q_m and q_s with $d_m(t) = 5.7 + 0.3\sin(t)$ and $d_s(t) = 8.4860 + 0.5\sin(t)$, while Fig. 3.3 shows the responses of the case $d_m(t) = 7.6860 + 0.3\sin(t)$ and $d_s(t) = 6.5 + 0.5\sin(t)$. From these figures, we can see that the closed-loop system is asymptotically stable.

3.5 Summary

The stability analysis problem for teleoperation systems with nonpassive forces and asymmetrical time-varying delays is studied in this chapter. The Lyapunov Krasovskii functional is used to derive the delay dependent stability criteria. The stability condition is given in the form of LMI, which can be used to optimize the allowable maximum delay values. Finally, numerical simulation results are presented to show the effectiveness of the main results.

References

1. W.R. Ferrell, Delayed force feedback. Hum. Factors **8**(5), 449–455 (1966)
2. N. Chopra, M.W. Spong, S. Hirche, M. Buss, Bilateral teleoperation over the internet: the time varying delay problem, in *Proceeding of American Control Conference*, pp. 155–160 (2003)
3. K. Gu, V.L. Kharitonov, J. Chen, *Stability of Time Delay Systems* (2003)
4. G. Niemeyer, J.J.E. Slotine, Telemanipulation with time delays. Int. J. Robot. Res. **23**(9), 873–890 (2004)
5. P. Arcara, C. Melchiorri, Control schemes for teleoperation with time delay: a comparative study. Robot. Auton. Syst. **38**(1), 49–64 (2002)
6. R.J. Anderson, M.W. Spong, Bilateral control of teleoperators with time delay. IEEE Trans. Autom. Control **34**(5), 494–501 (1989)
7. I.G. Polushin, P.X. Liu, C.H. Lung, A control scheme for stable force-reflecting teleoperation over ip networks. IEEE Trans. Syst. Man Cybern. B Cybern. **36**(4), 930–939 (2006)
8. I.G. Polushin, P.X. Liu, C.H. Lung, A force reflection algorithm for improved transparency in bilateral teleoperation with communication delay. IEEE/ASME Trans. Mechatron. **12**(3), 361–374 (2007)
9. I.G. Polushin, P.X. Liu, C.H. Lung, Projection-based force reflection algorithm for stable bilateral teleoperation over networks. IEEE Trans. Instrum. Measurement **57**(9), 1854–1865 (2008)
10. R. Lozano, N. Chora, M.W. Spong, Convergenc analysis of bilateral teleoperation with constant human input, in *Proceeding of American Control Conference*, pp. 1443–1448 (2007)

Chapter 4
Stability Analysis of Teleoperation Systems with Asymmetric Interval Time-Varying Delays

Abstract In this chapter, we investigate the synchronization problem of nonlinear teleoperators in the presence of variable time delay. Compared with previous works, we relax the restriction of the communication delay range. Although approaches to the interval time delay can be found in many literatures, they are not suitable for nonlinear teleoperator due to the properties of robotic systems. We firstly investigate the synchronization of teleoperation system with interval time delay and make full use of the information of both the lower bound and upper bound of delay. All the criteria are presented in form of LMI, which can be easily calculated by Matlab. Finally, numerical example and simulation are given to show the effectiveness of the main results.

4.1 Introduction

Bilateral teleoperation has been an active research topic over the last few decades [1–3]. Time delay, which exists in transmission channel between the master and the salve, has become an unavoidable factor affecting the stability of the closed-loop system [4, 5]. For teleoperation, Anderson and Spong firstly proposed a control law for teleoperator based on passivity and scattering theory in [6], in order to overcome instability caused by time delay. However, their control architecture could not deal with variable characteristics of the system transmission delay. Niemeyer and Slotine introduced the wave variables in teleoperation system in [7]. In this chapter, a passivity-based formalism was used, which was a sufficient condition for stability. Although stability can be achieved by scattering approach or wave variable formulations, they may cause the position drift. In [8], the model-free linear observer-based PD output-feedback control design was used to solve the output feedback stabilization for industrial robotic systems. A hybrid control strategy for the trajectory tracking control problem of robotic manipulators was presented in [9]. A robust adaptive control term was introduced in the design, in order to deal with the fuzzy approximation errors, external disturbances and other modeling uncertainties. Besides, we had given

C. Hua et al., *Analysis and Design for Networked Teleoperation System*,
https://doi.org/10.1007/978-981-13-7936-9_4

some criteria in form of LMI in [10, 11]. However, in most previous works, they did not take into account the information of the lower bound of delay.

In practice, a time-varying interval delay is often encountered, that is, the range of delay may vary in an interval. Therefore, it is significant to investigate the stability of system with interval time-varying delay. Many techniques have been reported in the literature to deal with such problem in time delay systems [1, 12–15]. Combined with the free-weighting matrix approach, He et al. presented a new method to estimate the upper bound of the variable delay in [15]. Shao improved the result with fewer matrix variables in [5]. However, the methods dealing with interval delay in time delay systems are not suitable in nonlinear teleoperators due to the properties of robotic systems.

In this chapter, the synchronization problem of nonlinear teleoperation systems in the presence of variable time delay is studied. We relax the restriction that the range of communication delay varies from zero. Information of both the lower bound and upper bound are fully used. Besides, we initially apply some significant conclusions in time delay systems to bilateral teleoperators with some modifications. Motivated by idea in this chapter, some advanced approaches in time delay systems can be adopted in teleoperation to make the system stable in a broader delay region. All of results are given in form of LMI, and thus, it can be calculated easily by Matlab. Finally, numerical example and simulation are given to demonstrate the effectiveness of our results. All of the theorems proposed in this chapter are verified.

4.2 Preliminaries and Problem Formulation

The dynamic behavior of master-slave bilateral teleoperation system is also reviewed in this chapter.

Consider a master-slave bilateral teleoperator given by the following model:

$$\begin{cases} M_m(q_m)\ddot{q}_m + C_m(q_m, \dot{q}_m)\dot{q}_m + G_m(q_m) = \tau_m + F_h \\ M_s(q_s)\ddot{q}_s + C_s(q_s, \dot{q}_s)\dot{q}_s + G_s(q_s) = \tau_s - F_e \end{cases} \tag{4.1}$$

The following lemmas and assumptions will be used in this chapter.

Lemma 4.1 ([16]) *For given positive integers n, m, a scalar α in the interval $(0, 1)$, a given $n \times n$ matrix $\mathcal{R} > 0$, two matrices W_1 and W_2 in $R^{n\times m}$. Define, for all vector ξ in R^m, the function $\Theta(\alpha, \mathcal{R})$ given by:*

$$\Theta(\alpha, \mathcal{R}) = \frac{1}{\alpha}\xi^T W_1^T R W_1 \xi + \frac{1}{1-\alpha}\xi^T W_2^T \mathcal{R} W_2 \xi$$

Then, if there exists a matrix X in $R^{n\times n}$ such that $\begin{bmatrix} \mathcal{R} & X \\ * & \mathcal{R} \end{bmatrix} > 0$, then the following inequality holds

$$\min_{\alpha\in(0,1)} \Theta(\alpha,\mathcal{R}) \geq \begin{bmatrix} W_1\xi \\ W_2\xi \end{bmatrix}^T \begin{bmatrix} \mathcal{R} & X \\ * & \mathcal{R} \end{bmatrix} \begin{bmatrix} W_1\xi \\ W_2\xi \end{bmatrix}$$

Lemma 4.2 ([17]) *Define $l_{\mathcal{R}}(\omega) = \int_a^b \omega^T(u)\mathcal{R}\omega(u)du$, for a given matrix $\mathcal{R} > 0$, the following inequality holds for all continuously differentiable function ω in $[a, b] \rightarrow R^n$:*

$$l_R(\dot{\omega}) \geq \frac{1}{b-a}(\omega(b) - \omega(a))^T\mathcal{R}(\omega(b) - \omega(a)) + \frac{3}{b-a}\tilde{\Omega}^T\mathcal{R}\tilde{\Omega}$$

where $\tilde{\Omega} = \omega(b) + \omega(a) - \frac{2}{b-a}\int_a^b \omega(u)du$.

Assumption 4.1 In this chapter, we focus on the time-varying and asymmetric delay case. $d_m(t)$ is defined as the forward (from master site to slave site) delay and $d_s(t)$ is defined as the backward (from slave site to master site) delay. They are both lower and upper bounded, i.e., there exist positive scalars d_{m1}, d_{m2}, d_{s1} and d_{s2} such that $0 < d_{i1} \leq d_i(t) \leq d_{i2} < \infty$ and $\mu_{i1} \leq \dot{d}_i(t) \leq \mu_{i2} < 1, i = m, s$.

Remark 4.1 The range of varying delay considered in many existing references is from zero to an upper bound. However, the time delays of practical systems are often in an interval; thus, it is reasonable to assume that there are lower bounds for the time delays. We relax the restriction in this chapter. Besides, the communication delays in network are time-varying and asymmetric, which is closer to real systems.

4.3 Main Results

In this section, we address the synchronization problems for nonlinear teleoperation systems. The delays vary in a range. Both passive and non-passive of human operator and environment are considered.

Firstly, assume that

$$F_h = k_0 - k_h q_m - \bar{k}_h \dot{q}_m, \; F_e = \bar{k}_e \dot{q}_s \tag{4.2}$$

where $k_0, k_h, \bar{k}_h$ and $\bar{k}_e$are positive scalars.

That is, the human operator is non-passive and the environment is only related to velocity of slave. PD control strategies with asymmetric time-varying delay are used, that is

$$\begin{aligned} \tau_m &= G_m(q_m) - k_{dm}(\dot{q}_m - \dot{q}_s(t - d_s(t))) - k_{pm}(q_m - q_s(t - d_s(t))) - \alpha_m \dot{q}_m \\ \tau_s &= G_s(q_s) + k_{ds}(\dot{q}_m(t - d_m(t)) - \dot{q}_s) + k_{ps}(q_m(t - d_m(t)) - q_s) - \alpha_s \dot{q}_s \end{aligned}$$

where k_{di}, k_{pi} are positive proportional and differential coefficients, respectively. α_i are positive damping coefficients, $i = m, s$.

We analyze the stability problem of the whole system in the following part.

Let $x = q_m - q_m^*$, $y = q_s - q_s^*$, where q_m^* and q_s^* are equilibrium points of master and slave positions, respectively. Then, with PD control law and (4.2), system (4.1) can be rewritten as

$$\begin{aligned} M_m(q_m)\ddot{x} + C_m(q_m, \dot{q}_m)\dot{x} &= -k_h x - \bar{k}_h \dot{x} - k_{dm}(\dot{x} - \dot{y}(t - d_s(t))) \\ &\quad - k_{pm}(x - y(t - d_s(t))) - \alpha_m \dot{x} \\ M_s(q_s)\ddot{y} + C_s(q_s, \dot{q}_s)\dot{y} &= k_{ds}(\dot{x}(t - d_m(t)) - \dot{y}) - \bar{k}_e \dot{y} \\ &\quad + k_{ps}(x(t - d_m(t)) - y) - \alpha_s \dot{y} \end{aligned} \tag{4.3}$$

For the static system, the equilibrium points of master and slave positions satisfy

$$\begin{aligned} 0 &= k_0 - k_{pm}(q_m^* - q_s^*) - k_h q_m^* \\ 0 &= k_{ps}(q_m^* - q_s^*) \end{aligned}$$

and then, $q_m^* = q_s^* = \frac{k_0}{k_h}$.

Theorem 4.1 *Consider the teleoperation error system* (4.3) *with non-passive human operator force and passive environment force, and assume that time delays satisfy* $d_r(t) = \{d_{r1}, d_{r2}\}$, $\dot{d}_r(t) = \{\mu_{r1}, \mu_{r2}\}$. *If there exist positive matrices* Q_r, R_r, H_r, P_r, $r = m, s$ *and matrices* X_r, N_r, M_r *with appropriate dimensions such that the following inequalities hold, then the error system is asymptotically stable.*

$$\begin{aligned} \Phi &= \begin{bmatrix} \Phi_{11} & \Phi_{12} \\ * & \Phi_{22} \end{bmatrix} < 0 \\ \Phi_{221} &= \begin{bmatrix} \tilde{R}_s & X_s \\ * & \tilde{R}_s \end{bmatrix} > 0 \\ \Phi_{111} &= \begin{bmatrix} \tilde{R}_m & X_m \\ * & \tilde{R}_m \end{bmatrix} > 0 \end{aligned} \tag{4.4}$$

where $\Phi_{11} = \Phi_{110} - \frac{1}{dm2}\Gamma_m^T \Phi_{111} \Gamma_m$

$$\Phi_{12} = \begin{bmatrix} 0 & k_{dm} & -k_{pm} & 0 & \dots & 0 \\ \frac{k_{pm}}{k_{ps}} k_{ds} & 0 & 0 & 0 & \dots & 0 \\ -k_{pm} & 0 & 0 & 0 & \dots & 0 \\ 0 & 0 & \ddots & \dots & \dots & 0 \\ \vdots & \vdots & \dots & \ddots & \dots & \vdots \\ 0 & 0 & \dots & \dots & \ddots & 0 \end{bmatrix}_{9n \times 9n}$$

$\Phi_{22} = \Phi_{220} - \frac{1}{ds2}\Gamma_s^T \Phi_{221} \Gamma_s$

$\Phi_{110} = \hat{H}_m + \hat{Q}_m + \hat{W}_m + \hat{N}_m + \hat{M}_m + \hat{N}_m^T + \hat{M}_m^T + J_{m0}^T P_m J_{m1} + J_{m1}^T P_m J_{m0}$

$\Phi_{220} = \hat{H}_s + \hat{Q}_s + \hat{W}_s + \hat{N}_s + \hat{M}_s + \hat{N}_s^T + \hat{M}_s^T + J_{s0}^T P_s J_{s1} + J_{s1}^T P_s J_{s0}$

$$\hat{H}_i = diag\{0_{5n}, (1-\dot{d}_i(t))H_i, -H_i, 0_{2n}\}$$
$$\hat{Q}_i = diag\{Q_i, -(1-\dot{d}_i(t))Q_i, 0_{7n}\}$$
$$\hat{W}_m = diag\{-2(\bar{k}_h+\alpha_m+k_{dm})I + d_{m2}R_m, 0_{8n}\}$$
$$\hat{W}_s = diag\{-2\tfrac{k_{pm}}{kps}(\bar{k}_e+\alpha_s+k_{ds})I + d_{s2}R_s, 0_{8n}\}$$
$$\hat{N}_i = N_i J_{i2}, \hat{M}_i = M_i J_{i3}$$
$$J_{i0} = \begin{bmatrix} 0\,0\,0\,0\,I\,0\,0 & 0 & 0 \\ 0\,0\,0\,0\,0\,I\,0 & 0 & 0 \\ 0\,0\,0\,0\,0\,0\,0 & d_i(t)I & 0 \\ 0\,0\,0\,0\,0\,0\,0 & 0 & (d_{i2}-d_i(t))I \end{bmatrix}$$
$$J_{i1} = \begin{bmatrix} I & 0 & 0\,0\,0 & 0 & 0\;\;0\,0 \\ 0 & (1-\dot{d}_i(t))I & 0\,0\,0 & 0 & 0\;\;0\,0 \\ 0 & 0 & 0\,0\,I & -(1-\dot{d}_i(t))I & 0\;\;0\,0 \\ 0 & 0 & 0\,0\,0 & (1-\dot{d}_i(t))I & -I\;\;0\,0 \end{bmatrix}$$
$$J_{i2} = \begin{bmatrix} 0\;0\;-I\;0\;I\;-I\;0\;0\;0 \end{bmatrix}, J_{i3} = \begin{bmatrix} 0\;0\;0\;-I\;0\;I\;-I\;0\;0 \end{bmatrix}$$
$$\Gamma_i^T = \begin{bmatrix} \Gamma_{i1}^T\;\Gamma_{i2}^T\;\Gamma_{i3}^T\;\Gamma_{i4}^T \end{bmatrix}^T$$
$$\Gamma_{i1} = \begin{bmatrix} 0\;0\;0\;I\;0\;0\;0\;0\;0 \end{bmatrix}, \Gamma_{i2} = \begin{bmatrix} 0\;0\;0\;0\;0\;I\;I\;0\;-2I \end{bmatrix}$$
$$\Gamma_{i3} = \begin{bmatrix} 0\;0\;I\;0\;0\;0\;0\;0\;0 \end{bmatrix}, \Gamma_{i4} = \begin{bmatrix} 0\;0\;0\;0\;I\;I\;0\;-2I\;0 \end{bmatrix}$$
$$\tilde{R}_i = diag\{R_i, 3R_i\}$$

Proof We choose the following Lyapunov functional

$$\begin{aligned}
V_1 &= \dot{x}^T M_m(q_m)\dot{x} + \frac{k_{pm}}{k_{ps}}\dot{y}^T M_s(q_s)\dot{y} \\
V_2 &= k_{pm}(x-y)^T(x-y) + k_h x^T x \\
V_3 &= \int_{-d_{m2}}^{0}\int_{t+\theta}^{t} \dot{x}^T(\xi) R_m \dot{x}(\xi)\, d\xi d\theta \\
&\quad + \int_{-d_{s2}}^{0}\int_{t+\theta}^{t} \dot{y}^T(\xi) R_s \dot{y}(\xi)\, d\xi d\theta
\end{aligned}$$

$$\begin{aligned}
V_4 &= \int_{t-d_{m2}}^{t-d_m(t)} x^T(\xi) H_m x(\xi)\, d\xi \\
&\quad + \int_{t-d_{s2}}^{t-d_s(t)} y^T(\xi) H_s y(\xi)\, d\xi \\
V_5 &= \int_{t-d_m(t)}^{t} \dot{x}^T(\xi) Q_m \dot{x}(\xi)\, d\xi \\
&\quad + \int_{t-d_s(t)}^{t} \dot{y}^T(\xi) Q_s \dot{y}(\xi)\, d\xi \\
V_6 &= \sum_{i=m,s} \theta_i^T P_i \theta_i
\end{aligned} \tag{4.5}$$

where $\theta_m^T = \left[x^T(t)\ x^T(t-d_m(t))\ \int_{t-d_m(t)}^{t} x^T(\xi)\,d\xi\ \int_{t-d_{m2}}^{t-d_m(t)} x^T(\xi)\,d\xi\right]$,
$\theta_s^T = \left[y^T(t)\ y^T(t-d_s(t))\ \int_{t-d_s(t)}^{t} y^T(\xi)\,d\xi\ \int_{t-d_{s2}}^{t-d_s(t)} y^T(\xi)\,d\xi\right]$

Obviously, $V = \sum_{i=1}^{6} V_i > 0$.

With Property 1.2 of system model shown in Chap. 1, the time derivative of $V_i, i = 1, 2, 3, 4, 5, 6$ is

$$\begin{aligned}
\dot{V}_1 &= 2\dot{x}^T[-k_h x - \bar{k}_h\dot{x} - k_{dm}(\dot{x} - \dot{y}(t-d_s(t))) - k_{pm}(x - y(t-d_s(t))) - \alpha_m\dot{x}] \\
&\quad + \frac{2k_{pm}}{k_{ps}}\dot{y}^T[k_{ds}(\dot{x}(t-d_m(t)) - \dot{y}) + k_{ps}(x(t-d_m(t)) - y) - \bar{k}_e\dot{y} - \alpha_s\dot{y}] \\
\dot{V}_2 &= 2k_{pm}(\dot{x}^T x - \dot{x}^T y - \dot{y}^T x + \dot{y}^T y) + 2k_h\dot{x}^T x \\
\dot{V}_3 &= d_{m2}\dot{x}^T R_m\dot{x} - \int_{t-d_{m2}}^{t} \dot{x}^T(\xi) R_m\dot{x}(\xi)\,d\xi \\
&\quad + d_{s2}\dot{y}^T R_s\dot{y} - \int_{t-d_{s2}}^{t} \dot{y}^T(\xi) R_s\dot{x}(\xi)\,d\xi \\
\dot{V}_4 &= (1-\dot{d}_m(t))x^T(t-d_m(t))H_m x(t-d_m(t)) - x^T(t-d_{m2})H_m x(t-d_{m2}) \\
&\quad + (1-\dot{d}_s(t))y^T(t-d_s(t))H_s y(t-d_s(t)) - y^T(t-d_{s2})H_s y(t-d_{s2}) \\
\dot{V}_5 &= \dot{x}^T(t)Q_m\dot{x}(t) - (1-\dot{d}_m(t))\dot{x}^T(t-d_m(t))Q_m\dot{x}(t-d_m(t)) \\
&\quad + \dot{y}^T(t)Q_s\dot{y}(t) - (1-\dot{d}_s(t))\dot{y}^T(t-d_s(t))Q_s\dot{y}(t-d_s(t)) \\
\dot{V}_6 &= \sum_{i=m,s} \dot{\theta}_i^T P_i\theta_i + \theta_i^T P_i\dot{\theta}_i
\end{aligned}$$

On the one hand, noticing that $\int_{t-d_{m2}}^{t} \dot{x}^T(\xi) R_m\dot{x}(\xi)\,d\xi$ in $\dot{V}_3$ can be divided into $\int_{t-d_{m2}}^{t-d_m(t)} \dot{x}^T(\xi) R_m\dot{x}(\xi)\,d\xi$ and $\int_{t-d_m(t)}^{t} \dot{x}^T(\xi) R_m\dot{x}(\xi)\,d\xi$. Invoking Lemma 4.2, we can obtain that

$$\begin{aligned}
-\int_{t-d_{m2}}^{t-d_m(t)} \dot{x}^T(\xi) R_m\dot{x}(\xi)\,d\xi &\le -\frac{1}{d_{m2}-d_m(t)}\left(\int_{t-d_{m2}}^{t-d_m(t)} \dot{x}(\xi)\,d\xi\right)^T R_m \\
&\quad \times \left(\int_{t-d_{m2}}^{t-d_m(t)} \dot{x}(\xi)\,d\xi\right) - \frac{1}{d_{m2}-d_m(t)}\epsilon_{m1}^T R_m\epsilon_{m1} \\
-\int_{t-d_m(t)}^{t} \dot{x}^T(\xi) R_m\dot{x}(\xi)\,d\xi &\le -\frac{1}{d_m(t)}\left(\int_{t-d_m(t)}^{t} \dot{x}(\xi)\,d\xi\right)^T R_m \\
&\quad \times \left(\int_{t-d_m(t)}^{t} \dot{x}(\xi)\,d\xi\right) - \frac{1}{d_m(t)}\epsilon_{m2}^T R_m\epsilon_{m2}
\end{aligned}$$

in which

$$\epsilon_{m1} = x(t-d_m(t)) + x(t-d_{m2}) - \frac{2}{d_{m2}-d_m(t)}\int_{t-d_{m2}}^{t-d_m(t)} x(\xi)\,d\xi$$

$$\epsilon_{m2} = x(t) + x(t - d_m(t)) - \frac{2}{d_m(t)} \int_{t-d_m(t)}^{t} x(\xi)\, d\xi$$

Further, Lemma 4.1 is applied to the above inequalities. It yields

$$-\int_{t-d_{m2}}^{t} \dot{x}^T(\xi) R_m \dot{x}(\xi)\, d\xi \le -\frac{1}{d_{m2}} \xi_m^T \Gamma_m^T \Phi_{111} \Gamma_m \xi_m$$

where

$$\xi_m^T = \Big[\dot{x}^T(t)\ \dot{x}^T(t - d_m(t))\ \int_{t-d_m(t)}^{t} \dot{x}^T(\xi)\, d\xi\ \int_{t-d_{m2}}^{t-d_m(t)} \dot{x}^T(\xi)\, d\xi$$
$$x^T(t)\ x^T(t - d_m(t))\ x^T(t - d_{m2})$$
$$\frac{1}{d_m(t)} \int_{t-d_m(t)}^{t} x(\xi)\, d\xi\ \frac{1}{d_{m2} - d_m(t)} \int_{t-d_{m2}}^{t-d_m(t)} x(\xi)\, d\xi \Big]$$

Similarly,

$$-\int_{t-d_{s2}}^{t} \dot{y}^T(\xi) R_s \dot{y}(\xi)\, d\xi \le -\frac{1}{d_{s2}} \xi_s^T \Gamma_s^T \Phi_{221} \Gamma_s \xi_s$$

where

$$\xi_s^T = \Big[\dot{y}^T(t)\ \dot{y}^T(t - d_s(t))\ \int_{t-d_s(t)}^{t} \dot{y}^T(\xi)\, d\xi\ \int_{t-d_{s2}}^{t-d_s(t)} \dot{y}^T(\xi)\, d\xi$$
$$y^T(t)\ y^T(t - d_s(t))\ y^T(t - d_{s2})$$
$$\frac{1}{d_s(t)} \int_{t-d_s(t)}^{t} y(\xi)\, d\xi\ \frac{1}{d_{s2} - d_s(t)} \int_{t-d_{s2}}^{t-d_s(t)} y(\xi)\, d\xi \Big]$$

On the other hand, $\theta_i = J_{i0}\xi_i$, $\dot{\theta}_i = J_{i1}\xi_i$, and thus, $\dot{V}_6$ can be written as

$$\sum_{i=m,s} \xi_i^T J_{i1}^T P_i J_{i0} \xi_i + .\xi_i^T J_{i0}^T P_i J_{i1} \xi_i$$

We introduce "free-weighting matrices" [18] into teleoperation system as follows

$$f_1 = 2\xi_m^T N_m [x(t) - x(t - d_m(t)) - \int_{t-d_m(t)}^{t} \dot{x}(\xi)\, d\xi] = 0$$
$$f_2 = 2\xi_m^T M_m [x(t - d_m(t)) - x(t - d_{m2}) - \int_{t-d_{m2}}^{t-d_m(t)} \dot{x}(\xi)\, d\xi] = 0$$
$$f_3 = 2\xi_s^T N_s [y(t) - y(t - d_s(t)) - \int_{t-d_s(t)}^{t} \dot{y}(\xi)\, d\xi] = 0$$
$$f_4 = 2\xi_s^T M_s [y(t - d_s(t)) - y(t - d_{s2}) - \int_{t-d_{s2}}^{t-d_s(t)} \dot{y}(\xi)\, d\xi] = 0$$

and therefore, one can obtain that

$$\dot{V} = \sum_{i=1}^{5} \dot{V}_i + \sum_{i=1}^{4} f_i$$
$$\le \xi^T \Phi \xi$$

with $\xi^T = \left[\xi_m^T \; \xi_s^T\right]$. Finally, $\dot{V}$ is negative definite if there matrices $X_{rij}, i, j = \{1, 2\}, N_r, M_r$ such that $\Phi_{221} > 0$ and $\Phi_{111} > 0$ and if $\Phi < 0$, for all $d_r(t) = \{d_{r1}, d_{r2}\}$ and $\dot{d}_r(t) = \{\mu_{r1}, \mu_{r2}\}$. Based on Lyapunov stability theory, we can obtain that $\dot{x}(t)$ and $\dot{y}(t)$ converge to zero asymptotically, $x(t)$ and $y(t)$ converge to constants as $t \to \infty$. Furthermore, (4.3) gives

$$
\begin{aligned}
0 &= -k_h x - k_{pm}(x - y) \\
0 &= k_{ps}(x - y)
\end{aligned}
$$

and thus, x and y converge to zero asymptotically as $t \to \infty$. Because $q_m \to q_m^*, q_s \to q_s^*$, we also obtain that $q_m \to q_s$. That completes the proof.

□

Next, we assume that both human operator and environment are passive. That is

$$
-\int_0^t \dot{q}_m^T(\sigma) F_h(\sigma) d\sigma \geq 0, \int_0^t \dot{q}_s^T(\sigma) F_e(\sigma) d\sigma \geq 0 \tag{4.6}
$$

Tracking performance with the controller below is analyzed in specific

$$
\begin{aligned}
\tau_m &= -k_m(q_m - q_s(t - d_s(t))) - \alpha_m \dot{q}_m \\
\tau_s &= k_s(q_m(t - d_m(t)) - q_s) - \alpha_s \dot{q}_s
\end{aligned}
$$

where positive scalars k_m, k_s are proportional coefficients and α_m, α_s are damping coefficients.

Theorem 4.2 *Consider the coupled master-slave closed loop teleoperator system along with passive inequalities. Time delays satisfy Assumption 4.1. If there exist positive matrices $H_r, P_r, R_r, r = m, s$ and matrices X_r, N_r, M_r with appropriate dimensions, such that*

$$
\begin{aligned}
\Phi' &= \begin{bmatrix} \Phi'_{11} & \Phi'_{12} \\ * & \Phi'_{22} \end{bmatrix} < 0 \\
\Phi_{221} &= \begin{bmatrix} \tilde{R}_s & X_s \\ * & \tilde{R}_s \end{bmatrix} > 0 \\
\Phi_{111} &= \begin{bmatrix} \tilde{R}_m & X_m \\ * & \tilde{R}_m \end{bmatrix} > 0
\end{aligned} \tag{4.7}
$$

where $\Phi'_{11} = \Phi'_{110} - \frac{1}{dm2}\Gamma_m'^T \Phi_{111} \Gamma'_m$

$$
\Phi'_{12} = \begin{bmatrix} 0 & -k_m & 0 & \dots & 0 \\ -k_m & 0 & 0 & \dots & 0 \\ \vdots & 0 & \ddots & \dots & 0 \\ 0 & 0 & \dots & \ddots & 0 \end{bmatrix}_{8n \times 8n}
$$

$$\Phi_{22}^{'} = \Phi_{220}^{'} - \frac{1}{ds2}\Gamma_{s}^{'T}\Phi_{221}\Gamma_{s}^{'}$$
$$\Phi_{110}^{'} = \hat{H}_{m}^{'} + \hat{W}_{m}^{'} + \hat{N}_{m}^{'} + \hat{M}_{m}^{'} + \hat{N}_{m}^{'T} + \hat{M}_{m}^{'T} + J_{m0}^{'T}P_{m}J_{m1}^{'} + J_{m1}^{'T}P_{m}J_{m0}^{'}$$
$$\Phi_{220}^{'} = \hat{H}_{s}^{'} + \hat{W}_{s}^{'} + \hat{N}_{s}^{'} + \hat{M}_{s}^{'} + \hat{N}_{s}^{'T} + \hat{M}_{s}^{'T} + J_{s0}^{'T}P_{s}J_{s1}^{'} + J_{s1}^{'T}P_{s}J_{s0}^{'}$$
$$\hat{H}_{i}^{'} = diag\{0_{4n}, (1-\dot{d}_i(t))H_i, -H_i, 0_{2n}\}$$
$$\hat{W}_{m}^{'} = diag\{-2\alpha_m I + d_{m2}R_m, 0_{7n}\}$$
$$\hat{W}_{s}^{'} = diag\{-2\frac{k_m\alpha_s}{ks}I + d_{s2}R_s, 0_{7n}\}$$
$$\hat{N}_{i}^{'} = N_i J_{i2}^{'}, \hat{M}_{i}^{'} = M_i J_{i3}^{'}$$
$$J_{i0}^{'} = \begin{bmatrix} 0\,0\,0\,I\,0\,0 & 0 & 0 \\ 0\,0\,0\,0\,0\,0 & d_i(t)I & 0 \\ 0\,0\,0\,0\,0\,0 & 0 & (d_{i2}-d_i(t))I \end{bmatrix}$$
$$J_{i1}^{'} = \begin{bmatrix} I\,0\,0\,0 & 0 & 0 & 0\,0 \\ 0\,0\,0\,I & -(1-\dot{d}_i(t))I & 0 & 0\,0 \\ 0\,0\,0\,0 & (1-\dot{d}_i(t))I & -I & 0\,0 \end{bmatrix}$$
$$J_{i2}^{'} = \begin{bmatrix}0 & -I & 0 & I & -I & 0 & 0 & 0\end{bmatrix}, J_{i3}^{'} = \begin{bmatrix}0 & 0 & -I & 0 & I & -I & 0 & 0\end{bmatrix}$$
$$\Gamma_{i}^{'T} = \begin{bmatrix}\Gamma_{i1}^{'T} & \Gamma_{i2}^{'T} & \Gamma_{i3}^{'T} & \Gamma_{i4}^{'T}\end{bmatrix}^{T}$$
$$\Gamma_{i1}^{'} = \begin{bmatrix}0 & 0 & I & 0 & 0 & 0 & 0 & 0\end{bmatrix}, \Gamma_{i2}^{'} = \begin{bmatrix}0 & 0 & 0 & 0 & I & I & 0 & -2I\end{bmatrix}$$
$$\Gamma_{i3}^{'} = \begin{bmatrix}0 & I & 0 & 0 & 0 & 0 & 0 & 0\end{bmatrix}, \Gamma_{i4}^{'} = \begin{bmatrix}0 & 0 & 0 & I & I & 0 & -2I & 0\end{bmatrix}$$
$$\tilde{R}_i = diag\{R_i, 3R_i\}$$

then the velocities of master and slave, and position error are bounded.

Proof Choose the following Lyapunov functional

$$\begin{aligned} V_1 =& \dot{q}_m^T M_m(q_m)\dot{q}_m + \frac{k_m}{k_s}\dot{q}_s^T M_s(q_s)\dot{q}_s + 2(U_m(q_m) - \beta_m) \\ &+ \frac{2k_m}{k_s}(U_s(q_s) - \beta_s) - 2\int_0^t (\dot{q}_m^T(\sigma)F_h(\sigma) - \frac{k_m}{k_s}\dot{q}_s^T(\sigma)F_e(\sigma))d\sigma \\ V_2 =& k_m(q_m - q_s)^T(q_m - q_s) \end{aligned}$$

$$\begin{aligned} V_3 &= \sum_{i=m,s}\int_{t-d_{i2}}^{t-d_i(t)} q_i^T(\xi)H_i q_i(\xi)d\xi \\ V_4 &= \sum_{i=m,s}\int_{-d_{i2}}^{0}\int_{t+\theta}^{t} \dot{q}_i^T(\xi)R_i\dot{q}_i(\xi)d\xi d\theta \\ V_5 &= \sum_{i=m,s}\theta_i^{'T}P_i\theta_i^{'} \end{aligned} \tag{4.8}$$

where $\theta_i^{'T} = \left[q_i^T(t)\ \int_{t-d_i(t)}^{t} q_i^T(\xi)d\xi\ \int_{t-d_{i2}}^{t-d_i(t)} q_i^T(\xi)d\xi\right]$

The time derivative of (4.8) along the system trajectories is given by

$$\begin{aligned}
\dot{V} = \sum_{i=1}^{5} \dot{V}_i = & - 2\alpha_m \dot{q}_m^T \dot{q}_m - \frac{2k_m\alpha_s}{k_s}\dot{q}_s^T \dot{q}_s \\
& - 2k_m \dot{q}_m^T \int_{t-d_s(t)}^{t} \dot{q}_s(\xi)d\xi - 2k_m \dot{q}_s^T \int_{t-d_m(t)}^{t} \dot{q}_m(\xi)d\xi \\
& + \sum_{i=m,s} (1-\dot{d}_i(t))q_i^T(t-d_i(t))H_i q_i(t-d_i(t)) - q_i^T(t-d_{i2})H_i q_i(t-d_{i2}) \\
& + \sum_{i=m,s} d_{i2}\dot{q}_i^T R_i \dot{q}_i - \int_{t-d_{i2}}^{t} \dot{q}_i^T(\xi) R_i \dot{q}_i(\xi)\, d\xi \\
& + \sum_{i=m,s} \dot{\theta}_i^{'T} P_i \theta_i^{'} + \theta_i^{'T} P_i \dot{\theta}_i^{'}
\end{aligned}$$

in which $-\int_{t-d_{i2}}^{t} \dot{q}_i^T(\xi) R_i \dot{q}_i(\xi)\, d\xi$ can be rewritten as $-\int_{t-d_{i2}}^{t-d_i(t)} \dot{q}_i^T(\xi) R_i \dot{q}_i(\xi)\, d\xi - \int_{t-d_i(t)}^{t} \dot{q}_i^T(\xi) R_i \dot{q}_i(\xi)\, d\xi$. The following inequalities hold with Lemma 4.2.

$$\begin{aligned}
-\int_{t-d_{i2}}^{t-d_i(t)} \dot{q}_i^T(\xi) R_i \dot{q}_i(\xi)\, d\xi \le & -\frac{1}{d_{i2}-d_i(t)}\left(\int_{t-d_{i2}}^{t-d_i(t)} \dot{q}_i(\xi)\, d\xi\right)^T R_i \\
& \times \left(\int_{t-d_{i2}}^{t-d_i(t)} \dot{q}_i(\xi)\, d\xi\right) - \frac{1}{d_{i2}-d_i(t)}\epsilon_{i1}^{'T} R_i \epsilon_{i1}^{'} \\
-\int_{t-d_i(t)}^{t} \dot{q}_i^T(\xi) R_i \dot{q}_i(\xi)\, d\xi \le & -\frac{1}{d_i(t)}\left(\int_{t-d_i(t)}^{t} \dot{q}_i(\xi)\, d\xi\right)^T R_i \\
& \times \left(\int_{t-d_i(t)}^{t} \dot{q}_i(\xi)\, d\xi\right) - \frac{1}{d_i(t)}\epsilon_{i2}^{'T} R_i \epsilon_{i2}^{'}
\end{aligned}$$

in which $\epsilon_{i1}^{'} = q_i(t-d_i(t)) + q_i(t-d_{i2}) - \dfrac{2}{d_{i2}-d_i(t)} \int_{t-d_{i2}}^{t-d_i(t)} q_i(\xi)\, d\xi$

$\epsilon_{i2}^{'} = q_i(t) + q_i(t-d_i(t)) - \dfrac{2}{d_i(t)} \int_{t-d_i(t)}^{t} q_i(\xi)\, d\xi$

The above inequalities can be further obtained

$$-\int_{t-d_{i2}}^{t} \dot{q}_i^T(\xi) R_i \dot{q}_i(\xi)\, d\xi \le -\frac{1}{d_{i2}}\xi_i^{'T} \Gamma_i^{'T} \Phi_{111} \Gamma_i^{'} \xi_i^{'}$$

where $\xi_i^{'T} = \Big[\dot{q}_i^T(t) \ \int_{t-d_i(t)}^{t} \dot{q}_i^T(\xi)\, d\xi \ \int_{t-d_{i2}}^{t-d_i(t)} \dot{q}_i^T(\xi)\, d\xi \ q_i^T(t) \ q_i^T(t-d_i(t))$
$q_i^T(t-d_{i2}) \ \dfrac{1}{d_i(t)} \int_{t-d_i(t)}^{t} q_i^T(\xi)\, d\xi \ \dfrac{1}{d_{i2}-d_i(t)} \int_{t-d_{i2}}^{t-d_i(t)} q_i^T(\xi)\, d\xi\Big]$

The last two terms in $\dot{V}$ can be rewritten as $\sum_{i=m,s} \xi_i^{'T} J_{i1}^{'T} P_i J_{i0}^{'} \xi_i^{'} + \xi_i^{'T} J_{i0}^{'T} P_i J_{i1}^{'} \xi_i^{'}$ because of $\theta_i^{'} = J_{i0}^{'}\xi_i^{'}$, $\dot{\theta}_i^{'} = J_{i1}^{'}\xi_i^{'}$.

The following free-weighting matrices are considered.

$$f_{i1}^{'} = 2\xi_i^{'T} N_i[q_i(t) - q_i(t - d_i(t)) - \int_{t-d_i(t)}^{t} \dot{q}_i\,(\xi)\,d\xi] = 0$$

$$f_{i2}^{'} = 2\xi_i^{'T} M_i[q_i(t - d_i(t)) - q_i(t - d_{i2}) - \int_{t-d_{i2}}^{t-d_i(t)} \dot{q}_i\,(\xi)\,d\xi] = 0$$

and therefore, one can obtain that

$$\begin{aligned}\dot{V} &= \sum_{i=1}^{5} \dot{V}_i + \sum_{\substack{j=1\\ i=m,s}}^{2} f_{ij}^{'} \\ &\leq \xi^{'T} \Phi^{'} \xi^{'}\end{aligned}$$

with $\xi^{'T} = \left[\xi_m^{'T}\ \xi_s^{'T}\right]$. Finally, $\dot{V}$ is negative definite if (4.7) satisfies, and thus, $\dot{q}_m, \dot{q}_s \in \ell_2$ and the velocities of master and slave, and position error are bounded. That completes the proof.

□

Theorem 4.3 *If human-input force F_h and environment force F_e are zero and inequality* (4.7) *holds, then the velocities $\dot{q}_m, \dot{q}_s$ and position error $q_m - q_s$ converge to zero asymptotically with the following controller*

$$\begin{aligned}\tau_m &= -k_m(q_m - q_s(t - d_s(t))) - \alpha_m \dot{q}_m + G_m(q_m) \\ \tau_s &= k_s(q_m(t - d_m(t)) - q_s) - \alpha_s \dot{q}_s + G_s(q_s)\end{aligned}$$

Proof Choose Lyapunov functional (4.8) without $(U_m(q_m) - \beta_m) + \frac{k_m}{k_s}(U_s(q_s) - \beta_s)$ in V_1. According to Theorem 4.2, we know that $\dot{q}_m, \dot{q}_s, q_m - q_s \in \ell_\infty$ and $\dot{q}_m, \dot{q}_s \in \ell_2$. With (4.1), we can obtain that

$$\ddot{q}_m = M_m^{-1}(q_m)[-C_m(q_m, \dot{q}_m)\dot{q}_m - k_m(q_m - q_s\,(t - d_s\,(t))) - \alpha_m \dot{q}_m]$$

By Properties 1.1 and 1.4 and $q_m - q_s\,(t - d_s\,(t)) = q_m - q_s + \int_{t-d_s(t)}^{t} \dot{q}_s\,(\xi)\,d\xi$, we have $\ddot{q}_m$ is bounded. Based on Barbalat's lemma, we know that $\dot{q}_m \to 0$ as $t \to \infty$, and similarly, $\dot{q}_s \to 0$ as $t \to \infty$.

Next, we will prove that the tracking error $q_m - q_s$ converges to zero asymptotically.

The time derivative of $\ddot{q}_m$ is

$$\begin{aligned}\dddot{q}_m &= \frac{dM_m^{-1}(q_m)}{dt}[-C_m(q_m, \dot{q}_m)\dot{q}_m - k_m(q_m - q_s\,(t - d_s\,(t))) - \alpha_m \dot{q}_m] \\ &\quad + M_m^{-1}(q_m)[-C_m(q_m, \dot{q}_m)\ddot{q}_m - \dot{C}_m(q_m, \dot{q}_m)\dot{q}_m \\ &\quad - k_m(\dot{q}_m - (1 - \dot{d}_s\,(t))\dot{q}_s\,(t - d_2\,(t))) - \alpha_m \ddot{q}_m]\end{aligned}$$

For the first term, we have
$\frac{dM_m^{-1}(q_m)}{dt} = -M_m^{-1}(q_m)\dot{M}_m(q_m)M_m^{-1}(q_m) = -M_m^{-1}(q_m)[C_m(q_m, \dot{q}_m) + \dot{C}_m$ $(q_m, \dot{q}_m)]M_m^{-1}(q_m)$, which is clearly bounded because of Properties 1.1 and 1.4. By Properties 1.1, 1.4 and 1.5, we have $\dddot{q}_m \in \ell_\infty$. Based on Barbalat's lemma, we know $\ddot{q}_m \to 0$, and further, $q_m - q_s \to 0$ as $t \to \infty$. That completes the proof.

□

Finally, with passive human operator and environment force, we consider the torque-feedback controller as follows

$$\begin{aligned} \tau_m &= -\tau_s(t - d_s(t)) - \alpha_m \dot{q}_m \\ \tau_s &= k_s(q_m(t - d_m(t)) - q_s) - \alpha_s \dot{q}_s \end{aligned} \tag{4.9}$$

where k_s is positive proportional parameter and α_m, α_s are positive damping parameters.

Theorem 4.4 *If there exist positive matrices $R, R_r, H, H_r, Q_s, P_r, r = m, s$ and matrices X, X_r, N_r, M_r, Y, G with appropriate dimensions, and assume that the human operator and environment force satisfy* (4.6), *and delays satisfy Assumption 4.1 such that*

$$\begin{aligned} \Phi^{''} &= \begin{bmatrix} \Phi_{11}^{''} & \Phi_{12}^{''} \\ * & \Phi_{22}^{''} \end{bmatrix} < 0 \\ \Phi_{221} &= \begin{bmatrix} \tilde{R}_s & X_s \\ * & \tilde{R}_s \end{bmatrix} > 0 \\ \Phi_{111} &= \begin{bmatrix} \tilde{R}_m & X_m \\ * & \tilde{R}_m \end{bmatrix} > 0 \\ \Phi_{311} &= \begin{bmatrix} \tilde{R} & X \\ * & \tilde{R} \end{bmatrix} > 0 \end{aligned} \tag{4.10}$$

where

$$\Phi_{11}^{''} = \Phi_{110}^{''} - \frac{1}{d_{m2}} \Gamma_m^{''T} \Phi_{111} \Gamma_m^{''} - \frac{1}{d_{m2} + d_{s2}} \Gamma_{mm}^T \Phi_{311} \Gamma_{mm}$$

$$\Phi_{12}^{''} = \begin{bmatrix} 0 & \alpha_s & -k_s & 0 & \dots & 0 \\ -k_s & 0 & 0 & 0 & \dots & 0 \\ 0 & 0 & \dots & \dots & \dots & 0 \\ \vdots & \vdots & \ddots & \ddots & \ddots & \vdots \\ 0 & 0 & \dots & \dots & \dots & 0 \end{bmatrix}_{14n \times 9n}$$

$$\Phi_{22}^{''} = \Phi_{220}^{''} - \frac{1}{d_{s2}} \Gamma_s^{''T} \Phi_{221} \Gamma_s^{''}$$

$$
\begin{aligned}
\Phi_{110}'' &= \hat{W}_m'' + \hat{H}_m'' + \hat{H} + \hat{N}_m'' + \hat{M}_m'' + \hat{N}_m''^T + \hat{M}_m''^T + J_{m0}''^T P_m J_{m1}'' \\
&\quad + J_{m1}''^T P_m J_{m0}'' + \hat{Y} + \hat{Y}^T + \hat{G} + \hat{G}^T \\
\hat{W}_m'' &= diag\{-2\alpha_m I + d_{m2} R_m + (d_{m2} + d_{s2})R, 0_{13n}\} \\
\hat{H}_m'' &= diag\{0_{6n}, (1 - \dot{d}_m(t))H_m, -H_m, 0_{6n}\} \\
\hat{H} &= diag\{0_{8n}, (1 - \dot{d}_m(t) - \dot{d}_s(t))H, -H, 0_{4n}\} \\
\hat{N}_m'' &= N_m J_{m2}'', \hat{M}_m = M_m J_{m3}'' \\
\hat{Y} &= Y J_{m4}, \hat{G} = G J_{m5}, \\
J_{m2}'' &= \begin{bmatrix} 0 & -I & 0 & 0 & 0 & I & -I & 0 & 0 & 0 & 0 & 0 & 0 & 0 \end{bmatrix} \\
J_{m3}'' &= \begin{bmatrix} 0 & 0 & -I & 0 & 0 & 0 & I & -I & 0 & 0 & 0 & 0 & 0 & 0 \end{bmatrix} \\
J_{m4} &= \begin{bmatrix} 0 & 0 & 0 & I & 0 & 0 & 0 & 0 & I & -I & 0 & 0 & 0 & 0 \end{bmatrix} \\
J_{m5} &= \begin{bmatrix} 0 & 0 & 0 & 0 & I & I & 0 & 0 & -I & 0 & 0 & 0 & 0 & 0 \end{bmatrix} \\
J_{m0}'' &= \begin{bmatrix}
0\,0\,0\,0\,0\,I\,0\,0\,0\,0 & 0 & 0 & 0 & 0 \\
0\,0\,0\,0\,0\,0\,0\,0\,0\,0 & d_m(t)I & 0 & 0 & 0 \\
0\,0\,0\,0\,0\,0\,0\,0\,0\,0 & 0 & (d_{m2} - d_m(t))I & 0 & 0 \\
0\,0\,0\,0\,0\,0\,0\,0\,0\,0 & 0 & 0 & d(t)I & 0 \\
0\,0\,0\,0\,0\,0\,0\,0\,0\,0 & 0 & 0 & 0 & (d_2 - d(t))I
\end{bmatrix} \\
J_{m1}'' &= \begin{bmatrix}
I\,0\,0\,0\,0\,0 & 0 & 0 & 0 & 0\;0\,0\,0\,0 \\
0\,0\,0\,0\,0\,I & -(1 - \dot{d}_m(t))I & 0 & 0 & 0\;0\,0\,0\,0 \\
0\,0\,0\,0\,0\,0 & (1 - \dot{d}_m(t))I & -I & 0 & 0\;0\,0\,0\,0 \\
0\,0\,0\,0\,0\,I & 0 & 0 & -(1 - \dot{d}(t))I & 0\;0\,0\,0\,0 \\
0\,0\,0\,0\,0\,0 & 0 & 0 & (1 - \dot{d}(t))I & -I\;0\,0\,0\,0
\end{bmatrix} \\
d(t) &= d_s(t) + d_m(t), d_2 = d_{s2} + d_{m2} \\
\Gamma_m''^T &= \begin{bmatrix} \Gamma_{m1}''^T & \Gamma_{m2}''^T & \Gamma_{m3}''^T & \Gamma_{m4}''^T \end{bmatrix}^T, \Gamma_{mm}^T = \begin{bmatrix} \Gamma_{mm1}^T & \Gamma_{mm2}^T & \Gamma_{mm3}^T & \Gamma_{mm4}^T \end{bmatrix} \\
\Gamma_{m1}'' &= \begin{bmatrix} 0 & 0 & I & 0 & 0 & 0 & 0 & 0 & 0 & 0 & 0 & 0 & 0 & 0 \end{bmatrix} \\
Gamma_{m2}'' &= \begin{bmatrix} 0 & 0 & 0 & 0 & 0 & 0 & I & I & 0 & 0 & 0 & -2I & 0 & 0 \end{bmatrix} \\
\Gamma_{m3}'' &= \begin{bmatrix} 0 & I & 0 & 0 & 0 & 0 & 0 & 0 & 0 & 0 & 0 & 0 & 0 & 0 \end{bmatrix} \\
\Gamma_{m4}'' &= \begin{bmatrix} 0 & 0 & 0 & 0 & 0 & I & I & 0 & 0 & 0 & -2I & 0 & 0 & 0 \end{bmatrix} \\
\Gamma_{mm1} &= \begin{bmatrix} 0 & 0 & 0 & I & 0 & 0 & 0 & 0 & 0 & 0 & 0 & 0 & 0 & 0 \end{bmatrix} \\
\Gamma_{mm2} &= \begin{bmatrix} 0 & 0 & 0 & 0 & 0 & 0 & 0 & 0 & I & I & 0 & 0 & 0 & -2I \end{bmatrix} \\
\Gamma_{mm3} &= \begin{bmatrix} 0 & 0 & 0 & 0 & I & 0 & 0 & 0 & 0 & 0 & 0 & 0 & 0 & 0 \end{bmatrix} \\
\Gamma_{mm4} &= \begin{bmatrix} 0 & 0 & 0 & 0 & 0 & I & 0 & 0 & I & 0 & 0 & 0 & -2I & 0 \end{bmatrix} \\
\Phi_{220}'' &= \hat{H}_s'' + \hat{Q}_s'' + \hat{W}_s'' + \hat{N}_s'' + \hat{M}_s'' + \hat{N}_s''^T + \hat{M}_s''^T + J_{s0}''^T P_s J_{s1}'' + J_{s1}''^T P_s J_{s0}'' \\
\hat{H}_s'' &= diag\{0_{5n}, (1 - \dot{d}_s(t))H_s, -H_s, 0_{2n}\}
\end{aligned}
$$

$$\hat{Q}_s'' = diag\{Q_s, -(1-\dot{d}_s(t))Q_s, 0_{7n}\}$$
$$\hat{W}_s = diag\{-2\alpha_s I + d_{s2}R_s, 0_{8n}\}$$
$$\hat{N}_s'' = N_s J_{s2}'', \hat{M}_s'' = M_s J_{s3}''$$
$$J_{s2}'' = \begin{bmatrix} 0 & 0 & -I & 0 & I & -I & 0 & 0 & 0 \end{bmatrix}$$
$$J_{s3}'' = \begin{bmatrix} 0 & 0 & 0 & -I & 0 & I & -I & 0 & 0 \end{bmatrix}$$
$$J_{s0}'' = \begin{bmatrix} 0\,0\,0\,0\,I\,0\,0 & 0 & 0 \\ 0\,0\,0\,0\,0\,I\,0 & 0 & 0 \\ 0\,0\,0\,0\,0\,0\,0 & d_s(t)I & 0 \\ 0\,0\,0\,0\,0\,0\,0 & 0 & (d_{s2}-d_s(t))I \end{bmatrix}$$
$$J_{s1}'' = \begin{bmatrix} I & 0 & 0\,0\,0 & 0 & 0\,0\,0 \\ 0 & (1-\dot{d}_s(t))I & 0\,0\,0 & 0 & 0\,0\,0 \\ 0 & 0 & 0\,0\,I & -(1-\dot{d}_s(t))I & 0\,0\,0 \\ 0 & 0 & 0\,0\,0 & (1-\dot{d}_s(t))I & -I\,0\,0 \end{bmatrix}$$
$$\Gamma_s^T = \begin{bmatrix} \Gamma_{s1}^T & \Gamma_{s2}^T & \Gamma_{s3}^T & \Gamma_{s4}^T \end{bmatrix}^T$$
$$\Gamma_{s1} = \begin{bmatrix} 0 & 0 & 0 & I & 0 & 0 & 0 & 0 & 0 \end{bmatrix}, \Gamma_{s2} = \begin{bmatrix} 0 & 0 & 0 & 0 & 0 & I & I & 0 & -2I \end{bmatrix}$$
$$\Gamma_{s3} = \begin{bmatrix} 0 & 0 & I & 0 & 0 & 0 & 0 & 0 & 0 \end{bmatrix}, \Gamma_{s4} = \begin{bmatrix} 0 & 0 & 0 & 0 & I & I & 0 & -2I & 0 \end{bmatrix}$$
$$\tilde{R}_i = diag\{R_i, 3R_i\}, \tilde{R} = diag\{R, 3R\}, i = m, s$$

then the controller (4.9) *renders velocities and position error are bounded;*

Proof Choose Lyapunov functional as follows

$$\begin{aligned} V_1 &= \dot{q}_m^T M_m(q_m)\dot{q}_m + \dot{q}_s^T M_s(q_s)\dot{q}_s + 2(U_m(q_m) - \beta_m) \\ &\quad + 2(U_s(q_s) - \beta_s) - 2\int_0^t (\dot{q}_m^T(\sigma)F_h(\sigma) - \dot{q}_s^T(\sigma)F_e(\sigma))d\sigma \\ V_2 &= k_s(q_m - q_s)^T(q_m - q_s) \\ V_3 &= \sum_{i=m,s}\int_{-d_{i2}}^0\int_{t+\theta}^t \dot{q}_i^T(\xi)R_i\dot{q}_i(\xi)d\xi d\theta + \int_{-d_2}^0\int_{t+\theta}^t \dot{q}_m^T(\xi)R\dot{q}_m(\xi)d\xi d\theta \\ V_4 &= \sum_{i=m,s}\int_{t-d_{i2}}^{t-d_i(t)} q_i^T(\xi)H_i q_i(\xi)d\xi + \int_{t-d_2}^{t-d(t)} q_m^T(\xi)Hq_m(\xi)d\xi \\ V_5 &= \int_{t-d_s(t)}^t \dot{q}_s^T(\xi)Q_s\dot{q}_s(\xi)d\xi \\ V_6 &= \sum_{i=m,s}\theta_i''^T P_i\theta_i'' \end{aligned} \tag{4.11}$$

where $\theta_m''^T = \Big[q_m^T(t)\ \int_{t-d_m(t)}^t q_m^T(\xi)d\xi\ \int_{t-d_{m2}}^{t-d_m(t)} q_m^T(\xi)d\xi\ \int_{t-d(t)}^t q_m^T(\xi)d\xi$
$\int_{t-d_2}^{t-d(t)} q_m^T(\xi)d\xi \Big]$

$$\theta_m^{''T} = \left[q_s^T(t)\ q_s^T(t-d_s(t))\ \int_{t-d_s(t)}^{t} q_s^T(\xi)\,d\xi\ \int_{t-d_{s2}}^{t-d_s(t)} q_s^T(\xi)\,d\xi \right]$$

The time derivative of Lyapunov function V_i satisfies

$$\begin{aligned}
\dot V_1 + \dot V_2 = & -2\alpha_m \dot q_m^T \dot q_m + 2\alpha_s \dot q_m^T \dot q_s(t-d_s(t)) \\
& + 2k_s \dot q_m^T \int_{t-d_m(t)-d_s(t)}^{t} \dot q_m(\xi)\,d\xi - 2k_s \dot q_m^T \int_{t-d_s(t)}^{t} \dot q_s(\xi)\,d\xi \\
& - 2\alpha_s \dot q_s^T \dot q_s - 2k_s \dot q_s^T \int_{t-d_m(t)}^{t} \dot q_m(\xi)\,d\xi \\
\dot V_3 = & \sum_{i=m,s} d_{i2} \dot q_i^T R_i \dot q_i - \int_{t-d_{i2}}^{t} \dot q_i^T(\xi) R_i \dot q_i(\xi)\,d\xi \\
& + d_2 \dot q_m^T R \dot q_m - \int_{t-d_2}^{t} \dot q_m^T(\xi) R \dot q_m(\xi)\,d\xi \\
\dot V_4 = & \sum_{i=m,s} (1-\dot d_i(t)) q_i^T(t-d_i(t)) H_i q_i(t-d_i(t)) - q_i^T(t-d_{i2}) H_i q_i(t-d_{i2}) \\
& + (1-\dot d(t)) q_m^T(t-d(t)) H q_m(t-d(t)) - q_m^T(t-d_2) H q_m(t-d_2) \\
\dot V_5 = & \dot q_s^T Q_s \dot q_s - (1-\dot d_s(t)) \dot q_s^T(t-d_s(t)) Q_s \dot q_s(t-d_s(t)) \\
\dot V_6 = & \sum_{i=m,s} \dot\theta_i^{''T} P_i \theta_i^{''} + \theta_i^{''T} P_i \dot\theta_i^{''}
\end{aligned}$$

Note that

$$\begin{aligned}
-\int_{t-d_{i2}}^{t} \dot q_i^T(\xi) R_i \dot q_i(\xi)\,d\xi &= -\int_{t-d_{i2}}^{t-d_i(t)} \dot q_i^T(\xi) R_i \dot q_i(\xi)\,d\xi - \int_{t-d_i(t)}^{t} \dot q_i^T(\xi) R_i \dot q_i(\xi)\,d\xi \\
-\int_{t-d_2}^{t} \dot q_m^T(\xi) R \dot q_m(\xi)\,d\xi &= -\int_{t-d_2}^{t-d(t)} \dot q_m^T(\xi) R \dot q_m(\xi)\,d\xi - \int_{t-d(t)}^{t} \dot q_m^T(\xi) R \dot q_m(\xi)\,d\xi
\end{aligned}$$

Employing Lemmas 4.2 and 4.1, we can finally obtain that

$$\begin{aligned}
-\int_{t-d_{i2}}^{t} \dot q_i^T(\xi) R_i \dot q_i(\xi)\,d\xi &\le -\frac{1}{d_{i2}} \xi_i^{''T} \Gamma_i^{''T} \Phi_{111} \Gamma_i^{''} \xi_i^{''}, \\
-\int_{t-d_2}^{t} \dot q_m^T(\xi) R \dot q_m(\xi)\,d\xi &\le -\frac{1}{d_2} \xi_m^{''T} \Gamma_{mm}^T \Phi_{311} \Gamma_{mm} \xi_m^{''}
\end{aligned}$$

where

$$\begin{aligned}
\xi_m^{''T} = \Big[& \dot q_m^T(t)\ \int_{t-d_m(t)}^{t} \dot q_m^T(\xi)\,d\xi\ \int_{t-d_{m2}}^{t-d_m(t)} \dot q_m^T(\xi)\,d\xi\ \int_{t-d_2}^{t-d(t)} \dot q_m^T(\xi)\,d\xi\ \int_{t-d(t)}^{t} \dot q_m^T(\xi)\,d\xi \\
& q_m^T(t)\ q_m^T(t-d_m(t))\ q_m^T(t-d_{m2})\ q_m^T(t-d(t))\ q_m^T(t-d_2) \\
& \frac{1}{d_m(t)} \int_{t-d_m(t)}^{t} q_m^T(\xi)\,d\xi\ \frac{1}{d_{m2}-d_m(t)} \int_{t-d_{m2}}^{t-d_m(t)} q_m^T(\xi)\,d\xi\ \frac{1}{d(t)} \int_{t-d(t)}^{t} q_m^T(\xi)\,d\xi \\
& \frac{1}{d_2-d(t)} \int_{t-d_2}^{t-d(t)} q_m^T(\xi)\,d\xi \Big]
\end{aligned}$$

and

$$\xi_s^{''T} = \Big[\dot{q}_s^T(t)\ \dot{q}_s^T(t-d_s(t))\ \int_{t-d_s(t)}^{t} \dot{q}_s^T(\xi)\,d\xi\ \int_{t-d_{s2}}^{t-d_s(t)} \dot{q}_s^T(\xi)\,d\xi\ q_s^T(t)\ q_s^T(t-d_s(t))$$
$$q_s^T(t-d_{s2})\ \tfrac{1}{d_s(t)} \int_{t-d_s(t)}^{t} q_s^T(\xi)\,d\xi\ \tfrac{1}{d_{s2}-d_s(t)} \int_{t-d_{s2}}^{t-d_s(t)} q_s^T(\xi)\,d\xi \Big]$$

The following free-weighting matrices are considered.

$$f_{i1}^{''} = 2\xi_i^{''T} N_i[q_i(t) - q_i(t-d_i(t)) - \int_{t-d_i(t)}^{t} \dot{q}_i(\xi)\,d\xi] = 0$$

$$f_{i2}^{''} = 2\xi_i^{''T} M_i[q_i(t-d_i(t)) - q_i(t-d_{i2}) - \int_{t-d_{i2}}^{t-d_i(t)} \dot{q}_i(\xi)\,d\xi] = 0$$

$$f_3^{'} = 2\xi_m^{''T} Y[q_m(t-d(t)) - q_m(t-d_2) - \int_{t-d_2}^{t-d(t)} \dot{q}_m(\xi)\,d\xi] = 0$$

$$f_4^{'} = 2\xi_m^{''T} G[q_m(t) - q_m(t-d(t)) - \int_{t-d(t)}^{t} \dot{q}_m(\xi)\,d\xi] = 0$$

and therefore, one can obtain that

$$\begin{aligned}\dot{V} &= \sum_{i=1}^{5} \dot{V}_i + \sum_{\substack{j=1\\ i=m,s}}^{2} f_{ij}^{''} + f_3^{'} + f_4^{'}\\ &\leq \xi^{''T} \Phi^{''} \xi^{''}\end{aligned}$$

with $\xi^{''T} = \left[\xi_m^{''T}\ \xi_s^{''T}\right]$. Finally, $\dot{V}$ is negative definite if (4.7) satisfies. Based on Lyapunov stability theory, velocities and position error are bounded and $\dot{q}_m, \dot{q}_s \in \ell_2$. That completes the proof.

Theorem 4.5 *When the human operator does not move the master robot and the slave robot does not contact with environment, the following controller*

$$\begin{aligned}\tau_m &= -\tau_s(t-d_s(t)) - \alpha_m \dot{q}_m + G_m(q_m)\\ \tau_s &= k_s(q_m(t-d_m(t)) - q_s) - \alpha_s \dot{q}_s + G_s(q_s)\end{aligned}$$

renders the velocities converge to zero and the synchronization of master/slave robots is achieved.

Proof V_1 is chosen as $V_1 = \dot{q}_m^T M_m(q_m)\dot{q}_m + \dot{q}_s^T M_s(q_s)\dot{q}_s$, others are the same with those in (4.11). With the system model, we have

$$\begin{aligned}\ddot{q}_m =& M_m^{-1}(q_m)[-C_m(q_m,\dot{q}_m)\dot{q}_m - k_s(q_m(t-d_m(t)-d_s(t)) - q_s(t-d_s(t)))\\ &+ \alpha_s \dot{q}_s(t-d_s(t)) - \alpha_m \dot{q}_m]\end{aligned}$$

According to the first part of the proof, we have known that $\dot{q}_m, \dot{q}_s, q_m - q_s \in \ell_\infty$ and $\dot{q}_m, \dot{q}_s \in \ell_2$. With Properties 1.1 and 1.4 and the fact that $q_m(t - d_m(t) - d_s(t)) - q_s(t - d_s(t)) = q_m - q_s - \int_{t-d_m(t)-d_s(t)}^{t} \dot{q}_m(\xi)\, d\xi + \int_{t-d_s(t)}^{t} \dot{q}_s(\xi)\, d\xi$ one can obtain that $\ddot{q}_m \in \ell_\infty$. Based on Barbalat's lemma, we know that $\dot{q}_m \to 0$ as $t \to \infty$, and similarly, $\dot{q}_s \to 0$ as $t \to \infty$.

The time derivative of $\ddot{q}_m$ is

$$
\begin{aligned}
\dddot{q}_m =& \frac{dM_m^{-1}(q_m)}{dt}[-C_m(q_m, \dot{q}_m)\dot{q}_m - k_s(q_m(t - d_m(t) - d_s(t)) - q_s(t - d_s(t))) \\
&+ \alpha_s \dot{q}_s(t - d_s(t)) - \alpha_m \dot{q}_m] + M_m^{-1}(q_m)[-C_m(q_m, \dot{q}_m)\ddot{q}_m - \dot{C}_m(q_m, \dot{q}_m)\dot{q}_m \\
&- k_s((1 - \dot{d}_m(t) - \dot{d}_s(t))\dot{q}_m(t - d_m(t) - d_s(t)) - (1 - \dot{d}_s(t))q_s(t - d_s(t))) \\
&+ \alpha_s(1 - \dot{d}_s(t))\ddot{q}_s(t - d_s(t)) - \alpha_m \ddot{q}_m]
\end{aligned}
$$

We have proved the boundness of $\frac{dM_m^{-1}(q_m)}{dt}$ in the proof of Theorem 4.3, and thus, $\dddot{q}_m \in Ł_\infty$ with Properties 1.1, 1.4 and 1.5. Based on Barbalat's lemma, we know $\ddot{q}_m \to 0$, and further, $q_m - q_s \to 0$ as $t \to \infty$, which implies the synchronization of master/slave robots is achieved. That completes the proof.

Remark 4.2 The results obtained in this chapter contain both the upper bound values and the lower bound values of the forward and backward time delays. In the inequalities, the maximum value of the delay can be solved by LMI tool box of Matlab when we fix another. In the next section, the upper bound of forward delay is known, and we can obtain the upper bound of backward delay. Compared with some existing results [5, 13, 19, 20], the criteria presented in this chapter are less conservative. Taking [5] for example, the authors used the fact $-2a^T(t)\int_{t-d(t)}^{t} b(\xi)\, d\xi - \int_{t-d(t)}^{t} b^T(\xi)\Upsilon b(\xi)\, d\xi \le \bar{d}a^T(t)\Upsilon^{-1}a(t)$ when dealing with $-\int_{t-\bar{d}_1}^{t} \dot{q}_m^T(\xi) Z \dot{q}_m(\xi)\, d\xi$ in $\dot{V}_3$, and finally got the inequality $\begin{bmatrix} -2\alpha_m I + \bar{d}_1 Z & \bar{d}_2 k_m & 0 & 0 \\ * & -\bar{d}_2 S & 0 & 0 \\ * & * & -\frac{2k_m\alpha_s}{k_s} I + \bar{d}_2 S & \bar{d}_1 k_m \\ * & * & * & -\bar{d}_1 Z \end{bmatrix} < 0$. However, in nature, Jensen inequality is employed during the process. $-\int_{t-\bar{d}_1}^{t} \dot{q}_m^T(\xi) Z \dot{q}_m(\xi)\, d\xi \le -\frac{1}{\bar{d}_1}(\int_{t-\bar{d}_1}^{t} \dot{q}_m(\xi)\, d\xi)^T Z(\int_{t-\bar{d}_1}^{t} \dot{q}_m(\xi)\, d\xi)$ and the final result can be obtained as $\begin{bmatrix} -2\alpha_m I + \bar{d}_1 Z & -k_m & 0 & 0 \\ * & -\frac{1}{\bar{d}_2} S & 0 & 0 \\ * & * & -\frac{2k_m\alpha_s}{k_s} I + \bar{d}_2 S & -k_m \\ * & * & * & -\frac{1}{\bar{d}_1} Z \end{bmatrix} < 0$, which is equivalent to the above inequality. Additionally, in [5], the derivative of $\int_{-\bar{d}_1}^{0}\int_{t+\theta}^{t} \dot{q}_m^T(\xi) Z \dot{q}_m(\xi)\, d\xi d\theta$

is estimated as $\bar{d}_1\dot{q}_m^T Z\dot{q}_m - \int_{t-d_1(t)}^{t} \dot{q}_m^T(\xi)Z\dot{q}_m(\xi)d\xi$ and the term $-\int_{t-\bar{d}_1}^{t-d_1(t)} \dot{q}_m^T(\xi) Z\dot{q}_m(\xi)\, d\xi$ is ignored, which may lead to considerable conservativeness. As is well known, Jensen inequality is commonly used but it is conservative. By contrast, Wirtinger inequality allows to consider a more accurate integral inequality which encompasses the Jensen one. Applying Wirtinger inequality, Seuret et. al. proposed a corollary (Lemma 4.2 in this chapter). It is worth while to note that the result can not be used directly to teleoperation due to the nonlinearity of the system. If we do so we find that the upper bound of delay can not be improved and the result is the same as the one in [5]. It is mainly because that some terms in inequality are restricted strongly. To solve the problem, we introduces some free-weighting matrices [18] to relax the restriction. Besides, the upper bounds increase with the increase of lower bounds. The result can be verified by necessary and sufficient condition. From the simulation part, one can see that the result obtained in this chapter is closer to the necessary and sufficient result.

Remark 4.3 As is well known, free-weighting matrices are widely used in time delay system. However, it can not be employed directly in teleoperator. There are four states in such a system, velocities and positions of master and slave robots. For example, in [5], if we apply free-weighting matrices $g_1 = 2\theta^T(t)M[q_i(t) - q_i(t - d_i(t)) - \int_{t-d_i(t)}^{t} \dot{q}_i(s)ds] = 0$, $g_2 = 2\theta^T(t)N[\dot{q}_i(t) - \dot{q}_i(t - d_i(t)) - \int_{t-d_i(t)}^{t} \ddot{q}_i(s)ds] = 0$, all of the terms in g_1 and g_2 can not be handled. In g_1, the negative quadratic terms can not be produced by $\dot{V}$ (If we add $V_5 = \int_{t-d_i(t)}^{t} q_i^T(s)H_i q_i(s)ds$ in [5], only $q_i^T H_i q_i > 0$ are arose in $\dot{V}_5$). In g_2, $\int_{t-d_i(t)}^{t} \ddot{q}_i(s)ds$ will not be found in $\dot{V}$ because $\ddot{q}_i$ stand for acceleration of robots. In this chapter, we employ free-weighting matrices similar with g_1, and the negative quadratic terms of $q_i(t)$ can be produced by the usage of Lemma 4.2.

4.4 Simulation Example

In this section, numerical example and simulation are given to demonstrate the effectiveness of our results. In Example 4.1, we relax the restriction that the lower bound of delay is zero and obtain larger maximal upper bound, and it is closer to the necessary and sufficient result. All of the theorems proposed in this chapter are verified. Figures in Example 4.2 show the tracking ability of teleoperation system. The simulations are performed on 2-degree-of-freedom manipulators.

Example 4.1 By Theorem 4.2, the maximal of communication delay can be found when the lower bound is known. First, we illustrate the validity of the theorem. That is, the upper bound of time delay obtained by our method cannot be greater than that obtained by the necessary and sufficient condition solved in the frequency domain. Choose the parameters as $\alpha_m = \alpha_s = 10$, $k_m = k_s = 100$. For the symmetric delay case, the result obtained by necessary and sufficient condition in frequency domain is 0.157. In [5, 13], the results are 0.0999. However, it can be improved to 0.1011 by Theorem 4.2 in this chapter. It is closer to the necessary and sufficient result.

Table 4.1 Calculated maximal upper bound of symmetric delay for $\alpha_m = \alpha_s = 10$, $k_m = k_s = 100$ of Example 4.1

References	d
Hua and Liu [5]	0.0999
Nuño et al. [13]	0.0999
Theorem 4.2	0.1011
Necessary and sufficient condition	0.157

Second, we illustrate the less conservation of the theorem. All of theorems in this chapter are checked (Table 4.1).

Take $\alpha_m = 1$, $k_m = 0.25$, $\alpha_s = 0.9$, $k_s = 0.25$, and fixed $d_{m1} = d_{s1} = 0$ and $d_{m2} = 1$, a summary of the maximal upper bound of backward delay obtained from Theorem 4.2 in this chapter and those in [5, 13, 20] is presented in Table 4.2. In [20], when d_{m2} is fixed at 1, we calculate that d_{s2} is 2.7947. That is, when $d_m(t)$ belongs to (0, 1), $d_s(t)$ belongs to (0, 2.7497). It was further improved in works [13] and [5]. The maximum allowable bound of forward delays were 5.2726 and 6.5894, respectively. On the other hand, by applying Theorem 4.2 in this chapter, the result is 6.6667. This implies that, for this example, the stability condition in this chapter is less conservative than those in [5, 13, 20].

Consider the teleoperation system (4.1), the controller parameters are taken from [19] $k_{dm} = k_{ds} = k_{pm} = k_{ps} = 0.25$, $\alpha_m = 1$, $\alpha_s = 0.9$, $\bar{k}_e = 1$, $\bar{k}_h = 1$. If the upper bound of forward delay is fixed as 6, stability is guaranteed for $d_{s2} = 8.9860$. However, from Table 4.3, one can see that an upper bound on the delay for stability of teleoperation system is obtained to be 9.9287 with Theorem 4.1 in this chapter.

For bilateral teleoperator with torque feedback controller, the parameters are chosen as $\alpha_m = \alpha_s = 100$, $k_m = k_s = 50$. The example was treated in [5]. It is assumed that the upper bound of forward delay is fixed as $d_{m2} = 0.2$. In [5], the obtained maximum backward delay bound is 0.6271. However, from Table 4.4, we can see that they are improved to 1.1380 by Theorems 4.4 and 4.5 in this chapter. It turns out that the criteria in this chapter outperform some existing results.

Besides, for interval time-varying delay of teleoperation system, there are four parameters of delay, the lower and upper bound of forward delay d_{m1} and d_{m2}, respectively and the lower and upper bound of backward delay d_{s1} and d_{s2}, respectively. We can obtain larger upper bound of delay by increasing the lower bound of delay. For example, when $\alpha_m = 1$, $k_m = k_s = 0.25$, $\alpha_s = 0.9$, if $d_m(t) \in [0, 1]$, we can calculate that $d_s(t) \in [0, 6.5956]$. But the result can be improved when the lower bound of delays are increased continuously and finally we obtain that $d_m(t) \in [0.9, 1]$ and $d_s(t) \in [6.5, 6.6667]$.

Example 4.2 To demonstrate the performance of our result, let us consider two degrees of freedom robotic manipulators, and we assume that there exists a pair of master and slave in teleoperation system.

In the simulation, choose $k_{dm} = k_{ds} = k_{pm} = k_{ps} = 0.25$, $\alpha_m = 1$, $\alpha_s = 0.9$, $\bar{k}_e = 1$, $\bar{k}_h = 1$, $k_h = 5$, $k_0 = \begin{bmatrix}20\ 10\end{bmatrix}^T$ and the parameters of the manipulators are

Table 4.2 Calculated maximal upper bound of backward delay for $d_{2m} = 1$ of Example 4.1

References	d
Ghorbanian et al. [20]	2.7947
Nuño et al. [13]	5.2726
Hua and Liu [5]	6.5894
Theorem 4.2	6.6667

Table 4.3 Calculated maximal upper bound of backward delay for $d_{2m} = 6$ of Example 4.1

Reference	d
Hua and Liu [19]	8.9860
Theorem 4.1	9.9287

Table 4.4 Calculated maximal upper bound of backward delay for $d_{2m} = 0.2$ of Example 4.1

Reference	d
Hua and Liu [5]	0.6271
Theorem 4.4	1.1380

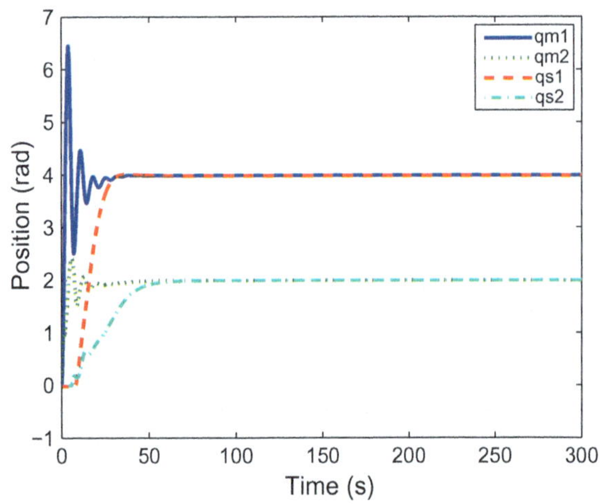

Fig. 4.1 Positions of master and slave

$m_1 = 10\,\text{kg}$, $m_2 = 5\,\text{kg}$, $l_1 = 0.7\,\text{m}$, $l_2 = 0.5\,\text{m}$, $g = 9.81\,\text{m/s}^2$. The initial values of velocities and positions are all zero. With the lower bounds of forward and backward delays $d_{1m} = 5.9$, $d_{1s} = 9.7$, calculating condition (4.4) in Theorem 4.1 gives the maximum allowable upper bounds are 6 and 9.9287, respectively. Figure 4.1 shows the positions of master and slave. Figure 4.2 shows the position error. From that, we can see that the closed-loop teleoperation system is asymptotically stable and the slave site tracks the movement of master site.

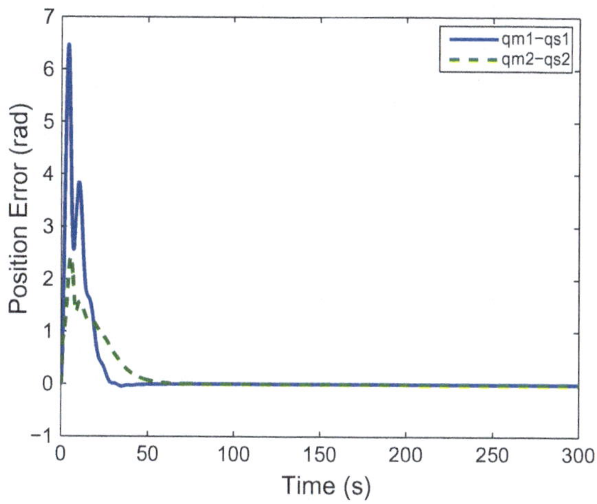

Fig. 4.2 Position error between two robotics

4.5 Summary

In this chapter, the synchronization problem of nonlinear teleoperation systems in the presence of variable time delay is studied. Different from some previous works, the range of varying delay is no longer from zero to an upper bound. Because delays in practical systems are often in an interval, we relax the restriction. Three classes of architectures are involved. As shown in simulation part, the criteria derived in this chapter improve over some previous ones in that the computed admissible upper bound of time delay is larger. In addition, all of results are given in form of LMI, which makes us calculate easily by Matlab.

References

1. Z.J. Li, X.Q. Cao, N. Ding, Adaptive fuzzy control for synchronization of nonlinear teleoperators with stochastic time-varying communication delays. IEEE Trans. Fuzzy Syst. **19**(4), 745–757 (2011)
2. C. Mitsantisuk, K. Ohishi, S. Katsura, Estimation of action/reaction forces for the bilateral control using kalman filter. IEEE Trans. Ind. Elecreon. **59**(11), 4383–4393 (2012)
3. I.G. Polushin, X.P. Liu, C.H. Lung, Stability of bilateral teleoperators with generalized projection-based force reflection algorithms. Automatica **48**(6), 1005–1016 (2012)
4. H.J. Yang, Y.Q. Xia, P. Shi, Stabilization of networked control systems with nonuniform random sampling periods. Int. J. Robust Nonlinear Control **21**(5), 501–526 (2011)
5. C.C. Hua, X.P. Liu, Delay-dependent stability criteria of teleoperation systems with asymmetric time-varying delays. IEEE Trans. Robot. **26**(5), 925–932 (2010)
6. R.J. Anderson, M.W. Spong, Bilateral control of teleoperators with time delay. IEEE Trans. Autom. Control **34**(5), 494–501 (1989)
7. G. Niemeyer, J.E. Slotine, Stable adaptive teleoperation. IEEE J. Ocean. Eng. **16**(1), 152–162 (1991)

8. S. Islam, P.X. Liu, Pd output feedback control design for industrial robotic manipulators. IEEE/ASME Trans. Mechatron. **16**(1), 187–197 (2011)
9. H.J. Gao, T.W. Chen, H-infinity estimation for uncertain systems with limited communication capacity. IEEE Trans. Autom. Control **52**(11), 2070–2084 (2007)
10. D. Lee, M.W. Spong, Passive bilateral teleoperation with constant time delay. IEEE Trans. Robot. Autom. **22**(2), 269–281 (2006)
11. K.W. Lee, H.K. Khalil, Adaptive output feedback control of robot manipulators using high-gain observer. Int. J. Control **67**(6), 869–886 (1997)
12. C.C. Hua, X.P. Liu, Teleoperation over the internet with/without velocity signal. IEEE Trans. Instrumen. Measurement **60**, 4–13 (2011)
13. E. Nuño, L. Basañez, R. Ortega, M.W. Spong, Position tracking for non-linear teleoperators with variable time delay. Int. J. Robot. Res. **28**(7), 895–910 (2009)
14. E. Nuño, R. Ortega, N. Barabanov, L. Basañez, A globally stable pd controller for bilateral teleoperators. IEEE Trans. Robot. **24**(3), 753–758 (2008)
15. E.J. Rodrigue-Seda, J.J. Troy et al., Bilateral teleoperation of multiple mobile agents: coordinated motion and collision avoidance. IEEE Trans. Control Syst. Technol. **18**(4), 984–992 (2010)
16. P.G. Park, J.W. Ko, C. Jeong, Reciprocally convex approach to stability of systems with time-varying delays. Automatica **47**(1), 235–238 (2011)
17. A. Seuret, F. Gouaisbaut, Wirtinger-based integral inequality: application to time-delay systems. Automatica **49**(9), 2860–2866 (2013)
18. Y. He, Q.G. Wang, L.H. Xie, C. Lin, Further improvement of free-weighting matrices technique for systems with time-varying delay. IEEE Trans. Autom. Control **52**(2), 293–299 (2007)
19. C.C. Hua, P.X. Liu, Convergence analysis of teleoperation systems with unsymmetric time-varying delays. IEEE Trans. Circuits Syst. II: Express Briefs **56**(3), 240–244 (2009)
20. A. Ghorbanian, S.M. Rezaei, A.R. Khoogar et al., A novel control framework for nonlinear time-delayed dual-master/single-slave teleoperation. ISA Trans. **52**(2), 268–277 (2013)

Chapter 5
Stability Analysis of Teleoperation Systems with Quantization

Abstract The stability problem is studied for teleoperation systems over general communication networks. Compared with previous work, both quantization and time delay issues are considered. The controller used in this chapter is in the form of proportional-derivative, and the output signals of master and slave systems are quantized before being transmitted. The stability criteria are presented to show that the controller can stabilize the master-slave system under quantization and variable time delay. Additionally, we propose a quantized measurement, which can decrease the quantization error. Finally, simulations and detailed analysis are given to show the effectiveness of the main results.

5.1 Introduction

A typical teleoperation system is composed of five elements: the human operator, the master robot, the communication channel, the slave robot and the environment. The human operator exerts torque on the master robot, which connected with the slave robot by a communication channel. At the same time, the slave robot interacts with the environment. In the bilateral teleoperation system, the master transmits its information to the slave, while the slave robot operates on the external environment and sends the slave's real-time information back to the master. In recent years, more and more researchers have focused on the teleoperation design owing to its wide range of applications, such as space and undersea exploration, handling of hazardous objects and robotic telesurgery [1–3].

The master and the slave sites exchange the information through communication networks, and then the time delays are unavoidable. Since time delays often lead to undesirable performance and even the instability, it is significant to consider it in system design. In the networks, another phenomenon is quantization. This is mainly due to the widespread use in control systems of digital computers that employ finite-precision arithmetic. The measurement and command signals should be quantized before being transmitted [4–6].

C. Hua et al., *Analysis and Design for Networked Teleoperation System*,
https://doi.org/10.1007/978-981-13-7936-9_5

There have been a great number of research results concerning bilateral teleoperation system with time delay. In [7], adaptive teleoperation controllers under time varying delay were designed to guarantee asymptotically stable. Reference [8] proposed a passive control framework for nonlinear robotic teleoperators with constant time delays. The controller designed satisfied demand of stability and transparency. Based on it, [9] extended the number of slave robots to N and considered collision avoidance at the same time. In [10], we addressed the problem of controller design for teleoperation when the velocity information was not available. The delay was asymmetric and time-varying. [11] presented a simple PD scheme with constant time delay, and it was extended in [12] with time-varying delay. Additionally, in [12], P-like controller, PD-like controller for variable time delay and scattering transformation were all designed. In [13], the adaptive fuzzy method was proposed for the teleoperation design with the consideration of the stochastic time-varying delays. However, in many previous works, they didn't consider the effect of quantization. Since it can not be neglected in networks, we address the stability problems associated with both time delays and quantization. The controller is similar to our work in [14], but the signals are quantized before they are transmitted to the networks. In the simulation section, we analyze the effect of quantization.

5.2 Problem Statement

Let us also consider a master-slave bilateral teleoperation system given by the following model:

$$\begin{cases} M_m(q_m)\ddot{q}_m + C_m(q_m, \dot{q}_m)\dot{q}_m + G_m(q_m) = \tau_m + F_h \\ M_s(q_s)\ddot{q}_s + C_s(q_s, \dot{q}_s)\dot{q}_s + G_s(q_s) = \tau_s - F_e \end{cases} \tag{5.1}$$

The following lemma and assumptions will be used in this chapter.

Lemma 5.1 ([5]) *For any vector signals x, y, any variable time delay $0 \leq d(t) \leq \bar{d} < \infty$ and any constant $\alpha > 0$ we have that*

$$-2\int_0^t x^T(\sigma)\int_{-d(\sigma)}^0 y(\sigma+\theta)\,d\theta d\sigma \leq \alpha\,\|x\|_2^2 + \frac{\bar{d}^2}{\alpha}\,\|y\|_2^2$$

Assumption 5.1 The human operator and the environment are passive, i.e.,

$$\int_0^t -\dot{q}_m^T(\sigma)\,F_h(\sigma)\,d\sigma \geq 0, \int_0^t \dot{q}_s^T(\sigma)\,F_e(\sigma)\,d\sigma \geq 0$$

Assumption 5.2 The time delays $d_m(t)$ and $d_s(t)$ are bounded, i.e., there exist positive scalars $\bar{d}_m$ and $\bar{d}_s$ such that $d_m(t) \leq \bar{d}_m$ and $d_s(t) \leq \bar{d}_s$.

The quantization strategies can be mainly categorized depending on whether the quantizer is static or dynamic. Static strategies presume that data quantization at time k is dependent on the data at time k only. Logarithmic quantizer is a typically static quantizer [15, 16]. The attraction of static quantizer is the simplicity of their coding/decoding schemes. But the main drawback is that they require an infinite number of quantization bits to ensure asymptotic stability. The second strategy is dynamic quantization strategy [17]. It has the advantage to scale the quantization levels dynamically in order to increase the region of attraction and to attenuate the steady state limit cycle. In some literatures, most dynamic quantizers employed are those use dynamic scaling in conjunction with static quantizers.

Let $z \in R^l$ be the variable being quantized. The quantizer $f(\cdot)$ is defined as a piecewise constant function $f : R^l \rightarrow \Im$, where $\Im$ is a finite subset of R^l. This leads to a partition of R^l into a finite number of quantization regions of the form $\{z \in R^l : f(z) = i\}, i \in \Im$. It is assumed that the quantizer $f(z)$ satisfies the following conditions:

$$\begin{aligned} &(1)\,\text{If } |z| \le M, \text{ then } |f(z) - z| \le \Delta \\ &(2)\,\text{If } |z| \ge M, \text{ then } |f(z)| \ge M - \Delta \end{aligned} \tag{5.2}$$

where M and Δ are the quantization range and quantization error of $f(\cdot)$, respectively. When z does not belong to the union of quantization regions of finite size, the *quantizer saturates*. Condition (1) gives a bound on the quantization error when the quantizer does not saturate. Condition (2) provides a way to detect the possibility of saturation. We use quantized measurements of the form

$$f_\mu(z) = \mu f\left(\frac{z}{\mu}\right) \tag{5.3}$$

where $\mu > 0$. The range of this quantizer is $M\mu$ and the quantization error is $\Delta\mu$. We can think of μ as the "zoom" variable: increasing μ corresponds to zooming out and essentially obtaining a new quantizer with larger range and quantization error, whereas decreasing μ corresponds to zooming in and obtaining a quantizer with a smaller range but also a smaller quantization error.

5.3 Main Results

In this chapter, we will consider the stability problem with the following controller (see Fig. 5.1):

$$\begin{aligned} \tau_m &= -k_m[q_m - f_{\mu_2}(q_s(t - d_s(t)))] - \alpha_m \dot{q}_m \\ \tau_s &= k_s[f_{\mu_1}(q_m(t - d_m(t))) - q_s] - \alpha_s \dot{q}_s \end{aligned} \tag{5.4}$$

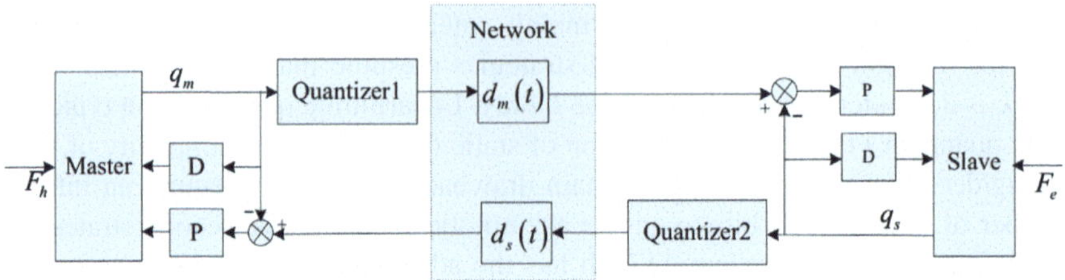

Fig. 5.1 The bilateral teleoperation system

where $d_m(t)$ is time delay from master to slave site and $d_s(t)$ is the opposite, k_m and k_s are proportional coefficients, and α_m and α_s are damping coefficients. $f_{\mu_2}(\cdot)$ and $f_{\mu_1}(\cdot)$ are quantization strategies.

With (5.3) and (5.4), we have the following main results.

Theorem 5.1 *If there exist positive scalars α_j, $\bar{d}_j$, $j = 1, 2$, k_i, α_i, $i = m, s$ and the quantization parameters M_j and Δ_j such that the following conditions hold:*

$$\begin{bmatrix} -\alpha_m + \frac{k_m\alpha_2}{2} & \sqrt{k_m}\bar{d}_m & \sqrt{k_m\Delta_2} \\ * & -2\alpha_1 & 0 \\ * & * & -M_2 \end{bmatrix} < 0 \tag{5.5}$$

$$\begin{bmatrix} -\frac{k_m\alpha_s}{k_s} + \frac{k_m\alpha_1}{2} & \sqrt{k_m}\bar{d}_s & \sqrt{k_m\Delta_1} \\ * & -2\alpha_2 & 0 \\ * & * & -M_1 \end{bmatrix} < 0 \tag{5.6}$$

Then the velocities $\dot{q}_m$, $\dot{q}_s$ and position error $q_m - q_s$ are bounded.

Proof We choose the following Lyapunov functional

$$\begin{aligned} V_1 =& \frac{1}{2}\dot{q}_m^T M_m(q_m)\dot{q}_m + \frac{k_m}{2k_s}\dot{q}_s^T M_s(q_s)\dot{q}_s \\ &+ (U_m(q_m) - \beta_m) + \frac{k_m}{k_s}(U_s(q_s) - \beta_s) \\ &- \int_0^t (\dot{q}_m^T(\sigma) F_h(\sigma) - \frac{k_m}{k_s}\dot{q}_s^T(\sigma) F_e(\sigma))d\sigma \\ V_2 =& \frac{k_m}{2}(q_m - q_s)^T(q_m - q_s) \end{aligned} \tag{5.7}$$

According to Assumption 5.1, we know that $V_1 > 0$, then, $V = V_1 + V_2 > 0$. With Properties 1.2 and 1.3, the time derivative of V is

$$\begin{aligned} \dot{V} =& \dot{q}_m^T\left(k_m f_{\mu_2}(q_s(t - d_s(t))) - \alpha_m\dot{q}_m\right) + \frac{k_m}{k_s}\dot{q}_s^T \\ &\times \left(k_s f_{\mu_1}(q_m(t - d_m(t))) - \alpha_s\dot{q}_s\right) - k_m\left(\dot{q}_m^T q_s + \dot{q}_s^T q_m\right) \end{aligned}$$

$$
\begin{aligned}
&= -\alpha_m \dot{q}_m^T \dot{q}_m - \frac{k_m \alpha_s}{k_s} \dot{q}_s^T \dot{q}_s + k_m \dot{q}_m^T [f_{\mu_2}(q_s(t - d_s(t))) - \\
&\quad q_s(t - d_s(t)) + q_s(t - d_s(t)) - q_s(t)] + k_m \dot{q}_s^T \\
&\quad \times [f_{\mu_1}(q_m(t - d_m(t))) - q_m(t - d_m(t)) \\
&\quad + q_m(t - d_m(t)) - q_m(t)]
\end{aligned}
$$

By (5.2) and (5.3), we can obtain that with $\left| \frac{q_s(t - d_s(t))}{\mu_2} \right| \le M_2$ and $\left| \frac{q_m(t - d_m(t))}{\mu_1} \right| \le M_1$

$\left| f_{\mu_2}(q_s(t - d_s(t))) - q_s(t - d_s(t)) \right| = \left| \mu_2 \left(f\left(\frac{q_s(t - d_s(t))}{\mu_2} \right) - \frac{q_s(t - d_s(t))}{\mu_2} \right) \right|$ $\le \mu_2 \Delta_2$, similarly, $\left| f_{\mu_1}(q_m(t - d_m(t))) - q_m(t - d_m(t)) \right| \le \mu_1 \Delta_1$.

Because $q_s(t - d_s(t)) - q_s(t) = -\int_{-d_s(t)}^{0} \dot{q}_s(t + \theta)\, d\theta$ and $q_m(t - d_m(t)) - q_m(t) = -\int_{-d_m(t)}^{0} \dot{q}_m(t + \theta)\, d\theta$,

$$
\begin{aligned}
\dot{V} &\le -\alpha_m \dot{q}_m^T \dot{q}_m - \frac{k_m \alpha_s}{k_s} \dot{q}_s^T \dot{q}_s + k_m |\dot{q}_m| \mu_2 \Delta_2 \\
&\quad - k_m \dot{q}_m^T \int_{-d_s(t)}^{0} \dot{q}_s(t + \theta)\, d\theta \\
&\quad + k_m |\dot{q}_s| \mu_1 \Delta_1 - k_m \dot{q}_s^T \int_{-d_m(t)}^{0} \dot{q}_m(t + \theta)\, d\theta
\end{aligned}
$$

Integrating from zero to t with Lemma 5.1, and its derivative is

$$
\begin{aligned}
\dot{V} \le &- \left(\alpha_m - \frac{k_m \alpha_2}{2} - \frac{k_m \bar{d}_m^2}{2\alpha_1} \right) |\dot{q}_m| \\
&\times \left(|\dot{q}_m| - \frac{k_m \mu_2 \Delta_2}{\alpha_m - \frac{k_m \alpha_2}{2} - \frac{k_m \bar{d}_m^2}{2\alpha_1}} \right) \\
&- \left(\frac{k_m \alpha_s}{k_s} - \frac{k_m \bar{d}_s^2}{2\alpha_2} - \frac{k_m \alpha_1}{2} \right) |\dot{q}_s| \\
&\times \left(|\dot{q}_s| - \frac{k_m \mu_1 \Delta_1}{\frac{k_m \alpha_s}{k_s} - \frac{k_m \bar{d}_s^2}{2\alpha_2} - \frac{k_m \alpha_1}{2}} \right)
\end{aligned}
$$

From (5.5) and (5.6), we can always find scalars $\varepsilon_1, \varepsilon_2 \in (0, 1)$, such that

$$M_2 > \frac{k_m \Delta_2}{\alpha_m - \frac{k_m \alpha_2}{2} - \frac{k_m \bar{d}_m^2}{2\alpha_1}} \times \frac{1}{1-\varepsilon_2},$$

$$M_1 > \frac{k_m \Delta_1}{\frac{k_m \alpha_s}{k_s} - \frac{k_m \bar{d}_s^2}{2\alpha_2} - \frac{k_m \alpha_1}{2}} \times \frac{1}{1-\varepsilon_1}$$

Therefore, for any nonzero $\dot{q}_m$ and $\dot{q}_s$, we can find positive scalars μ_1 and μ_2 such that

$$M_2\mu_2 > |\dot{q}_m| > \frac{k_m \Delta_2 \mu_2}{\alpha_m - \frac{k_m \alpha_2}{2} - \frac{k_m \bar{d}_m^2}{2\alpha_1}} \times \frac{1}{1-\varepsilon_2} \tag{5.8}$$

$$M_1\mu_1 > |\dot{q}_s| > \frac{k_m \Delta_1 \mu_1}{\frac{k_m \alpha_s}{k_s} - \frac{k_m \bar{d}_s^2}{2\alpha_2} - \frac{k_m \alpha_1}{2}} \times \frac{1}{1-\varepsilon_1} \tag{5.9}$$

Notice that $1 = \frac{1}{1-\varepsilon_2} - \frac{\varepsilon_2}{1-\varepsilon_2}$.
Furthermore, we get

$$|\dot{q}_m| - \frac{k_m \mu_2 \Delta_2}{\alpha_m - \frac{k_m \alpha_2}{2} - \frac{k_m \bar{d}_m^2}{2\alpha_1}} > \varepsilon_2 |\dot{q}_m| \tag{5.10}$$

Similarly, $|\dot{q}_s| - \frac{k_m \mu_1 \Delta_1}{\frac{k_m \alpha_s}{k_s} - \frac{k_m \bar{d}_s^2}{2\alpha_2} - \frac{k_m \alpha_1}{2}} > \varepsilon_1 |\dot{q}_s|$.
As mentioned above,

$$\begin{aligned} V(t) - V(0) \leq & -\varepsilon_2 \left(\alpha_m - \frac{k_m \alpha_2}{2} - \frac{k_m \bar{d}_m^2}{2\alpha_1} \right) \|\dot{q}_m\|_2^2 \\ & -\varepsilon_1 \left(\frac{k_m \alpha_s}{k_s} - \frac{k_m \bar{d}_s^2}{2\alpha_2} - \frac{k_m \alpha_1}{2} \right) \|\dot{q}_s\|_2^2 \end{aligned}$$

Conditions (5.5) and (5.6) imply that $\alpha_m - \frac{k_m \alpha_2}{2} - \frac{k_m \bar{d}_m^2}{2\alpha_1} > 0$ and $\frac{k_m \alpha_s}{k_s} - \frac{k_m \bar{d}_s^2}{2\alpha_2} - \frac{k_m \alpha_1}{2} > 0$, we can obtain $V(t) \leq V(0)$. That completes the proof.

Theorem 5.2 *If human-input force F_h and environment force F_e are zero, the following controllers*

$$\tau_m = -k_m[q_m - f_{\mu_2}(q_s(t-d_s(t)))] - \alpha_m\dot{q}_m + G_m(q_m)$$
$$\tau_s = k_s[f_{\mu_1}(q_m(t-d_m(t))) - q_s] - \alpha_s\dot{q}_s + G_s(q_s)$$

render velocities $\dot{q}_m, \dot{q}_s$ and position error $q_m - q_s$ converge to zero asymptotically.

Proof Choose Lyapunov functional (5.7) without $(U_m(q_m) - \beta_m) + \frac{k_m}{k_s}(U_s(q_s) - \beta_s)$ in V_1. According to Theorem 5.1, we know that $\dot{q}_m, \dot{q}_s, q_m - q_s \in \ell_\infty$ and $\dot{q}_m, \dot{q}_s \in \ell_2$. With (5.1), we can obtain that

$$\begin{aligned}\ddot{q}_m &= M_m^{-1}(q_m)[-C_m(q_m, \dot{q}_m)\dot{q}_m - k_m(q_m - \mu_2 f\left(\frac{q_s(t-d_s(t))}{\mu_2}\right)) - \alpha_m\dot{q}_m] \\ &= M_m^{-1}(q_m)[-C_m(q_m, \dot{q}_m)\dot{q}_m - k_m(q_m - q_s(t-d_s(t))) \\ &+ k_m(\mu_2 f\left(\frac{q_s(t-d_s(t))}{\mu_2}\right) - q_s(t-d_s(t))) - \alpha_m\dot{q}_m]\end{aligned}$$

By Properties 1.1 and 1.4 and $q_m - q_s(t-d_s(t)) = q_m - q_s + \int_{t-d_s(t)}^{t}\dot{q}_s(\xi)\,d\xi$, $\left|\mu_2 f\left(\frac{q_s(t-d_s(t))}{\mu_2}\right) - q_s(t-d_s(t))\right| \le \mu_2\Delta_2$, we have $\ddot{q}_m$ is bounded. Based on Barbalat's lemma, we know that $\dot{q}_m \to 0$ as $t \to \infty$, and similarly, $\dot{q}_s \to 0$ as $t \to \infty$.

Next, we will prove that the tracking error $q_m - q_s$ converges to zero asymptotically.

The time derivative of $\ddot{q}_m$ is

$$\begin{aligned}\dddot{q}_m =& \frac{dM_m^{-1}(q_m)}{dt}[-C_m(q_m, \dot{q}_m)\dot{q}_m - k_m(q_m - \mu_2 f\left(\frac{q_s(t-d_s(t))}{\mu_2}\right)) - \alpha_m\dot{q}_m] \\ &+ M_m^{-1}(q_m)[-C_m(q_m, \dot{q}_m)\ddot{q}_m - \dot{C}_m(q_m, \dot{q}_m)\dot{q}_m \\ &- k_m(\dot{q}_m - \frac{d\left(\mu_2 f\left(\frac{q_s(t-d_s(t))}{\mu_2}\right)\right)}{dt}) - \alpha_m\ddot{q}_m]\end{aligned}$$

For the first term, we have

$$\begin{aligned}\frac{dM_m^{-1}(q_m)}{dt} &= -M_m^{-1}(q_m)\dot{M}_m(q_m)M_m^{-1}(q_m) \\ &= -M_m^{-1}(q_m)[C_m(q_m, \dot{q}_m) + C_m^T(q_m, \dot{q}_m)]M_m^{-1}(q_m)\end{aligned}$$

which is clearly bounded because of Properties 1.1 and 1.4. Furthermore, by Properties 1.1, 1.4 and 1.5, we have $\dddot{q}_m \in \ell_\infty$. Based on Barbalat's lemma, we know $\ddot{q}_m \to 0$, and further, $q_m - q_s \to 0$ as $t \to \infty$. That completes the proof.

We give detailed proof on how to get (5.10). The key transformation is $1 = \frac{1}{1-\varepsilon_2} - \frac{\varepsilon_2}{1-\varepsilon_2}$. By (5.8), we have $|\dot{q}_m| - \frac{k_m \Delta_2 \mu_2}{\alpha_m - \frac{k_m \alpha_2}{2} - \frac{k_m \bar{d}_m^2}{2\alpha_1}} \times \frac{1}{1-\varepsilon_2} > 0$, which is equivalent to $\frac{1}{1-\varepsilon_2}(|\dot{q}_m| - \frac{k_m \Delta_2 \mu_2}{\alpha_m - \frac{k_m \alpha_2}{2} - \frac{k_m \bar{d}_m^2}{2\alpha_1}}) - \frac{\varepsilon_2}{1-\varepsilon_2}|\dot{q}_m| > 0$, and further, we get (5.10).

The quantization effect in the communication channel for teleoperation systems is considered in this chapter. That is important in real world application. Figure 5.1 shows that the command signals q_m, q_s have been quantized before they are transmitted to network. Because of the advantage of dynamic quantizer, we choose the dynamic quantizer.

Conditions (5.5) and (5.6) are equal to $\frac{k_m \Delta_2}{M_2} < \alpha_m - \frac{k_m \alpha_2}{2} - \frac{k_m \bar{d}_m^2}{2\alpha_1}$ and $\frac{k_m \Delta_1}{M_1} < \frac{k_m \alpha_s}{k_s} - \frac{k_m \bar{d}_s^2}{2\alpha_2} - \frac{k_m \alpha_1}{2}$. We can calculate $\bar{d}_m$ if and only if all the parameters were given. However, they are presented in the form of LMIs. When $\alpha_i, M_i, \Delta_i, \bar{d}_i, i = m, s$ are all unknown, we can also solve them by the numerically efficient Matlab LMI toolbox.

5.4 Simulation Example

In this section, some simulations are showed to verify the effectiveness of our results.

The simulations are performed on 2-degree-of-freedom manipulators. We apply a human force F to the master site in the Y-direction (see Fig. 5.2). In the slave site,

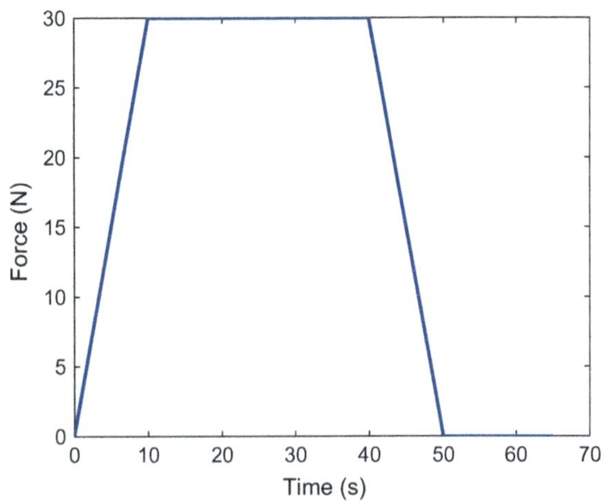

Fig. 5.2 Human-force input F

there is a wall at $y = 0.5m$, when the slave joint reaches the wall and moves further, the feedback force is $10000 * (y - 0.5)\ N$. The torque $F_h = -J_m^T * [0\ 1]^T * F$ and $F_e = -J_s^T * [0\ 1]^T * 10000 * (y - 0.5)$, where J_m and J_s are the Jacobian matrices of the master and slave manipulators, respectively.

The parameters are chosen as $m_1 = 10\,\text{kg}$, $m_2 = 5\,\text{kg}$, $L_1 = 0.7\,\text{m}$, $L_2 = 0.5m$, $g = 9.81\,\text{m/s}^2$, $\alpha_m = k_m = k_s = 100$, $\alpha_s = 200$, $M_i = 5$, $\Delta_i = 0.001$, $\alpha_i = 1$, $i = 1, 2$ and initial conditions are $q_{m1} = q_{m2} = \dot{q}_{m1} = \dot{q}_{m2} = 0$, $q_{s1} = q_{s2} = \dot{q}_{s1} = \dot{q}_{s2} = 0$. When the quantizer doesn't saturate, we have the following measurement

$$\mu q\left(\frac{z}{\mu}\right) = \mu\Delta\left[\frac{z}{\mu\Delta}\right], -(M+0.5)\Delta \leq \frac{z}{\mu} \leq (M+0.5)\Delta \tag{5.11}$$

$$\mu = \begin{cases} 8, -(M+0.5)\Delta \leq \frac{z}{8} \leq (M+0.5)\Delta \\ \left\{\frac{z}{(M+0.5)\Delta}\right\}, \text{ others} \end{cases} \tag{5.12}$$

where $[\cdot]$ is the operation of round, $\left\{\frac{z}{(M+0.5)\Delta}\right\}$ means the round number which is greater than or equal to $\frac{z}{(M+0.5)\Delta}$. The basic idea of the measurement is, when $\frac{z}{\mu} > (M+0.5)\,\Delta$ or $\frac{z}{\mu} < -(M+0.5)\,\Delta$, we adjust μ dynamically in order to let $\frac{z}{\mu}$ in the interval $[-(M+0.5)\,\Delta, (M+0.5)\,\Delta]$. That improves the performance of quantizer and decreases the quantization error.

Now we do the simulation to verify the performance of system.

The human operator exerts torque on the master site during the first ten seconds. The force is 0 at 0s and then increases to 30 N at 10 s. As a result, Y-Position of

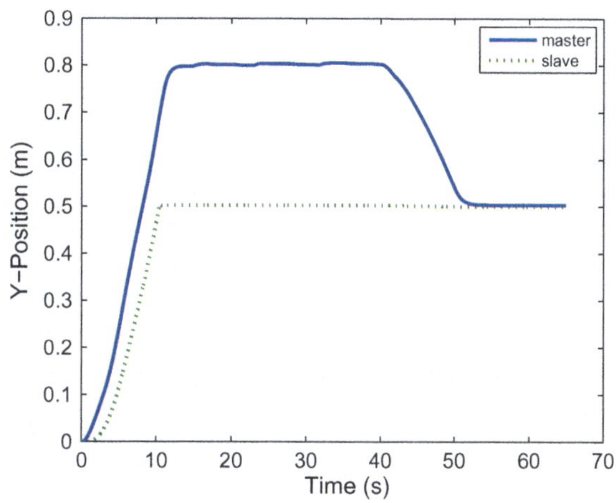

Fig. 5.3 Y-Position of master and slave with $\Delta = 0.001$

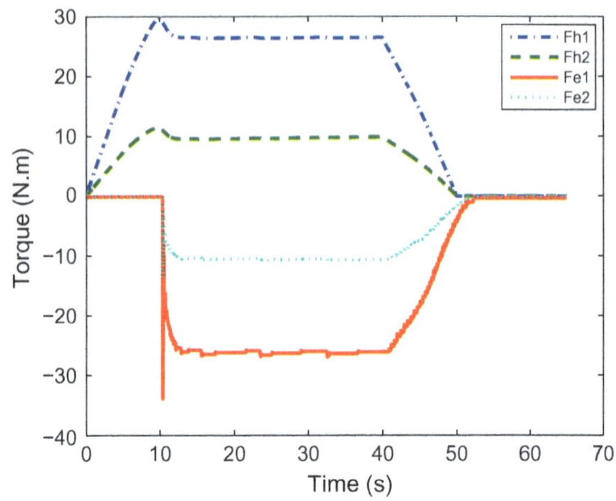

Fig. 5.4 Torque of F_h and F_e with $\Delta = 0.001$

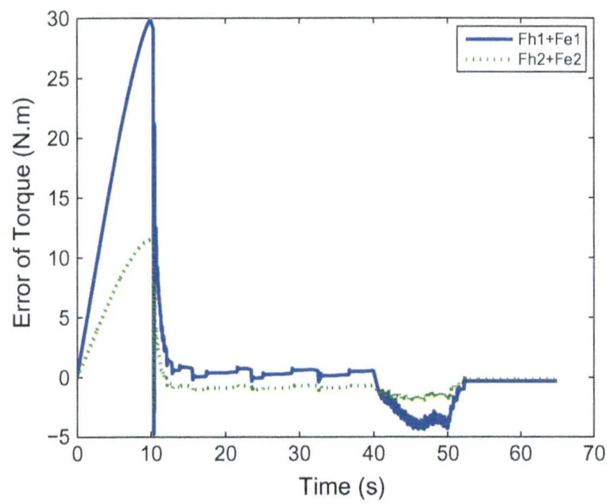

Fig. 5.5 Torque error between master and slave with $\Delta = 0.001$

master moves from 0 to 0.6542 m and Y-Position of slave moves from 0 to 0.4661 m. We can see that the slave joint tracks the master joint. From 10 s to 40 s, the human force keeps constant and the master remains the former values. The human force decreases to zero from 40 to 50 s and Y-Position of master decreases also. After 50 s, the human force is 0N and the error between master site and slave site is zero (see Fig. 5.3).

From 0 to 10.38s, the slave joint doesn't touch the obstacle and $F_e = 0$. In 10.38 s, it reaches the wall and a large environmental-torque pulse appears. In Fig. 5.4, we can see that the numbers of F_h and F_e are equal but the orientations are opposite. At last, they all converge to zero. The errors between them are presented in Fig. 5.5. Figure 5.6 also shows tracking performance and stability of the system.

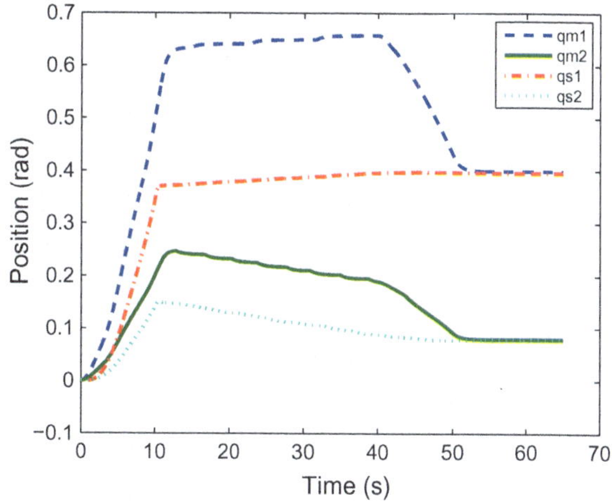

Fig. 5.6 Position of master and slave with $\Delta = 0.001$

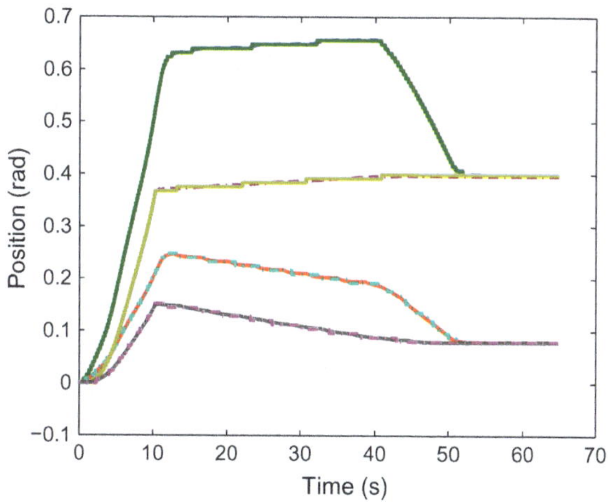

Fig. 5.7 Quantized signals and original signals of q_m and q_s with $\Delta = 0.001$

Next, we analyze the effect of quantization. Compared with our previous work [14], we can find that the figures are similar. It is because the quantization error is small. But the advantage couldn't be presented in the simulation. From Figs. 5.3, 5.4, 5.5 and 5.6, we can see that introducing quantization doesn't do harm to the performance of system. Figure 5.7 shows the quantized signals and original signals of q_m and q_s. In order to see the effect clearly, we enlarge the figure in Fig. 5.8 (from 0 to 5 s), Fig. 5.9 (from 40 to 50 s) and Fig. 5.10 (from 50 to 55 s). Additionally, from Fig. 5.10, we can see that the error between original signal and quantized signal converges to zero with time increasing.

What is the relationship between quantization and stability? In our previous work, we have calculated $\bar{d}_s = 1.4999$ s when $\bar{d}_m = 0.5$ s. With the same parameters, we obtain $\bar{d}_s = 0.9606$ s by (5.5)–(5.6). It is less than 1.4999. According to (5.6), we can

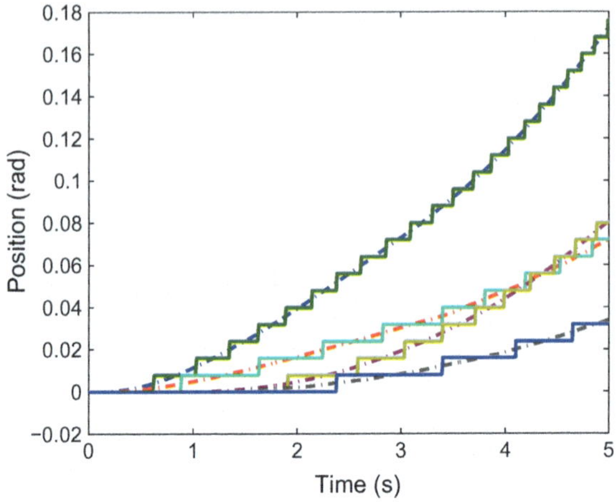

Fig. 5.8 Enlarged figure from 0 to 5 s

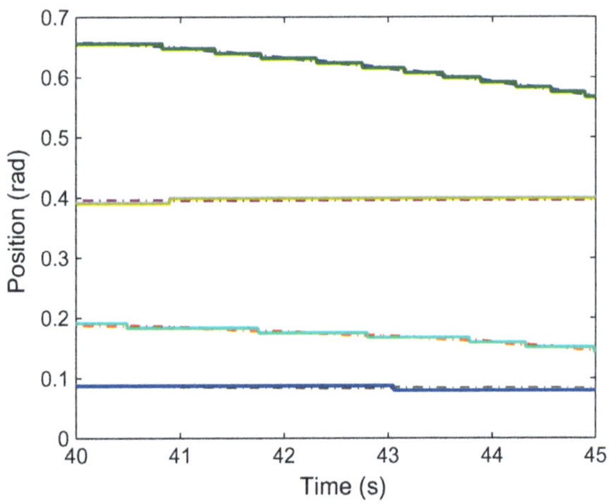

Fig. 5.9 Enlarged figure from 40 to 45 s

find that: increasing Δ_1 leads to larger quantization error and essentially obtaining smaller maximal upper bound of delay, whereas decreasing Δ_1 leads to smaller quantization error and essentially obtaining larger maximal upper bound.

How about choosing larger Δ? If we choose $\Delta = 0.05$, we can obtain the following data: $q_{m1} = 0.6712$, $f(q_{m1}) = 0.8\,(0.6720)$, $q_{m2} = 0.107$, $f(q_{m2}) = 0\,(0.1040)$, $q_{s1} = 0.4833$, $f(q_{s1}) = 0.4\,(0.4800)$, $q_{s2} = -0.1202$, $f(q_{s2}) = 0\,(-0.1200)$ at t=16.135s; $q_{m1} = 0.5647$, $f(q_{m1}) = 0.4\,(0.5680)$, $q_{m2} = 0.06632$, $f(q_{m2}) = 0\,(0.0640)$, $q_{s1} = 0.449$, $f(q_{s1}) = 0.4\,(0.4480)$, $q_{s2} = -0.07911$, $f(q_{s2}) = 0\,(-0.0800)$ at t = 45.362 s, where $f(q(\cdot))$ is the quantized number when $\Delta = 0.05$ and $(\cdot)$ is the one when $\Delta = 0.001$. Figures 5.11, 5.12, 5.13 and 5.14 show the performance of system when $\Delta = 0.05$. How to choose Δ depends on the precision of the data. From above, we can find that q_m and q_s are small. Most of them will be quantized to 0, such as 0.06632. However, if we choose $\Delta = 0.001$, the quantized

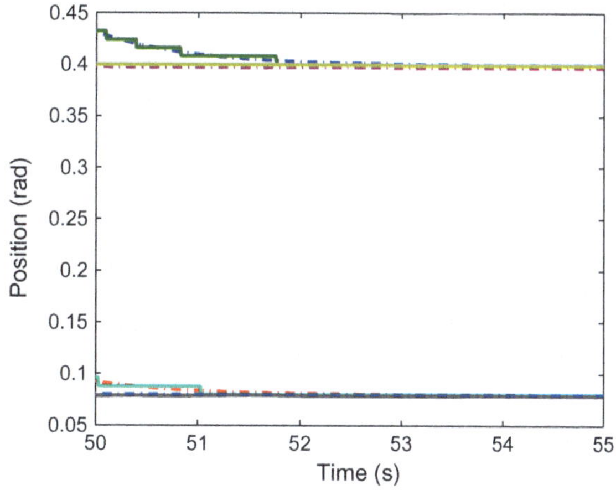

Fig. 5.10 Enlarged figure from 50 to 55 s

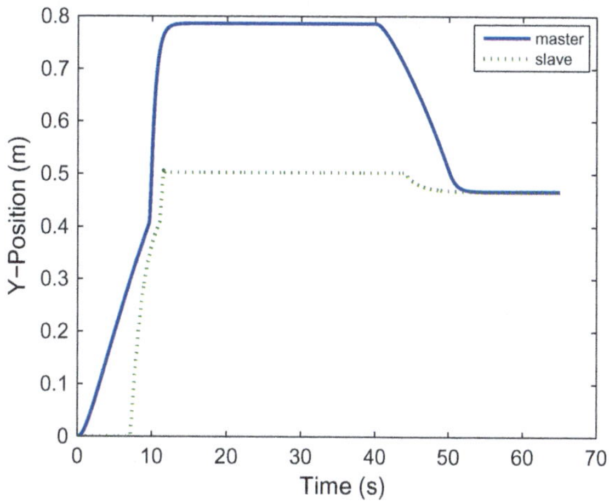

Fig. 5.11 Y-Position of master and slave with $\Delta = 0.05$

number will be 0.0640. From Fig. 5.11, we can see that slave joint doesn't track the master joint from 0 to 7.071 s. What's worse, if we choose $\Delta = 0.5$, slave joint doesn't track the master joint totally. Why? When $\Delta = 0.5$, $q_{m1} = 0.05695$, $f(q_{m1}) = 0\,(0.0560)$, $q_{m2} = 0.02387$, $f(q_{m2}) = 0\,(0.0240)$ at t = 2.546s; $q_{m1} = 0.3325$, $f(q_{m1}) = 0\,(0.3360)$, $q_{m2} = 0.134$, $f(q_{m1}) = 0\,(0.1360)$ at t = 19.748 s; $q_{m1} = 0.2557$, $f(q_{m1}) = 0\,(0.2560)$, $q_{m2} = 0.1039$, $f(q_{m1}) = 0\,(0.1040)$ at t = 43.469 s. Because q_m is small and Δ is relatively large, the quantized number is 0. Consequently, q_s is 0 and slave joint doesn't track the master joint. As mentioned above, we conclude that $\Delta = 0.001$ leads to better performance.

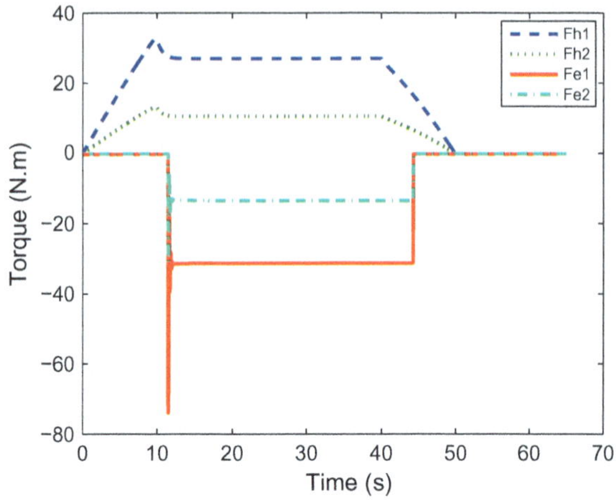

Fig. 5.12 Torque of F_h and F_e with $\Delta = 0.05$

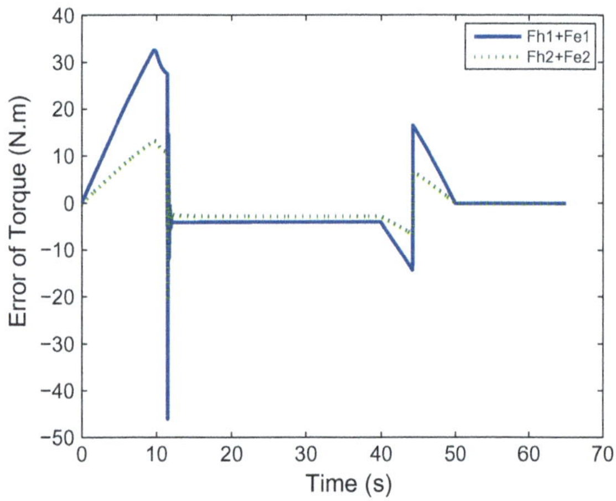

Fig. 5.13 Torque error between master and slave with $\Delta = 0.05$

5.5 Summary

In this chapter, the stability problem is studied for teleoperation systems in the presence of variable time delay and quantization. The framework uses the simple proportional-derivative control, i.e., the master and slave robots are directly connected via spring and damper over the delayed communication channels. The output signals of master and slave systems are quantized before transmission. New conditions are presented to show that the controller can stabilize the master-slave system under time-varying delay and quantization. Besides, we propose a quantized measurement, which decreases the quantization error. Simulations are presented to verify the effectiveness of the main results. In the future work, the model uncertainties will

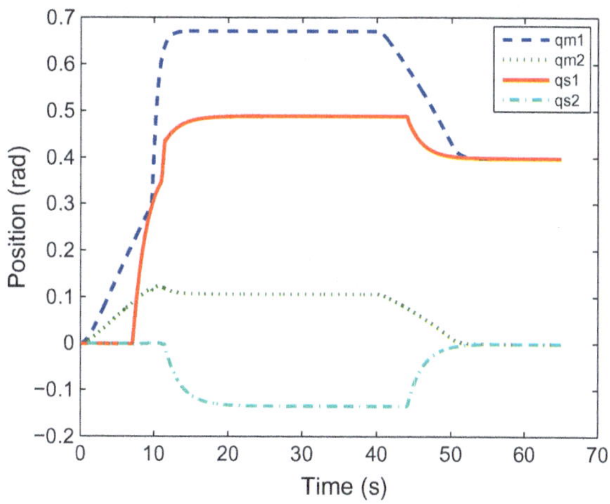

Fig. 5.14 Position of master and slave with $\Delta = 0.05$

be considered for the teleoperation design, and more communication effects will be investigated for the general network transmission channels.

References

1. S. Islam, P.X. Liu, Robust adaptive fuzzy output feedback control system for robot manipulators. IEEE/ASME Trans. Mechatron. **16**(2), 288–296 (2011)
2. I.G. Polushin, P.X. Liu, C.H. Lung et al., Position-error based schemes for bilateral teleoperation with time delay: theory and experiments. J. Dyn. Syst. Measurement Control **132**(3), 1–11 (2010)
3. I.G. Polushin, X.P. Liu, C.H. Lung, Stability of bilateral teleoperators with generalized projection-based force reflection algorithms. Automatica **48**(6), 1005–1016 (2012)
4. H.J. Gao, T.W. Chen, A new approach to quantized feedback control systems. Automatica **44**, 534–542 (2008)
5. H.J. Gao, T.W. Chen, J. Lam, A new delay system approach to network-based control. Automatica **44**, 39–52 (2008)
6. H.J. Gao, T.W. Chen, H-infinity estimation for uncertain systems with limited communication capacity. IEEE Trans. Autom. Control **52**(11), 2070–2084 (2007)
7. S. Islam, P.X. Liu, A.E.I. Saddik, Bilateral teleoperation system with time varying communication delay: stability and convergence. Auton. Intell. Syst. **6752**(2), 156–166 (2011)
8. D. Lee, M.W. Spong, Passive bilateral teleoperation with constant time delay. IEEE Trans. Robot. Autom. **22**(2), 269–281 (2006)
9. E.J. Rodrigue-Seda, J.J. Troy et al., Bilateral teleoperation of multiple mobile agents: coordinated motion and collision avoidance. IEEE Trans. Control Syst. Technol. **18**(4), 984–992 (2010)
10. C.C. Hua, X.P. Liu, Teleoperation over the internet with/without velocity signal. IEEE Trans. Instrum. Measurement **60**, 4–13 (2011)
11. E. Nuño, R. Ortega, N. Barabanov, L. Basañez, A globally stable pd controller for bilateral teleoperators. IEEE Trans. Robot. **24**(3), 753–758 (2008)
12. E. Nuño, L. Basañez, R. Ortega, M.W. Spong, Position tracking for non-linear teleoperators with variable time delay. Int. J. Robot. Res. **28**(7), 895–910 (2009)

13. Z.J. Li, Y.Q. Xia, F.C. Sun, Adaptive fuzzy control for multilateral cooperative teleoperation of multiple robotic manipulators under random network-induced delays. IEEE Trans. Fuzzy Syst. **22**(2), 437–450 (2014)
14. C.C. Hua, X.P. Liu, Delay-dependent stability criteria of teleoperation systems with asymmetric time-varying delays. IEEE Trans. Robot. **26**(5), 925–932 (2010)
15. M.Y. Fu, L.H. Xie, The sector bound approach to quantized feedback control. IEEE Trans. Autom. Control **50**, 1698–1711 (2005)
16. B. Zhou, G.R. Duan, J. Lam, On the absolute stability approach to quantized feedback control. Automatica **46**, 337–346 (2010)
17. G.N. Nair, R.J. Evans, Exponential stabilizability of finite-dimensional linear systems with limited data rates. Automatica **39**, 585–593 (2003)

Part II
High-Performance Controller Design for Teleoperation System

Chapter 6
Directed Force Feedback Control Design with Asymmetric and Time-Varying Delays

Abstract This chapter proposes a new control design method for nonlinear teleoperation systems with asymmetric and time-varying delays. Compared with previous P+d controller or the PD+d controller, delayed torque feedback information is applied to design the new control scheme, which renders the entire system stable with relatively good steady-state and transient-state performances. Proper Lyapunov Krasovskii functional are chosen to show the master-slave teleoperation system is stable under specific LMI conditions. Finally, simulation and experiments are both given to demonstrate the effectiveness of the designed control approach.

6.1 Introduction

For the design of a teleoperation system, the performances of transparency [1–4] and stability [5–9] were extensively investigated. For the enhancement of the transparency performance, the slave force feedback was often employed. For the stability analysis and control design of an Internet-based teleoperation system, the time delay is a crucial issue. With the introduction of computer communication networks into teleoperation systems, time delays are unavoidable when data are exchanged between the master and slave sites [8, 10, 11]. Many methods have been presented for dealing with this problem.

This chapter addresses specifically the stability analysis problem for a class of teleoperation systems with slave force feedback signal, in particular for network-based teleoperation systems. The forward and backward transmission time delays are also assumed to be time-varying and asymmetric. By employing Lyapunov Krasovskii functional, we establish the relationships among the controller design parameters and the upper bounds of the time-varying delays. The stability criteria are given in the form of linear matrix inequalities (LMIs), which can be used to compute the allowed maximum values of forward and backward delays. Finally, simulations and experiments are performed to show the effectiveness of the proposed main results.

C. Hua et al., *Analysis and Design for Networked Teleoperation System*,
https://doi.org/10.1007/978-981-13-7936-9_6

6.2 System Formulation and Assumptions

In this chapter, we also consider the following model of master-slave bilateral teleoperation system given by

$$\begin{cases} M_m(q_m)\ddot{q}_m + C_m(q_m, \dot{q}_m)\dot{q}_m + G_m(q_m) = \tau_m + F_h \\ M_s(q_s)\ddot{q}_s + C_s(q_s, \dot{q}_s)\dot{q}_s + G_s(q_s) = \tau_s - F_e \end{cases} \tag{6.1}$$

In this chapter, we will consider the stability problem of system (6.1) with proportional position errors and delayed slave torque feedback plus damping injection (see Fig. 6.1)

$$\begin{cases} \tau_m = -\tau_s(t - d_s(t)) - \alpha_m \dot{q}_m, \\ \tau_s = k_s(q_m(t - d_m(t)) - q_s) - \alpha_s \dot{q}_s. \end{cases} \tag{6.2}$$

where $d_m(t)$ and $d_s(t)$ are forward time delay (from the master site to slave site) and backward time delay (from the slave site to master site), respectively; k_s is proportional coefficient; α_m and α_s are damping coefficients.

The assumptions on the human operator and environment insert forces is same as the Chap. 2.

For the sake of completeness, the following lemma is presented as follows.

Lemma 6.1 *For a positive definite matrix* Υ*, the following inequality holds*

$$\begin{aligned} &-2a^T(t)\int_{t-d(t)}^{t} b(\xi)\,d\xi - \int_{t-d(t)}^{t} b^T(\xi)\Upsilon b_m(\xi)\,d\xi \\ &\le \bar{d}a^T(t)\Upsilon^{-1}a(t), \end{aligned}$$

where $a(\cdot)$ *and* $b(\cdot)$ *are vector functions,* $d(t)$ *is a time-varying scalar with* $0 \le d(t) \le \bar{d}$.

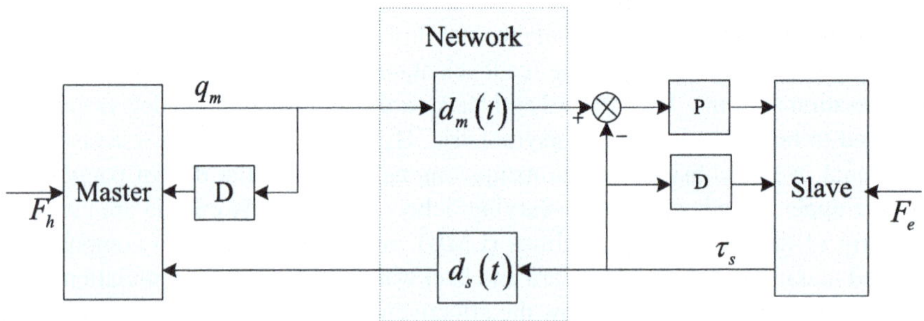

Fig. 6.1 Master with slave torque feedback plus damping and slave with positive error plus damping

6.3 Main Results

In this section, the stability problems are analyzed for system (6.1) with controller (6.2). Before proposing the stability criteria, we need the following assumption on $d_s(t)$.

Assumption 6.1 The backward time-varying delay $d_s(t)$ of the system satisfies $\dot{d}_s(t) \le \mu < 1$.

Remark 6.1 For any communication channel/communication protocol where a received signal is used by the teleoperation system (i.e. not discarded) until a fresher (more recently sent) packet arrives, time delays cannot grow faster than the time. In other words, the time derivative of the delay is less than 1. We can see that Assumption 6.1 is actually always satisfied.

With (6.1) and (6.2), we have the following system

$$\begin{cases} M_m(q_m)\ddot{q}_m + C_m(q_m, \dot{q}_m)\dot{q}_m + G_m(q_m) \\ = F_h - \tau_s(t - d_s(t)) - \alpha_m \dot{q}_m, \\ M_s(q_s)\ddot{q}_s + C_s(q_s, \dot{q}_s)\dot{q}_s + G_s(q_s) \\ = k_s((q_m(t - d_m(t))) - q_s) - \alpha_s \dot{q}_s - F_e. \end{cases}$$

Choose the Lyapunov functional $V = V_1 + V_2 + V_3 + V_4$ with

$$\begin{aligned} V_1 =& \dot{q}_m^T M_m(q_m)\dot{q}_m + \dot{q}_s^T M_s(q_s)\dot{q}_s \\ &+ 2U_m(q_m) + 2U_s(q_s) - 2\beta_m - 2\beta_s \\ &- 2\int_0^t \left(\dot{q}_m^T(\sigma) F_h(\sigma) - \dot{q}_s^T(\sigma) F_e(\sigma)\right) d\sigma, \\ V_2 =& \int_{-\bar{d}_m - \bar{d}_s}^{0} \int_{t+\theta}^{t} \dot{q}_m(\xi)^T Z_1 \dot{q}_m(\xi)\, d\xi d\theta \\ &+ \int_{-\bar{d}_m}^{0} \int_{t+\theta}^{t} \dot{q}_m(\xi)^T Z_2 \dot{q}_m(\xi)\, d\xi d\theta \\ &+ \int_{-\bar{d}_s}^{0} \int_{t+\theta}^{t} \dot{q}_s(\xi)^T S \dot{q}_s(\xi)\, d\xi d\theta, \\ V_3 =& k_s(q_m - q_s)^T (q_m - q_s), \\ V_4 =& \int_{t-d_s(t)}^{t} \dot{q}_s(\xi)^T N \dot{q}_s(\xi)\, d\xi. \end{aligned}$$

where Z_1, Z_2, S and N are positive definite matrices.

The time derivatives of V_1, V_2, V_3 and V_4 are as follows

$$\begin{aligned}\dot{V}_1 =& 2\dot{q}_m^T\left(-\tau_s\left(t-d_s\left(t\right)\right)-\alpha_m\dot{q}_m\right)\\&+2\dot{q}_s^T\left(k_s\left(q_m\left(t-d_m\left(t\right)\right)-q_s\right)-\alpha_s\dot{q}_s\right),\end{aligned} \tag{6.3}$$

$$\begin{aligned}\dot{V}_2 \le& \left(\overline{d}_m+\overline{d}_s\right)\dot{q}_m^T Z_1\dot{q}_m+\overline{d}_m\dot{q}_m^T Z_2\dot{q}_m\\&-\int_{t-d_m(t)-d_s(t)}^{t}\dot{q}_m\left(\xi\right)^T Z_1\dot{q}_m\left(\xi\right)d\xi\\&-\int_{t-d_m(t)}^{t}\dot{q}_m\left(\xi\right)^T Z_2\dot{q}_m\left(\xi\right)d\xi d\theta\\&+\overline{d}_s\dot{q}_s^T S\dot{q}_s-\int_{t-d_s(t)}^{t}\dot{q}_s\left(\xi\right)^T S\dot{q}_s\left(\xi\right)d\xi,\end{aligned} \tag{6.4}$$

$$\dot{V}_3 = 2k_s\left(\dot{q}_m-\dot{q}_s\right)^T\left(q_m-q_s\right), \tag{6.5}$$

and

$$\begin{aligned}\dot{V}_4 \le& \dot{q}_s^T N\dot{q}_s\\&-\left(1-\mu\right)\dot{q}_s\left(t-d_s\left(t\right)\right)^T N\dot{q}_s\left(t-d_s\left(t\right)\right).\end{aligned} \tag{6.6}$$

By verification, we have the following equations

$$\begin{aligned}&2\dot{q}_m^T\left(-\tau_s\left(t-d_s\left(t\right)\right)-\alpha_m\dot{q}_m\right)\\&=-2\alpha_m\dot{q}_m^T\dot{q}_m+2\alpha_s\dot{q}_m^T\dot{q}_s\left(t-d_s\left(t\right)\right)\\&\quad-2\dot{q}_m^T\left(k_s\left(q_m\left(t-d_m\left(t\right)-d_s\left(t\right)\right)-q_s\left(t-d_s\left(t\right)\right)\right)\right)\\&=-2\alpha_m\dot{q}_m^T\dot{q}_m+2\alpha_s\dot{q}_m^T\dot{q}_s\left(t-d_s\left(t\right)\right)\\&\quad-2\dot{q}_m^T k_s(q_m-q_s-q_m+q_s\\&\quad+q_m\left(t-d_m\left(t\right)-d_s\left(t\right)\right)-q_s\left(t-d_s\left(t\right)\right))\\&=-2\alpha_m\dot{q}_m^T\dot{q}_m+2\alpha_s\dot{q}_m^T\dot{q}_s\left(t-d_s\left(t\right)\right)\\&\quad-2\dot{q}_m^T k_s\left(q_m-q_s\right)+2\dot{q}_m^T k_s\int_{t-d_m(t)-d_s(t)}^{t}\dot{q}_m\left(\xi\right)d\xi\\&\quad-2\dot{q}_m^T k_s\int_{t-d_s(t)}^{t}\dot{q}_s\left(\xi\right)d\xi,\end{aligned} \tag{6.7}$$

and

$$
\begin{aligned}
&2\dot{q}_s^T\left(k_s\left(q_m\left(t-d_m\left(t\right)\right)-q_s\right)-\alpha_s\dot{q}_s\right)\\
&=-2\alpha_s\dot{q}_s^T\dot{q}_s-2k_s\dot{q}_s^T\left(q_s-q_m\right)\\
&\quad-2k_s\dot{q}_s^T\int_{t-d_m(t)}^{t}\dot{q}_m\left(\xi\right)d\xi.
\end{aligned}\tag{6.8}
$$

With Lemma 6.1, it is easy to verify that

$$
\begin{aligned}
&2\dot{q}_m^Tk_s\int_{t-d_m(t)-d_s(t)}^{t}\dot{q}_m\left(\xi\right)d\xi-\int_{t-d_m(t)-d_s(t)}^{t}\dot{q}_m^T\left(\xi\right)Z_1\dot{q}_m\left(\xi\right)d\xi\\
&\leq\left(\overline{d}_m+\overline{d}_s\right)k_s^2\dot{q}_m^TZ_1^{-1}\dot{q}_m,
\end{aligned}\tag{6.9}
$$

$$
\begin{aligned}
&-2k_s\dot{q}_s^T\int_{t-d_m(t)}^{t}\dot{q}_m\left(\xi\right)d\xi-\int_{t-d_m(t)}^{t}\dot{q}_m^T\left(\xi\right)Z_2\dot{q}_m\left(\xi\right)d\xi\\
&\leq\overline{d}_mk_s^2\dot{q}_s^TZ_2^{-1}\dot{q}_s,
\end{aligned}\tag{6.10}
$$

and

$$
\begin{aligned}
&-2k_s\dot{q}_m^T\int_{t-d_s(t)}^{t}\dot{q}_s\left(\xi\right)d\xi-\int_{t-d_s(t)}^{t}\dot{q}_s^T\left(\xi\right)S\dot{q}_s\left(\xi\right)d\xi\\
&\leq\overline{d}_sk_s^2\dot{q}_m^TS^{-1}\dot{q}_m.
\end{aligned}\tag{6.11}
$$

With (6.7)–(6.11), the time derivative of V satisfies

$$
\begin{aligned}
\dot{V}=&\dot{V}_1+\dot{V}_2+\dot{V}_3+\dot{V}_4\\
&\leq-2\alpha_m\dot{q}_m^T\dot{q}_m+2\alpha_s\dot{q}_m^T\dot{q}_s\left(t-d_s\left(t\right)\right)\\
&-2\alpha_s\dot{q}_s^T\dot{q}_s+\dot{q}_s^T\left(t\right)N\dot{q}_s\left(t\right)+\overline{d}_mk_s^2\dot{q}_s^TZ_2^{-1}\dot{q}_s\\
&+\left(\overline{d}_m+\overline{d}_s\right)k_s^2\dot{q}_m^TZ_1^{-1}\dot{q}_m+\left(\overline{d}_m+\overline{d}_s\right)\dot{q}_m^TZ_1\dot{q}_m\\
&+\overline{d}_m\dot{q}_m^TZ_2\dot{q}_m+\overline{d}_sk_s^2\dot{q}_m^TS^{-1}\dot{q}_m+\overline{d}_s\dot{q}_s^TS\dot{q}_s\\
&-\left(1-\mu\right)\dot{q}_s\left(t-d_s\left(t\right)\right)^TN\dot{q}_s\left(t-d_s\left(t\right)\right)\\
&=\eta^T\Psi\eta,
\end{aligned}\tag{6.12}
$$

where $\eta=\left[\dot{q}_m^T\ \dot{q}_s^T\ \dot{q}_s^T\left(t-d_s\left(t\right)\right)\right]^T$ and

$$
\Psi=\begin{bmatrix}\Psi_{(1,1)} & 0 & \alpha_sI\\ * & \Psi_{(2,2)} & 0\\ * & * & -\left(1-\mu\right)N\end{bmatrix},
$$

in which $\Psi_{(1,1)}=-2\alpha_mI+\overline{d}_sk_s^2S^{-1}+\left(\overline{d}_m+\overline{d}_s\right)k_s^2Z_1^{-1}+\left(\overline{d}_m+\overline{d}_s\right)Z_1+\overline{d}_mZ_2$, $\Psi_{(2,2)}=-2\alpha_sI+N+\overline{d}_mk_s^2Z_2^{-1}+\overline{d}_sS$.

Theorem 6.1 *For system (6.1) with controller (6.2), if there exist positive definite matrices Z_2, S and N such that the following LMI holds*

$$\overline{\Psi} = \begin{bmatrix} \overline{\Psi}_{(1,1)} & 0 & \alpha_s I & 0 & \overline{d}_s k_s \\ * & \overline{\Psi}_{(2,2)} & 0 & \overline{d}_m k_s & 0 \\ * & * & \overline{\Psi}_{(3,3)} & 0 & 0 \\ * & * & * & -\overline{d}_m Z_2 & 0 \\ * & * & * & 0 & -\overline{d}_s S \end{bmatrix} < 0, \tag{6.13}$$

where $\overline{\Psi}_{(1,1)} = -2\alpha_m I + 2\left(\overline{d}_m + \overline{d}_s\right) k_s I + \overline{d}_m Z_2$, $\overline{\Psi}_{(2,2)} = -2\alpha_s + N + \overline{d}_s S$ *and* $\overline{\Psi}_{(3,3)} = -(1-\mu) N$, *then the velocities* $\dot{q}_m, \dot{q}_s$ *and the position error* $q_m - q_s$ *are bounded.*

Proof One knows that the minimum value of $(\overline{d}_m + \overline{d}_s)k_s^2 Z_1^{-1} + (\overline{d}_m + \overline{d}_s)Z_1$ is $2(\overline{d}_m + \overline{d}_s)k_s I$. Then using $2(\overline{d}_m + \overline{d}_s)k_s I$ instead of $(\overline{d}_m + \overline{d}_s)k_s^2 Z_1^{-1} + (\overline{d}_m + \overline{d}_s)Z_1$ in $\Psi_{(1,1)}$, we can find that $\overline{\Psi} < 0$ is equivalent to $\Psi < 0$. By Lyapunov stability theory, the velocities $\dot{q}_i$ and position error $q_m - q_s$ are bounded. The proof is completed.

Furthermore, we have the following result on the asymptotical stability of the closed-loop system.

Theorem 6.2 *For system (6.1) with controller (6.2), if there exist positive definite matrices Z_2, S and N such that the LMI (6.13) holds, and human input F_h and environment force F_e are zero, then the following controller*

$$\begin{cases} \tau_m = -k_s\left(q_m\left(t - d_m(t) - d_s(t)\right) - q_s\left(t - d_s(t)\right)\right) \\ +\alpha_s \dot{q}_s\left(t - d_s(t)\right) - \alpha_m \dot{q}_m + G_m(q_m), \\ \tau_s = k_s\left(q_m\left(t - d_m(t)\right) - q_s\right) - \alpha_s \dot{q}_s + G_s(q_s). \end{cases} \tag{6.14}$$

renders velocities $\dot{q}_m, \dot{q}_s$ and position error $q_m - q_s$ converge to zero asymptotically.

The proof is similar to that of Theorem 6.1, thus it is omitted here.

Remark 6.2 Compared with the P+d controller, with the directed force feedback signals, in this chapter Theorems 6.1 and 6.2 need the information μ. The reason is as follows: The feedback controller (6.2) contains $\dot{q}_s\left(t - d_s(t)\right)$ resulted by $\tau_s\left(t - d_s(t)\right)$ (It does not appear for controller (2.2) in Chap. 2), we should choose the Lyapunov functional V_4 to compensate for it, then this condition $d_s(t) \leq \mu$ is needed. For the parameters μ, α_s, k_s and one of delay bound parameters known, we can obtain the allowable maximum value of the other delay by using the LMI optimization tool box of Matlab. For example, if μ, k_s and $\overline{d}_m$ are known, we can obtain the allowable maximum value $\overline{d}_s$.

Remark 6.3 It was assumed in [9, 12] that the forward and backward time delays are symmetric and constant. The corresponding stability conditions were developed for system (6.1). Obviously, with $\mu = 0$ and $\overline{d}_m = \overline{d}_s$ the LMI condition is identical to the stability condition in [9, 12], thus the result presented in this paper actually includes the symmetric and constant time delay as a special case.

Remark 6.4 In this chapter, we employ a new Lyapunov Krasovskii functional to derive the stability condition. The method is applied to a new directed force feedback control scheme. From the transparency aspect, it is obvious that the torque reflection controller (6.2) results in better performance than P+d controller (2.2) designed in Chap. 2. Nevertheless, stability and transparency are both important properties for teleoperation systems, but they are often contradictive from the design point view. Therefore, the stability performance of the closed-loop system with P+d is better than that of the closed-loop system with controller (6.2). We can see this from the following fact: with given the same parameters α_m, α_s, k_m, k_s and μ for LMI (2.11) in Chap. 2 and LMI (6.13), the allowable maximum time delay via P+d controller is much bigger than that via controller (6.2) designed in this chapter. This can be also seen by the example of the next section.

Remark 6.5 Although in this chapter we investigate the case that the control parameters k_s, α_m and α_s are scalars, the method is also applicable to the case that they are matrices. This method proposed in this chapter can be used for the stability analysis for other controllers by constructing the similar Lyapunov Krasovskii functional.

6.4 Simulation and Experiment

Similar with the Chap. 2, in this section, simulation results are first given to verify the effectiveness of the main result. Then, the experiments performed on a teleoperation system consisting of two Phantom Premium 1.5 A robot manipulators are presented.

The simulations are also performed on 2-DOF manipulators:

$$\begin{aligned} M_m(q_m)\ddot{q}_m + C_m(q_m, \dot{q}_m)\dot{q}_m + G_m(q_m) &= \tau_m + F_h \\ M_s(q_s)\ddot{q}_s + C_s(q_s, \dot{q}_s)\dot{q}_s + G_s(q_s) &= \tau_s - F_e \end{aligned}.$$

The specific system model parameters are chosen as same as Chap. 2. Firstly, We apply a human force F to the master site on the Y-direction, which is shown in Fig. 6.2. The human input is zero at 0 s, and then increases to 30 N at 10 s. The force is kept constant from 10 to 40 s, and it decreases to zero from 40 to 50 s. In the slave site, there is also a wall at $y = 0.5$ m. When the slave joint reaches the wall and moves further, the feedback force is $10000 * (y - 0.5)\,N$. With the human force input, we know the torque $F_h = -J_m^T * \begin{bmatrix} 0 & 1 \end{bmatrix}^T * F$, where J_m is the Jacobian matrix of the master manipulator. The torque F_e is $F_e = -J_s^T * \begin{bmatrix} 0 & 1 \end{bmatrix}^T * 10000N * (y - 0.5)$.

The slave torque reflection control (6.14) is employed. With the parameters $\alpha_m = \alpha_s = 100$, $k_m = k_s = 50$, $\mu = 0.2$ and $\overline{d}_m = 0.2$ s, solving LMI (6.13) gives the allowable maximum delay value $\overline{d}_s = 0.6271s$ based on Theorem 6.2. In simulation,

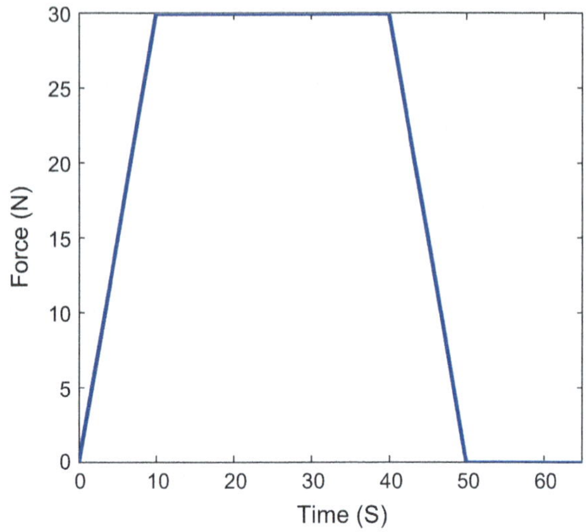

Fig. 6.2 The human force input F

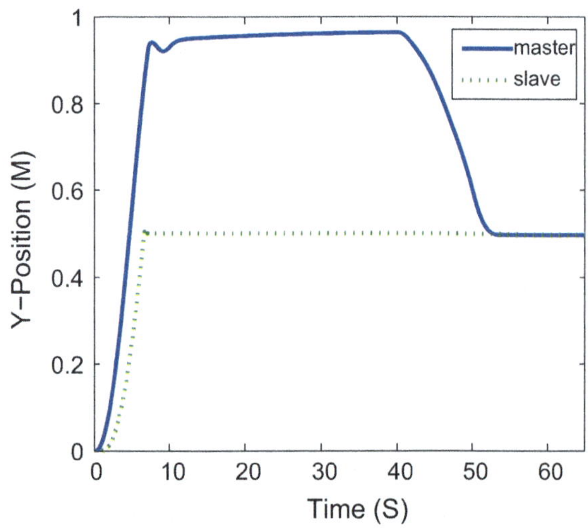

Fig. 6.3 The Y-position of master and slave with control (6.14)

we choose $q_m(0) = \dot{q}_m(0) = q_s(0) = \dot{q}_s(0) = \left[0\ 0\right]^T$, and the delays $d_m(t) = 0.1 + 0.1\sin(t)$ and $d_s(t) = 0.4271 + 0.2\sin(t)$, the simulation results are shown in Figs. 6.3, 6.4, 6.5 and 6.6. Figure 6.3 shows the Y-position of the master and the slave joints, and Fig. 6.4 shows the human torque F_h and environment force F_e (the wall feedback torque), and the error between the two torques is shown in Fig. 6.5. Figure 6.6 shows the positions of the q_m and q_s. When the human force increases, the Y-positions of master and slave joints move forward, which can be seen from Fig. 6.3. The torque F_h increases with the augment of F, while $F_e = 0$ at the beginning for no touch of the obstacle. In 9.5 s, the slave joint reaches the wall at $y = 0.5$ m and

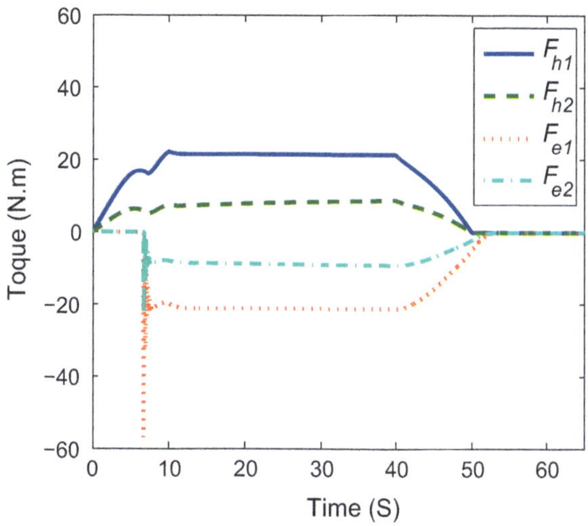

Fig. 6.4 The torque of F_h and F_e with control (6.14)

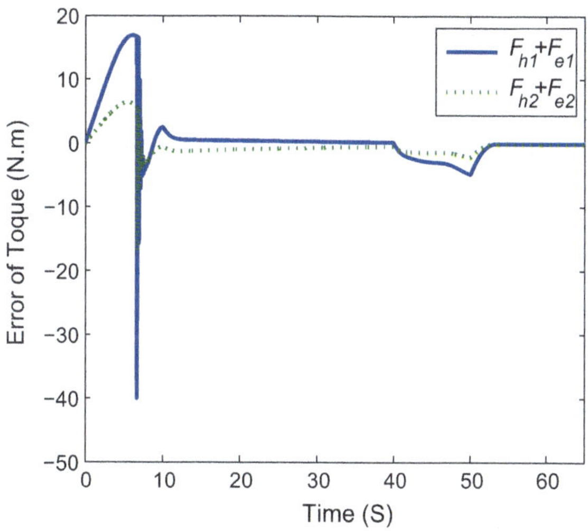

Fig. 6.5 The torque error between master and slave with control (6.14)

could not move any more, while the master joint moves further because of the human input force, and finally it reaches $y \approx 0.8\,\mathrm{m}$ in 10 s, see Fig. 6.3. In Fig. 6.4, we can see that a large environmental torque pulse appears in about 9.5 s for the touch of the wall. From 9.5 s to 10 s, we can see that environment torque F_e increases quickly, and $F_e = -F_h$ at 10 s. With the force F keeps 30 N from 10 s to 40 s, we find that torques F_h and F_e also keep the former values, and the master joint and slave joint do not move any more on the Y-direction during this time interval. With the decrease of F from 40 s, F_h and F_e also reduces, and the Y-position of master moves backward, which can be seen from Figs. 6.3 and 6.5. When $F = 0$ at $t = 50\,\mathrm{s}$, $F_h = 0$ and $F_e = 0$, and the master reaches $y = 0.5\,\mathrm{m}$, and the positions of the two systems synchronize.

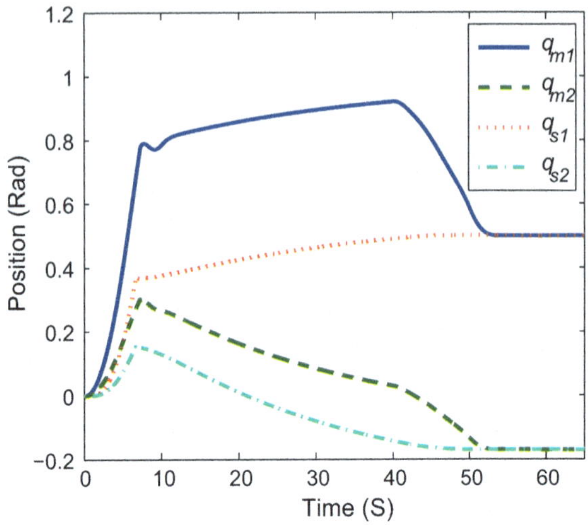

Fig. 6.6 The position of q_m and q_s with control (6.14)

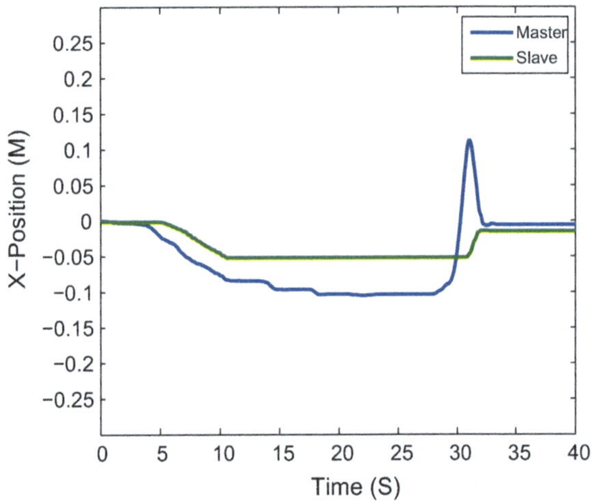

Fig. 6.7 The X-position of master and slave with control (6.14)

For the controller is with the gravity compensator and there is no negative force to push the master back, the positions of the two joints will not move any more. From these figures we can see that the designed controller is effective.

The experiments on two Phantom Premium 1.5 A robots (SensAble Technologies Inc.) are also conducted. The experiment platform is same as Chap. 2, see Fig. 2.7. Similar with Chap. 2 we also put a wood box at −0.06 m on the X-direction of the slave manipulator, which represents a stiff wall. We also apply the human input to the master manipulator and push the last joint to move on the X-direction. The slave joint will follows the movement of the master joint, and then the slave joint touches the stiff box and could not move any more. We move the master joint further and can feel the feedback force, then move the master joint to the original position. Our

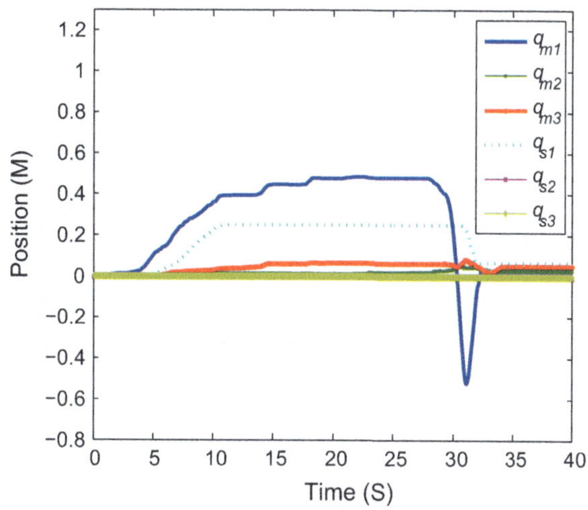

Fig. 6.8 The position of master and slave with control (6.14)

objective here is to verify the effectiveness of the proposed controller (6.14) and check whether the slave joint follows the movement of the master joint and whether they can return eventually to the original position.

With the use of the proposed controller (6.14) and same parameters above, the movement of the master and slave joints on the X-direction is shown in Fig. 6.7. In Fig. 6.8, the positions of q_m and q_s are shown. We can see that the slave joint reaches -0.06 m and could not move any more. At the master site, we can feel the feedback force. With the master joint back to the origin, the slave manipulator follows the movement and reaches the origin. The experimental results further show the effectiveness and correctness of the proposed main results.

6.5 Summary

In this chapter, the control problem of nonlinear teleoperation system with asymmetric and time-varying delays is studied. We consider a class of the slave torque feedback plus damping injection controller to improve the transparency of closed-loop master-slave system. The Lyapunov Krasovskii functional is also used to derive the delay-dependent stability criteria, which are given in the form of LMIs. The maximum allowable time delay values can be obtained by solving the LMIs. The simulations and experiments performed verify well the theoretical results and demonstrate their effectiveness.

References

1. D.A. Lawrence, Stability and transparency in bilateral teleoperation. IEEE Trans. Rob. Autom. **9**(5), 624–637 (1993)
2. K. Hashtrudi, S.E. Salcudean, Transparency in time-delayed systems and the effect of local force feedback for transparent teleoperation. IEEE Trans. Rob. Autom. **18**(1), 108–114 (2002)
3. I.G. Polushin, P.X. Liu, C.H. Lung, A control scheme for stable force-reflecting teleoperation over ip networks. IEEE Trans. Syst. Man Cybern. B Cybern. **36**(4), 930–939 (2006)
4. I.G. Polushin, P.X. Liu, C.H. Lung, A force reflection algorithm for improved transparency in bilateral teleoperation with communication delay. IEEE/ASME Trans. Mechatron. **12**(3), 361–374 (2007)
5. J.H. Ryu, D.S. Kwon, B. Hannaford, Stable teleoperation with time-domain passivity control. IEEE Trans. Rob. Autom. **20**(2), 365–373 (2004)
6. W.H. Zhu, S.E. Salcudean, Stability guaranteed teleoperation: an adaptive motion/force control approach. IEEE Trans. Autom. Control **45**(11), 1951–1969 (2000)
7. C.C. Hua, X.P. Liu, Delay-dependent stability criteria of teleoperation systems with asymmetric time-varying delays. IEEE Trans. Rob. **26**(5), 925–932 (2010)
8. R.J. Anderson, M.W. Spong, Bilateral control of teleoperators with time delay. IEEE Trans. Autom. Control **34**(5), 494–501 (1989)
9. D. Lee, M.W. Spong, Passive bilateral teleoperation with constant time delay. IEEE Trans. Rob. Autom. **22**(2), 269–281 (2006)
10. W.R. Ferrell, Delayed force feedback. Hum. Factors **8**(5), 449–455 (1966)
11. N. Chopra, M.W. Spong, S. Hirche, M. Buss, Bilateral teleoperation over the internet: the time varying delay problem, in *Proceeding of American Control Conference* (2003), pp. 155–160
12. E. Nuño, R. Ortega, N. Barabanov, L. Basañez, A globally stable pd controller for bilateral teleoperators. IEEE Trans. Rob. **24**(3), 753–758 (2008)

Chapter 7
Teleoperation over the Internet with Exponential Convergence

Abstract This chapter addresses the controller design problem for teleoperation over networks such as the Internet with exponential convergence. The forward and backward network transmission time delays are also assumed to be asymmetric and time-varying, which is the case for computer network communications. A novel master-slave controller, which renders the entire system stable with relatively good steady-state and transient-state performances is proposed. The relations among the parameters of the controller and the allowable maximum time delays are built in the form of linear matrix inequality. The stability of the closed-loop system is proved, strictly. Both simulations and experiments are performed to verify the correctness and effectiveness of the proposed method.

7.1 Introduction

In above chapters, we designed the P+d [1], PD+d [2] and the directed force feedback controllers [3] for the nonlinear teleoperation system. Moreover, the asymptotic stability of the closed-loop teleoperation are achieved [4, 5]. It is well known that compared with the asymptotic stability, the exponent stability provide some superior convergence performances [5–8].

Based on the above considerations, this chapter addresses the controller design problem for a class of teleoperation systems providing exponent stability. The forward and the backward transmission time delays between the master and the slave are time-varying and asymmetric. We proposed a control scheme for general Internet (or network) based teleoperation systems. With the proposed control scheme, we prove the stability of the closed-loop system and show that the synchronization error converges exponentially to a bounded region. The stability of the closed-loop system is also guaranteed by choosing a Lyapunov Krasovskii functional. The controller design parameters are determined by solving linear matrix inequalities. Finally, simulations and experiments are performed to show the effectiveness of the proposed main results.

C. Hua et al., *Analysis and Design for Networked Teleoperation System*,
https://doi.org/10.1007/978-981-13-7936-9_7

7.2 Preliminaries and Problem Formulation

Consider a master-slave system described by

$$\begin{cases} M_m(q_m)\ddot{q}_m + C_m(q_m, \dot{q}_m)\dot{q}_m = \tau_m + F_h \\ M_s(q_s)\ddot{q}_s + C_s(q_s, \dot{q}_s)\dot{q}_s = \tau_s - F_e \end{cases} \tag{7.1}$$

It has assumed that the gravity torque has been compensated in advance.

For system (7.1), we impose the following assumption:

Assumption 7.1 The time delays $d_m(t)$ and $d_s(t)$ are bounded, that is, there exist positive scalars $\overline{d}_m$ and $\overline{d}_s$ such that $d_m(t) \leq \overline{d}_m$ and $d_s(t) \leq \overline{d}_s$.

In this chapter, we will propose a control scheme for the considered system (7.1). The new controller guarantees that the closed-loop system is exponentially stable.

7.3 Main Results

In this section, we consider the controller design problem for system (7.1). We propose the master and slave controller as follows

$$\begin{cases} \tau_m = -k_m(q_m - q_s(t - d_s(t))) \\ -\alpha_m\dot{q}_m - \beta_m\|\dot{q}_m\|^2\dot{q}_m, \\ \tau_s = k_s(q_m(t - d_m(t)) - q_s) \\ -\alpha_s\dot{q}_s - \beta_s\|\dot{q}_s\|^2\dot{q}_s. \end{cases} \tag{7.2}$$

where $d_m(t)$ and $d_s(t)$ are forward time delay (from the master site to slave site) and backward time delay (from the slave site to master site), respectively; k_m and k_s are proportional coefficients; α_m, α_s, β_m and β_s are positive parameters. The control scheme is shown in Fig. 7.1, in which the function $\phi_i(\dot{q}_i) = -\alpha_i\dot{q}_i - \beta_i\|\dot{q}_i\|^2\dot{q}_i$. Different from the classic PD controller, the new controller (7.2) includes the nonlinear parts $\beta_i\|\dot{q}_i\|^2\dot{q}_i$. With the new parts, we will show that the controller (7.2) can realize the whole system exponentially stable.

With (7.1) and (7.2), the controlled system is as follows

$$\begin{cases} M_m(q_m)\ddot{q}_m + C_m(q_m, \dot{q}_m)\dot{q}_m \\ \quad = -k_m(q_m - q_s(t - d_s(t))) \\ \quad -\alpha_m\dot{q}_i - \beta_m\|\dot{q}_m\|^2\dot{q}_m + F_h, \\ M_s(q_s)\ddot{q}_s + C_s(q_s, \dot{q}_s)\dot{q}_s \\ \quad = k_s(q_m(t - d_m(t)) - q_s) \\ \quad -\alpha_s\dot{q}_s - \beta_s\|\dot{q}_s\|^2\dot{q}_s - F_e. \end{cases} \tag{7.3}$$

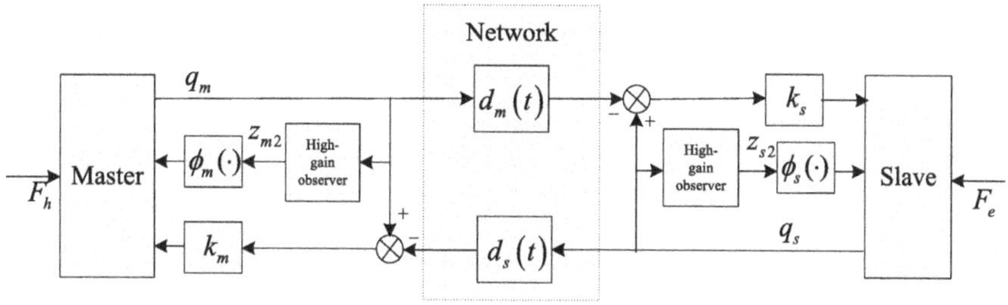

Fig. 7.1 The control scheme for teleoperation system

Now let us analyze the stability of the closed-loop system (7.3). Choose the Lyapunov functional $V = V_1 + V_2$ with

$$\begin{aligned}
V_1 =& a_m \dot{q}_m^T M_m (q_m) \dot{q}_m + b_m \left(\dot{q}_m^T M_m (q_m) \dot{q}_m\right)^2 \\
&+ a_s \left(\dot{q}_s^T M_s (q_s) \dot{q}_s\right) + b_s \left(\dot{q}_s^T M_s (q_s) \dot{q}_s\right)^2 \\
&+ 2\epsilon_m (q_m - q_s)^T M_m (q_m) \dot{q}_m \\
&- 2\epsilon_s (q_m - q_s)^T M_s (q_s) \dot{q}_s \\
&+ k (q_m - q_s)^T (q_m - q_s), \\
V_2 =& \int_{-\bar{d}_m}^{0} \int_{t+\theta}^{t} e^{-\gamma(t-\xi)} \dot{q}_m (\xi)^T Z \dot{q}_m (\xi) \, d\xi d\theta \\
&+ \int_{-\bar{d}_s}^{0} \int_{t+\theta}^{t} e^{-\gamma(t-\xi)} \dot{q}_s (\xi)^T S \dot{q}_s (\xi) \, d\xi d\theta,
\end{aligned} \tag{7.4}$$

where a_i, b_i and ϵ_i are positive scalars for $i = m, s$, Z and S are positive definite matrices, k is a positive scalar satisfying $k > a_m^{-1} \mu_{m1}^{-1} (\mu_{m2}\epsilon_m)^2 + a_s^{-1} \mu_{s1}^{-1} (\mu_{s2}\epsilon_s)^2$.

Remark 7.1 Different from [2], the Lyapunov function V_1 contains a new part $2\epsilon_m (q_m - q_s) M_m (q_m) \dot{q}_m - 2\epsilon_s (q_m - q_s) M_s (q_s) \dot{q}_s$, which is necessary to guarantee the boundedness of the position tracking error. Using Property 1.1 of robotic system shown in Chap. 1 and $\pm 2a^T b \leq a^T a + b^T b$, we have

$$\begin{aligned}
&2\epsilon_m (q_m - q_s)^T M_m (q_m) \dot{q}_m + a_m \dot{q}_m^T M_m (q_m) \dot{q}_m \\
&\geq -a_m^{-1} \mu_{m1}^{-1} (\mu_{m2}\epsilon_m)^2 \|q_m - q_s\|^2
\end{aligned}$$

and

$$\begin{aligned}
&- 2\epsilon_s (q_m - q_s)^T M_s (q_s) \dot{q}_s + a_s \dot{q}_s^T M_s (q_s) \dot{q}_s \\
&\geq -a_s^{-1} \mu_{s1}^{-1} (\mu_{s2}\epsilon_s)^2 \|q_m - q_s\|^2 .
\end{aligned}$$

With the chosen k, functional V_1 is positive definite.

From the Properties 1.1 and 1.2, we know that the time derivative of V_1 along system (7.3) is

$$\begin{aligned}
\dot{V}_1 =& \rho_m(-2k_m\dot{q}_m^T(q_m - q_s(t-d_s(t)))) \\
&+ 2\rho_m\dot{q}_m^T(-\alpha_m\dot{q}_i - \beta_m\|\dot{q}_m\|^2\dot{q}_m + F_h) \\
&+ \rho_s\left(2k_s\dot{q}_s^T(q_m(t-d_m(t)) - q_s)\right) \\
&+ 2\rho_s\dot{q}_s^T(-\alpha_s\dot{q}_s - \beta_s\|\dot{q}_s\|^2\dot{q}_s - F_e) \\
&+ 2\epsilon_m(q_m - q_s)^T(-C_m\dot{q}_m - \alpha_m\dot{q}_m + \dot{M}_m\dot{q}_m \\
&- \beta_m\|\dot{q}_m\|^2\dot{q}_m - k_m(q_m - q_s(t-d_s(t))) \\
&+ F_h) + 2\epsilon_m(\dot{q}_m - \dot{q}_s)^T M_m\dot{q}_m \\
&- 2\epsilon_s(\dot{q}_m - \dot{q}_s)^T M_s\dot{q}_s \\
&- 2\epsilon_s(q_m - q_s)^T(k_s(q_m(t-d_m(t)) - q_s) \\
&- C_s\dot{q}_s - \alpha_s\dot{q}_s + \dot{M}_s\dot{q}_s - \beta_s\|\dot{q}_s\|^2\dot{q}_s \\
&- F_e) + 2k(q_m - q_s)^T(\dot{q}_m - \dot{q}_s)
\end{aligned} \tag{7.5}$$

where $\rho_i = a_i + 2b_i\dot{q}_i^T M_i(q_i)\dot{q}_i$. With Property 1.1, $\rho_i = \overline{\rho}_i + b_i\delta_{\mu_i}\|\dot{q}_i\|^2$ where $\overline{\rho}_i = a_i + (\mu_{i1} + \mu_{i2})b_i\|\dot{q}_i\|^2$ and $|\delta_{\mu_i}| \le \overline{\delta}_{\mu_i} = \mu_{i2} - \mu_{i1}$. Here we decompose ρ_i into the certain part $\overline{\rho}_i$ and uncertain part $b_i\delta_{\mu_i}\|\dot{q}_i\|^2$ with the new parameter δ_{μ_i} as bounded disturbance.

Using $q_i(t) - q_i(t-d(t)) = \int_{t-d(t)}^{t}\dot{q}_i(\xi)d\xi$, one has

$$\begin{aligned}
&-2\rho_m k_m\dot{q}_m^T(q_m - q_s(t-d_s(t))) + 2\rho_s k_s\dot{q}_s^T(q_m(t-d_m(t)) - q_s) \\
=& -2k_m\rho_m\dot{q}_m^T(q_m - q_s) - 2k_m\rho_m\dot{q}_m^T\int_{t-d_s(t)}^{t}\dot{q}_s(\xi)d\xi \\
&+ 2k_s\rho_s\dot{q}_s^T(q_m - q_s) - 2k_s\rho_s\dot{q}_s^T\int_{t-d_m(t)}^{t}\dot{q}_m(\xi)d\xi.
\end{aligned} \tag{7.6}$$

With the decompositions of ρ_i, we have

$$
\begin{aligned}
& -2k_m\rho_m\dot{q}_m^T(q_m-q_s)+2\rho_s k_s\dot{q}_s^T(q_m-q_s) \\
= & -2k_m(a_m+(\mu_{m1}+\mu_{m2})b_m\|\dot{q}_m\|^2 \\
& +b_m\delta_{\mu_m}\|\dot{q}_m\|^2)\dot{q}_m^T(q_m-q_s) \\
& +2k_s(a_s+(\mu_{s1}+\mu_{s2})b_s\|\dot{q}_s\|^2 \\
& +b_s\delta_{\mu_s}\|\dot{q}_s\|^2)\dot{q}_s^T(q_m-q_s) \\
\leq & -2k_m\left(a_m+(\mu_{m1}+\mu_{m2})b_m\|\dot{q}_m\|^2\right)\dot{q}_m^T(q_m-q_s) \\
& +2k_s\left(a_s+(\mu_{s1}+\mu_{s2})b_s\|\dot{q}_s\|^2\right)\dot{q}_s^T(q_m-q_s) \\
& +\eta_1 k_m^2\overline{\delta}_{\mu_m}^2 b_m^2\|\dot{q}_m\|^4\dot{q}_m^T\dot{q}_m+\eta_1^{-1}(q_m-q_s)^T(q_m-q_s) \\
& +\eta_2 k_s^2\overline{\delta}_{\mu_s}^2 b_s^2\|\dot{q}_s\|^4\dot{q}_s^T\dot{q}_s+\eta_2^{-1}(q_m-q_s)^T(q_m-q_s)
\end{aligned} \tag{7.7}
$$

where η_1 and η_2 are positive scalars.

In addition, using the Properties 1.1–1.3 gives

$$
\begin{aligned}
& 2\epsilon_m(q_m-q_s)^T(\dot{M}_m-C_m)\dot{q}_m-2\epsilon_s(q_m-q_s)^T(\dot{M}_s-C_s)\dot{q}_s \\
\leq & 2h_m^{-1}\epsilon_m^2(q_m-q_s)^T(q_m-q_s)+5h_m K_{cm}^2\|\dot{q}_m\|^4 \\
& +2h_s^{-1}\epsilon_s^2(q_m-q_s)^T(q_m-q_s)+5h_s K_{cs}^2\|\dot{q}_s\|^4,
\end{aligned} \tag{7.8}
$$

$$
\begin{aligned}
& 2\epsilon_m(\dot{q}_m-\dot{q}_s)^T M_m\dot{q}_m-2\epsilon_s(\dot{q}_m-\dot{q}_s)^T M_s\dot{q}_s \\
\leq & 2\mu_{m2}\epsilon_m\dot{q}_m^T\dot{q}_m+\omega_1\epsilon_m^2\mu_{m2}^2\dot{q}_s^T\dot{q}_s+\omega_1^{-1}\dot{q}_m^T\dot{q}_m \\
& +2\mu_{s2}\epsilon_s\dot{q}_s^T\dot{q}_s+\omega_2\epsilon_s^2\mu_{s2}^2\dot{q}_s^T\dot{q}_s+\omega_2^{-1}\dot{q}_m^T\dot{q}_m,
\end{aligned} \tag{7.9}
$$

and

$$
\begin{aligned}
& 2\rho_m\dot{q}_m^T F_h-2\rho_s\dot{q}_s^T F_e+2\epsilon_m(q_m-q_s)^T(F_h)+2\epsilon_s(q_m-q_s)^T F_e \\
\leq & \vartheta_1\rho_m^2\dot{q}_m^T\dot{q}_m+\vartheta_1^{-1}\overline{F}_h^2+\vartheta_2\rho_s^2\dot{q}_s^T\dot{q}_s+\vartheta_2^{-1}\overline{F}_e^2+\vartheta_3^{-1}\overline{F}_h^2 \\
& +\vartheta_3\epsilon_m^2(q_m-q_s)^T(q_m-q_s)+\vartheta_4\epsilon_s^2\|q_m-q_s\|^2+\vartheta_4^{-1}\overline{F}_e^2,
\end{aligned} \tag{7.10}
$$

where $h_m, h_s, \omega_1, \omega_2, \vartheta_1, \vartheta_2, \vartheta_3$ and ϑ_4 are positive scalars, $\overline{F}_h \geq \|F_h\|$ and $\overline{F}_e \geq \|F_e\|$.

For the delayed signal, one has

$$
\begin{aligned}
& 2\epsilon_m(q_m-q_s)^T(-k_m(q_m-q_s(t-d_s(t)))) \\
= & -2k_m\epsilon_m\|q_m-q_s\|^2-2k_m\epsilon_m(q_m-q_s)^T\int_{t-d_s(t)}^{t}\dot{q}_s(\xi)\,d\xi,
\end{aligned} \tag{7.11}
$$

and

$$\begin{aligned}
&- 2\epsilon_s (q_m - q_s)^T (k_s (q_m (t - d_m (t)) - q_s)) \\
&= - 2k_s\epsilon_s \|q_m - q_s\|^2 + 2k_s\epsilon_s (q_m - q_s)^T \int_{t-d_m(t)}^{t} \dot{q}_m (\xi) d\xi.
\end{aligned} \tag{7.12}$$

The time derivative of V_2 is

$$\begin{aligned}
\dot{V}_2 \leq & - \gamma V_2 + \overline{d}_m\dot{q}_m^T Z\dot{q}_m - e^{-\gamma\overline{d}_m} \int_{t-\overline{d}_m}^{t} \dot{q}_m^T (\xi) Z\dot{q}_m (\xi) d\xi \\
& + \overline{d}_s\dot{q}_s^T S\dot{q}_s - e^{-\gamma\overline{d}_s} \int_{t-\overline{d}_s}^{t} \dot{q}_s^T (\xi) S\dot{q}_s (\xi) d\xi \\
\leq & - \gamma V_2 + \overline{d}_m\dot{q}_m^T Z\dot{q}_m - \int_{t-d_m(t)}^{t} \dot{q}_m^T (\xi) \overline{Z}\dot{q}_m (\xi) d\xi \\
& + \overline{d}_s\dot{q}_s^T S\dot{q}_s - \int_{t-d_s(t)}^{t} \dot{q}_s^T (\xi) \overline{S}\dot{q}_s (\xi) d\xi.
\end{aligned} \tag{7.13}$$

where $\overline{S} = e^{-\gamma\overline{d}_s} S$ and $\overline{Z} = e^{-\gamma\overline{d}_m} Z$.

We deal with the delay parts as follows

$$\begin{aligned}
&(-2k_m\rho_m\dot{q}_m - 2k_m\epsilon_m (q_m - q_s))^T \int_{t-d_s(t)}^{t} \dot{q}_s (\xi) d\xi - \int_{t-d_s(t)}^{t} \dot{q}_s^T (\xi) \overline{S}\dot{q}_s (\xi) d\xi \\
&\leq \overline{d}_s k_m^2((\overline{\rho}_m + b_m\delta_{\mu_m} \|\dot{q}_m\|^2)\dot{q}_m + \epsilon_m (q_m - q_s))^T\overline{S}^{-1} * \\
&\quad ((\overline{\rho}_m + b_m\delta_{\mu_m} \|\dot{q}_m\|^2)\dot{q}_m + \epsilon_m (q_m - q_s)) \\
&\leq \overline{d}_s \left(k_m\overline{\rho}_m\dot{q}_m + k_m\epsilon_m (q_m - q_s)\right)^T \overline{S}^{-1} \left(k_m\overline{\rho}_m\dot{q}_m + k_m\epsilon_m (q_m - q_s)\right) \\
&\quad + 2\overline{d}_s \left(k_m\overline{\rho}_m\dot{q}_m + k_m\epsilon_m (q_m - q_s)\right)^T \overline{S}^{-1} \left(k_m b_m\delta_{\mu_m} \|\dot{q}_m\|^2\right) \dot{q}_m, \\
&\quad + \overline{d}_s \left(k_m b_m\overline{\delta}_{\mu_m} \|\dot{q}_m\|^2\right)^2 \dot{q}_m^T\overline{S}^{-1}\dot{q}_m
\end{aligned} \tag{7.14}$$

and

$$\begin{aligned}
&(-2k_s\rho_s\dot{q}_s + 2k_s\epsilon_s (q_m - q_s))^T \int_{t-d_m(t)}^{t} \dot{q}_m (\xi) d\xi - \int_{t-d_m(t)}^{t} \dot{q}_m^T (\xi) \overline{Z}\dot{q}_m (\xi) d\xi \\
&\leq \overline{d}_m \left(-k_s \left(\overline{\rho}_s + b_s\delta_{\mu_s} \|\dot{q}_s\|^2\right) \dot{q}_s + k_s\epsilon_s (q_m - q_s)\right)^T \overline{Z}^{-1} * \\
&\quad \left(-k_s \left(\overline{\rho}_s + b_s\delta_{\mu_s} \|\dot{q}_s\|^2\right) \dot{q}_s + k_s\epsilon_s (q_m - q_s)\right) \\
&\leq \overline{d}_m \left(-k_s\overline{\rho}_s\dot{q}_s + k_s\epsilon_s (q_m - q_s)\right)^T \overline{Z}^{-1} \left(-k_s\overline{\rho}_s\dot{q}_s + k_s\epsilon_s (q_m - q_s)\right) \\
&\quad + 2\overline{d}_m \left(-k_s\overline{\rho}_s\dot{q}_s + k_s\epsilon_s (q_m - q_s)\right)^T \overline{Z}^{-1} \left(-k_s b_s\delta_{\mu_s} \|\dot{q}_s\|^2\right) \dot{q}_s, \\
&\quad + \overline{d}_m k_s^2 \left(b_s\overline{\delta}_{\mu_s} \|\dot{q}_s\|^2\right)^2 \dot{q}_s^T\overline{Z}^{-1}\dot{q}_s,
\end{aligned} \tag{7.15}$$

Now let us deal with the uncertain parameters δ_{μ_m} and δ_{μ_s}, and we can obtain

$$
\begin{aligned}
&2\overline{d}_s\left(k_m\overline{\rho}_m\dot{q}_m + k_m\epsilon_m\left(q_m - q_s\right)\right)^T\overline{S}^{-1}\left(k_m b_m\delta_{\mu_m}\left\|\dot{q}_m\right\|^2\right)\dot{q}_m\\
&= 2\overline{d}_s k_m\overline{\rho}_m\dot{q}_m^T\overline{S}^{-1}\left(b_m k_m\delta_{\mu_m}\left\|\dot{q}_m\right\|^2\right)\dot{q}_m\\
&\quad + 2\overline{d}_s k_m^2\epsilon_m b_m\delta_{\mu_m}\left\|\dot{q}_m\right\|^2\left(q_m - q_s\right)^T\overline{S}^{-1}\dot{q}_m\\
&\leq 2\overline{d}_s k_m^2\overline{\rho}_m\left(b_m\overline{\delta}_{\mu_m}\left\|\dot{q}_m\right\|^2\right)\dot{q}_m^T\overline{S}^{-1}\dot{q}_m\\
&\quad + \eta_3^{-1}\left(k_m\overline{d}_s\epsilon_m\right)^2\left(q_m - q_s\right)^T\overline{S}^{-2}\left(q_m - q_s\right)\\
&\quad + \eta_3\left(k_m b_m\overline{\delta}_{\mu_m}\left\|\dot{q}_m\right\|^2\right)^2\dot{q}_m^T\dot{q}_m.
\end{aligned}
\tag{7.16}
$$

and

$$
\begin{aligned}
&2\overline{d}_m\left(-k_s\overline{\rho}_s\dot{q}_s + k_s\epsilon_s\left(q_m - q_s\right)\right)^T\overline{Z}^{-1}\left(-k_s b_s\delta_{\mu_s}\left\|\dot{q}_s\right\|^2\right)\dot{q}_s,\\
&\leq 2\overline{d}_m k_s\overline{\rho}_s\left(k_s b_s\overline{\delta}_{\mu_s}\left\|\dot{q}_s\right\|^2\right)\dot{q}_s^T\overline{Z}^{-1}\dot{q}_s\\
&\quad + \eta_4^{-1}\left(\overline{d}_m k_s\epsilon_s\right)^2\left(q_m - q_s\right)^T\overline{Z}^{-2}\left(q_m - q_s\right)\\
&\quad + \eta_4\left(k_s b_s\overline{\delta}_{\mu_s}\left\|\dot{q}_s\right\|^2\right)^2\dot{q}_s^T\dot{q}_s,
\end{aligned}
\tag{7.17}
$$

where η_3 and η_4 are positive scalars.

It can be verified that

$$
\begin{aligned}
&2\rho_i\dot{q}_i^T\left(-\alpha_i\dot{q}_i - \beta_i\left\|\dot{q}_i\right\|^2\dot{q}_i\right)\\
&\leq 2\left(a_i + 2\mu_{i1}b_i\left\|\dot{q}_i\right\|^2\right)\dot{q}_i^T\left(-\alpha_i\dot{q}_i - \beta_i\left\|\dot{q}_i\right\|^2\dot{q}_i\right)\\
&\leq -2a_i\alpha_i\left\|\dot{q}_i\right\|^2 - 4\mu_{i1}\beta_i b_i\left\|\dot{q}_i\right\|^6\\
&\quad - \left(2a_i\beta_i + 4\alpha_i\mu_{i1}b_i\right)\left\|\dot{q}_i\right\|^4.
\end{aligned}
\tag{7.18}
$$

With the defined Lyapunov functional (7.4), there exist positive scalars l_{m1}, l_{m2}, l_{s1}, l_{s2} and l_3 such that

$$
\begin{aligned}
V_1 \leq& l_{m1}\left\|\dot{q}_m\right\|^2 + 2l_{m2}\left\|\dot{q}_m\right\|^4 + l_{s1}\left\|\dot{q}_s\right\|^2\\
&+ 2l_{s2}\left\|\dot{q}_s\right\|^4 + l_3\left(q_m - q_s\right)^T\left(q_m - q_s\right).
\end{aligned}
\tag{7.19}
$$

With (7.5)–(7.19) and $c_i\left\|\dot{q}_i\right\|^4 \geq 0$, the time derivative of Lyapunov functional V satisfies

$$
\dot{V} \leq -\gamma V + \Theta^T\Phi\Theta + \left(\vartheta_1^{-1} + \vartheta_3^{-1}\right)\overline{F}_h^2 + \left(\vartheta_2^{-1} + \vartheta_4^{-1}\right)\overline{F}_e^2
\tag{7.20}
$$

where $\Theta^T = \left[\, q_m^T - q_s^T \;\; \dot{q}_m^T \;\; \dot{q}_s^T \;\; \left\|\dot{q}_m\right\|^2\dot{q}_m^T \;\; \left\|\dot{q}_s\right\|^2\dot{q}_s^T \,\right]$ and

$$
\Phi = \begin{bmatrix} \Phi_{(1,1)} & \Phi_{(1,2)} & \Phi_{(1,3)} & \Phi_{(1,4)} & \Phi_{(1,5)} \\ * & \Phi_{(2,2)} & 0 & \Phi_{(2,4)} & 0 \\ * & * & \Phi_{(3,3)} & 0 & \Phi_{(3,5)} \\ * & * & * & \Phi_{(4,4)} & 0 \\ * & * & * & * & \Phi_{(5,5)} \end{bmatrix}
$$

where $*$ is the transpose of the corresponding matrix, and the elements are shown below:

$$
\begin{aligned}
\Phi_{(1,1)} =& \eta_1^{-1} I + \eta_2^{-1} I - 2k_m \epsilon_m I - 2k_s \epsilon_s I + 2h_m^{-1} \epsilon_m^2 I + 2h_s^{-1} \epsilon_s^2 I + \overline{d}_s k_m^2 \epsilon_m^2 \overline{S}^{-1} \\
& + \overline{d}_m k_s^2 \epsilon_s^2 \overline{Z}^{-1} + \eta_3^{-1} \left(\overline{d}_s k_m \epsilon_m\right)^2 \overline{S}^{-2} + \eta_4^{-1} \left(\overline{d}_m k_s \epsilon_s\right)^2 \overline{Z}^{-2} \\
& + \gamma l_3 I + \vartheta_3 \epsilon_m^2 I + \vartheta_4 \epsilon_s^2 I, \\
\Phi_{(1,2)} =& - k_m a_m I + \overline{d}_s k_m^2 a_m \epsilon_m \overline{S}^{-1} + kI, \\
\Phi_{(1,3)} =& k_s a_s I - \overline{d}_m k_s^2 \epsilon_s a_s \overline{Z}^{-1} - kI, \\
\Phi_{(1,4)} =& - (\mu_{m1} + \mu_{m2}) k_m b_m I + \overline{d}_s b_m k_m^2 \epsilon_m (\mu_{m1} + \mu_{m2}) \overline{S}^{-1}, \\
\Phi_{(1,5)} =& (\mu_{s1} + \mu_{s2}) b_s k_s I - \overline{d}_m b_s k_s^2 \epsilon_s (\mu_{s1} + \mu_{s2}) \overline{Z}^{-1}, \\
\Phi_{(2,2)} =& - 2a_m \alpha_m I + 2\mu_{m2} \epsilon_m I + \omega_1^{-1} I + \omega_2^{-1} I + \overline{d}_m Z + \overline{d}_s k_m^2 a_m^2 \overline{S}^{-1} \\
& + \gamma l_{m1} I + \vartheta_1 a_m^2, \\
\Phi_{(2,4)} =& 2.5 h_m K_{cm}^2 I + \overline{d}_s k_m^2 a_m b_m (\mu_{m1} + \mu_{m2}) \overline{S}^{-1} + \overline{d}_s k_m^2 a_m b_m \overline{\delta}_{\mu_m} \overline{S}^{-1} \\
& - a_m \beta_m I - 2\alpha_m \mu_{m1} b_m I + c_m I + \gamma l_{m2} I + 2\vartheta_1 \mu_m a_m b_m I, \\
\Phi_{(3,3)} =& - 2\alpha_s a_s I + 2\mu_{s2} \epsilon_s I + \omega_1 \epsilon_m^2 \mu_{m2}^2 I + \omega_2 \epsilon_s^2 \mu_{s2}^2 I + \overline{d}_s S + \overline{d}_m k_s^2 a_s^2 \overline{Z}^{-1} \\
& + \gamma l_{s1} I + \vartheta_2 a_s^2, \\
\Phi_{(3,5)} =& 2.5 h_s K_{cs}^2 I + \overline{d}_m (\mu_{s1} + \mu_{s2}) k_s^2 a_s b_s \overline{Z}^{-1} + \overline{d}_m k_s^2 a_s b_s \overline{\delta}_{\mu_s} \overline{Z}^{-1} \\
& - a_s \beta_s I - 2\alpha_s \mu_{s1} b_s I + c_s I + \gamma l_{s2} I + 2\vartheta_2 \mu_s a_s b_s I, \\
\Phi_{(4,4)} =& \eta_1 \overline{\delta}_{\mu_m}^2 b_m^2 k_m^2 I + \overline{d}_s k_m^2 (\mu_{m1} + \mu_{m2})^2 b_m^2 \overline{S}^{-1} + \overline{d}_s \left(k_m b_m \overline{\delta}_{\mu_m}\right)^2 \overline{S}^{-1} \\
& + 2\overline{d}_s k_m^2 (\mu_{m1} + \mu_{m2}) b_m^2 \overline{\delta}_{\mu_m} \overline{S}^{-1} + \eta_3 \left(k_m b_m \overline{\delta}_{\mu_m}\right)^2 I - 4b_m \mu_{m1} \beta_m I \\
& + 4\vartheta_1 b_m^2 \mu_m^2 I, \\
\Phi_{(5,5)} =& \eta_2 \overline{\delta}_{\mu_s}^2 b_s^2 k_s^2 I + \overline{d}_m k_s^2 (\mu_{s1} + \mu_{s2})^2 b_s^2 \overline{Z}^{-1} + \overline{d}_m k_s^2 \left(b_s \overline{\delta}_{\mu_s}\right)^2 \overline{Z}^{-1} \\
& + 2\overline{d}_m k_s^2 (\mu_{s1} + \mu_{s2}) b_s^2 \overline{\delta}_{\mu_s} \overline{Z}^{-1} + \eta_4 \left(k_s b_s \overline{\delta}_{\mu_s}\right)^2 I \\
& - 4b_s \mu_{s1} \beta_s I + 4\vartheta_2 b_s^2 \mu_s^2 I.
\end{aligned}
$$

With the above analysis, we have the main result:

Theorem 7.1 *For the teleoperation system (7.1) with the control scheme (7.2), if there exist positive scalars $\omega_i, \eta_i, h_m, h_s, \vartheta_i$ and positive definite Z, S such that matrix $\Phi < 0$, then the velocities $\dot{q}_m, \dot{q}_s$ and position tracking error $q_m - q_s$ converge to small bounded regions exponentially and their convergence regions are determined by $\overline{F}_h$ and $\overline{F}_e$.*

Proof With (7.20) and $\Phi < 0$, we have $\dot{V} \le -\gamma V + \nu$ where $\nu = \left(\vartheta_1^{-1} + \vartheta_3^{-1}\right) \overline{F}_h^2 + \left(\vartheta_2^{-1} + \vartheta_4^{-1}\right) \overline{F}_e^2$. Then, one further has $V(t) \le e^{-\gamma t} V(0) + \frac{1}{\gamma}\nu$. With the defined V in (7.4), there exist positive scalars $\overline{a}_m$, $\overline{a}_s$ and $\overline{k}$ such that $V(t) \ge \overline{a}_m \mu_{m1} \dot{q}_m^T \dot{q}_m + \overline{a}_s \mu_{s1} \dot{q}_s^T \dot{q}_s + \overline{k}(q_m - q_s)^T (q_m - q_s)$. Therefore, we have

$$\begin{cases} \|\dot{q}_m\|^2 \le \dfrac{1}{\overline{a}_m \mu_{m1}} \left(e^{-\gamma t} V(0) + \dfrac{1}{\gamma}\nu\right), \\ \|\dot{q}_s\|^2 \le \dfrac{1}{\overline{a}_s \mu_{s1}} \left(e^{-\gamma t} V(0) + \dfrac{1}{\gamma}\nu\right), \\ \|q_m - q_s\|^2 \le \dfrac{1}{\overline{k}} \left(e^{-\gamma t} V(0) + \dfrac{1}{\gamma}\nu\right). \end{cases} \tag{7.21}$$

Then $\dot{q}_m, \dot{q}_s$ and $q_m - q_s$ converge to the bounded regions exponentially, and their bounds are determined by ν. The proof is completed.

Remark 7.2 By using the Schur complement lemma, we can convert the inequality $\Phi < 0$ into a strict linear matrix inequality $\overline{\Phi} < 0$ where

$$\begin{aligned}
\overline{\Phi}_{(1,1)} =& -2k_m\epsilon_m I - 2k_s\epsilon_s I + \overline{d}_m k_s^2 \epsilon_s^2 \overline{Z}^{-1} + \overline{d}_s k_m^2 \epsilon_m^2 \overline{S}^{-1} + \gamma l_3 I + \vartheta_3 \epsilon_m^2 I \\
& + \vartheta_4 \epsilon_s^2 I, \\
\overline{\Phi}_{(1,2)} =& -k_m a_m I + \overline{d}_s k_m^2 a_m \epsilon_m \overline{S}^{-1} + kI, \\
\overline{\Phi}_{(1,3)} =& k_s a_s I - \overline{d}_m k_s^2 \epsilon_s a_s \overline{Z}^{-1} - kI, \\
\overline{\Phi}_{(1,4)} =& -(\mu_{m1} + \mu_{m2}) k_m b_m I + \overline{d}_s b_m k_m^2 \epsilon_m (\mu_{m1} + \mu_{m2}) \overline{S}^{-1}, \\
\overline{\Phi}_{(1,5)} =& (\mu_{s1} + \mu_{s2}) b_s k_s I - \overline{d}_m b_s k_s^2 \epsilon_s (\mu_{s1} + \mu_{s2}) \overline{Z}^{-1}, \\
\overline{\Phi}_{(2,2)} =& -2a_m\alpha_m I + 2\mu_{m2}\epsilon_m I + \overline{d}_s k_m^2 a_m^2 \overline{S}^{-1} + \gamma l_{m1} I + \vartheta_1 a_m^2, \\
\overline{\Phi}_{(2,4)} =& 2.5h_m K_{cm}^2 I + \overline{d}_s k_m^2 a_m b_m (\mu_{m1} + \mu_{m2}) \overline{S}^{-1} + \overline{d}_s k_m^2 a_m b_m \overline{\delta}_{\mu_m} \overline{S}^{-1} \\
& - a_m \beta_m I - 2\alpha_m \mu_{m1} b_m I + c_m I + \gamma l_{m2} I + 2\vartheta_1 \mu_m a_m b_m, \\
\overline{\Phi}_{(3,3)} =& -2\alpha_s a_s I + 2\mu_{s2}\epsilon_s I + \omega_1 \epsilon_m^2 \mu_{m2}^2 I + \omega_2 \epsilon_s^2 \mu_{s2}^2 I + \overline{d}_m k_s^2 a_s^2 \overline{Z}^{-1} \\
& + \gamma l_{s1} I + \vartheta_2 a_s^2, \\
\overline{\Phi}_{(3,5)} =& 2.5h_s K_{cs}^2 I + \overline{d}_m (\mu_{s1} + \mu_{s2}) k_s^2 a_s b_s \overline{Z}^{-1} + \overline{d}_m k_s^2 a_s b_s \overline{\delta}_{\mu_s} \overline{Z}^{-1} - a_s \beta_s I \\
& - 2\alpha_s \mu_{s1} b_s I + c_s I + \gamma l_{s2} I + 2\vartheta_2 \mu_s a_s b_s, \\
\overline{\Phi}_{(4,4)} =& \eta_1 \overline{\delta}_{\mu_m}^2 b_m^2 k_m^2 I + \overline{d}_s k_m^2 (\mu_{m1} + \mu_{m2})^2 b_m^2 \overline{S}^{-1} + \overline{d}_s \left(k_m b_m \overline{\delta}_{\mu_m}\right)^2 \overline{S}^{-1}
\end{aligned}$$

$$+ 2\overline{d}_s k_m^2 (\mu_{m1} + \mu_{m2}) b_m^2 \overline{\delta}_{\mu_m} \overline{S}^{-1} + \eta_3 \left(k_m b_m \overline{\delta}_{\mu_m}\right)^2 - 4 b_m \mu_{m1} \beta_m I$$
$$+ 4\vartheta_1 b_m^2 \mu_m^2,$$

$$\overline{\Phi}_{(5,5)} = \eta_2 \overline{\delta}_{\mu_s}^2 b_s^2 k_s^2 I + \overline{d}_m k_s^2 (\mu_{s1} + \mu_{s2})^2 b_s^2 \overline{Z}^{-1} + \overline{d}_m k_s^2 \left(b_s \overline{\delta}_{\mu_s}\right)^2 \overline{Z}^{-1}$$
$$+ 2\overline{d}_m k_s^2 (\mu_{s1} + \mu_{s2}) b_s^2 \overline{\delta}_{\mu_s} \overline{Z}^{-1} + \eta_4 \left(k_s b_s \overline{\delta}_{\mu_s}\right)^2 - 4 b_s \mu_{s1} \beta_s I + 4\vartheta_2 b_s^2 \mu_s^2.$$

$$\overline{\Phi}_{(1,6)} = I, \Phi_{(1,7)} = I, \overline{\Phi}_{(6,6)} = -\eta_1 I, \Phi_{(7,7)} = -\eta_2 I,$$
$$\overline{\Phi}_{(1,8)} = \overline{d}_s k_m \epsilon_m \overline{S}^{-1}, \overline{\Phi}_{(8,8)} = -\eta_3 I,$$
$$\overline{\Phi}_{(1,9)} = \overline{d}_m k_s \epsilon_s \overline{Z}^{-1}, \overline{\Phi}_{(9,9)} = -\eta_4 I,$$
$$\overline{\Phi}_{(1,10)} = \epsilon_m I, \overline{\Phi}_{(10,10)} = -0.5 h_m I,$$
$$\overline{\Phi}_{(1,11)} = \epsilon_s I, \overline{\Phi}_{(11,11)} = -0.5 h_s I$$
$$\overline{\Phi}_{(2,12)} = \overline{d}_m I, \overline{\Phi}_{(12,12)} = -\overline{d}_m Z^{-1} = -\overline{d}_m e^{-\gamma \overline{d}_m} \overline{Z}^{-1},$$
$$\overline{\Phi}_{(2,13)} = I, \overline{\Phi}_{(13,13)} = -\omega_1 I;$$
$$\overline{\Phi}_{(2,14)} = I, \overline{\Phi}_{(14,14)} = -\omega_2 I;$$
$$\overline{\Phi}_{(3,15)} = \overline{d}_s I, \overline{\Phi}_{(15,15)} = -\overline{d}_s S^{-1} = -\overline{d}_s e^{-\gamma \overline{d}_s} \overline{S}^{-1}.$$

Then we may use the LMI tool box of Matlab to solve it.

Remark 7.3 The parameters $\omega_1, \omega_2, \eta_1, \eta_2, \eta_3, \eta_4, h_m, h_s, \vartheta_1, \vartheta_2, \vartheta_3$ and ϑ_4 are used to deal with the uncertainties in inequalities (7.7)–(7.10), (7.16) and (7.17), which brings the freedom for solving $\Phi < 0$. The optimal values of them can be obtained by solving the transformed LMI $\overline{\Phi} < 0$. For $c_i \|\dot{q}_i\|^4 \geq 0$, the free parameters c_i are adopted to reduce the conservativeness of the matrix inequality. The LMI $\overline{\Phi} < 0$ contains the controller design parameters $k_m, k_s, \alpha_m, \alpha_s, \beta_m$ and β_s, and the bound values $\overline{d}_m, \overline{d}_s$ of the forward and backward time delays. With the given delay parameters $\overline{d}_m$ and $\overline{d}_s$, we can design the parameters $k_m, k_s, \alpha_m, \alpha_s, \beta_m$ and β_s based on the given LMI condition, then the stability of the closed-loop system is guaranteed. On the other hand, for the given set of control parameters and the delay bound of one path, we can get the allowable maximal values of the delay of the other path by using the LMI optimization tool box provided with Matlab. For example, if $\overline{d}_s$ is known, we can obtain the allowable maximal value $\overline{d}_m$.

7.4 Simulations and Experiments

In this section, simulations are first presented to verify the effectiveness of the main result. Then, experiments performed on a teleoperation system consisting of two Phantom Premium 1.5 A robot manipulators are also shown.

7.4.1 Simulation Results

The simulations assumed a teleoperation system consisting of 2-DOF manipulators (master and slave) with the model shown below:

$$\begin{aligned} M_m(q_m)\ddot{q}_m + C_m(q_m, \dot{q}_m)\dot{q}_m &= \tau_m + F_h \\ M_s(q_s)\ddot{q}_s + C_s(q_s, \dot{q}_s)\dot{q}_s &= \tau_s - F_e \end{aligned},$$

where

$$M_m(q) = M_s(q) = \begin{bmatrix} M_{11} & M_{12} \\ * & M_{22} \end{bmatrix},$$

$$C_m(q, \dot{q}) = C_s(q, \dot{q}) = \begin{bmatrix} C_{11} & C_{12} \\ C_{21} & C_{22} \end{bmatrix},$$

in which

$$\begin{aligned} M_{11} &= (2L_1 \cos q_2 + L_2) L_2 m_2 + L_1^2 (m_1 + m_2) \\ M_{12} &= L_2^2 m_2 + L_1 L_2 m_2 \cos q_2, M_{22} = L_2^2 m_2, \\ C_{11} &= -L_1 L_2 m_2 \sin(q_2)\dot{q}_2, C_{21} = L_1 l_2 m_2 \sin(q_2)\dot{q}_1, \\ C_{12} &= -L_1 L_2 m_2 \sin(q_2)(\dot{q}_1 + \dot{q}_2), C_{22} = 0, \end{aligned}$$

The parameters used for simulation are: $m_1 = 1\,\text{kg}$, $m_2 = 0.5\,\text{kg}$, $L_1 = 0.5\,\text{m}$, $L_2 = 0.5\,\text{m}$, $g = 9.81\,\text{m/s}^2$. The user exerted force F to the master robot on the Y-direction, which is shown in Fig. 7.2. The force that the user applied was zero at 0 s, and then was increased to 5 N at 10 s. It was kept constant from time 10 to 40 s, and decreased to zero from time 40 to 50 s. At the slave side, there was a wall at $y = 0.5$ m.

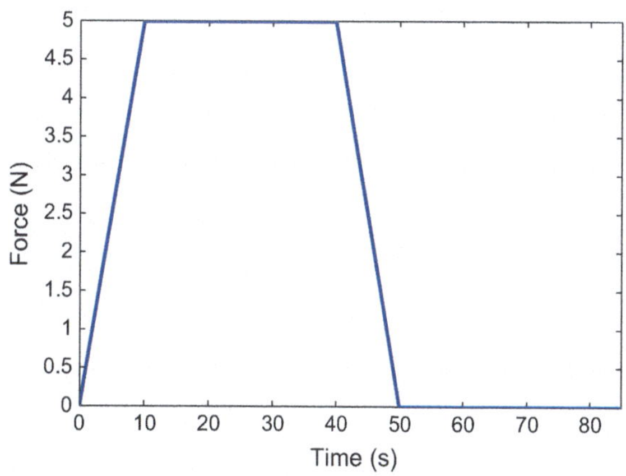

Fig. 7.2 The human force input

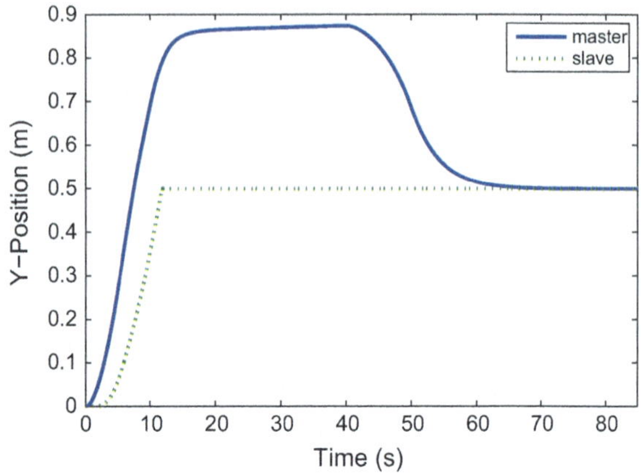

Fig. 7.3 The Y-position of the master and the slave with controller (7.2)

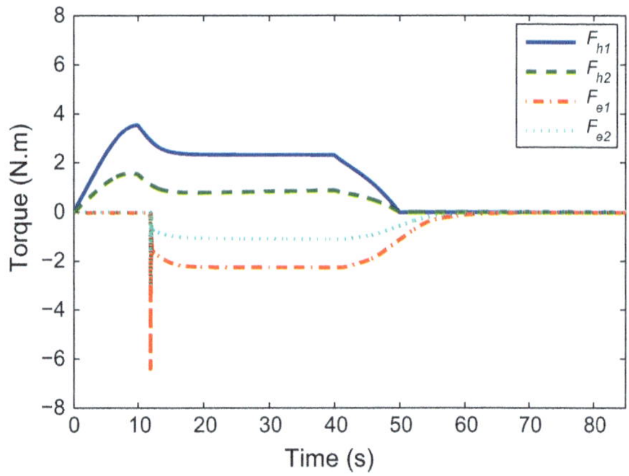

Fig. 7.4 The torques of F_h and F_e with controller (7.2)

When the slave joint reached the wall and moved further, the feedback force was $10000 * (y - 0.5)\, N$. We can see that the wall is quite stiff.

With the human force input, we know the torque $F_h = -J_m^T * \begin{bmatrix} 0 & 1 \end{bmatrix}^T * F$, where J_m is the Jacobian matrix of the master manipulator. The torque F_e is $F_e = -J_s^T * \begin{bmatrix} 0 & 1 \end{bmatrix}^T * 10000N * (y - 0.5)$. The simulation was used to verify the following: (i) When the user moves the master robot, does the slave follow the movement of the master? (ii) When the slave contacts the wall, is the torque feedback F_e from the wall close to the human input F_h? (iii) When the human input force is changed to zero, does the error between the master and slave positions disappear? For (i) and (iii), our aim is to show the stability of the closed-loop system. For (ii), it is used to show whether the slave torque changes with the changes of the master torque. If we see F_e as the environment force, the change of F_e can be used to show the transparency performance. To show the effectiveness of the proposed delay-dependent controller,

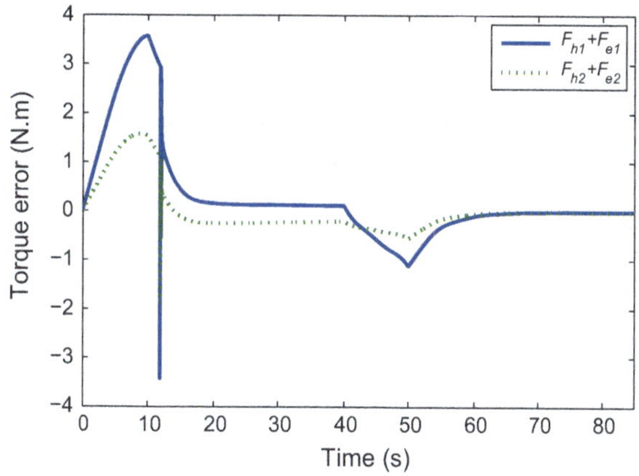

Fig. 7.5 The error between F_h and F_e with controller (7.2)

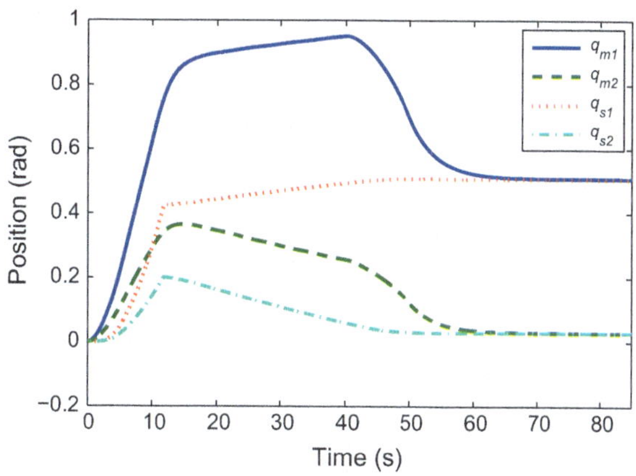

Fig. 7.6 The positions of the master and slave joints with controller (7.2)

all the above questions will be checked with the computed maximum allowable time delays.

Based on the properties, we have $\mu_{i1} = 0.1250$, $\mu_{i2} = 0.2500$, $K_{ci} = 0.3$. Based on the values of the physical parameters, we chose $\epsilon_i = \epsilon_s = 0.1, k = 1, l_{m1} = l_{m2} = l_{s1} = l_{s2} = 0.5, l_3 = 1$. Now we use Theorem 7.1 for the controller design.

With Theorem 7.1, the control design parameters are given as $k_i = 5, \alpha_i = \beta_i = 20, \gamma = 0.1$. Choose the forward delay parameter $\overline{d}_m = 1$, solving the LMI gives the maximum allowable backward time delay $\overline{d}_s = 2.7084$. The time delays chosen for simulations were $d_m(t) = 0.8 + 0.2\sin(t)$ and $d_s(t) = 2 + 0.7084\sin(t)$. Now we employ Theorem 7.1 for the controller design.

With the controller (7.2), the responses of y-position are shown in Fig. 7.3. From Fig. 7.3, we can see that the master and slave joints move forward with the increase of human force F. The slave joint meets the stiff wall at position $y = 0.5$ and could not move anymore, while the master moves further because of the force applied by the user. With the same force maintained from time 10 to 40 s, the Y positions of the master and slave joints keep unchanged. With the decrease of F from 40 s, the master joint moves backward and reaches $y = 0.5$. Because there is no negative force to move backward, the master and the slave joints stop at $y = 0.5$. The applied master and slave torques are shown in Fig. 7.4, and the error between them is shown in Fig. 7.5. If we see F_e as the force reflection of the input F_h, we can see that good transparency performance is achieved. The positions of q_m and q_s are shown in Fig. 7.6, from which we can see that the closed-loop system shows good steady-state and transient performances.

7.4.2 Experimental Results

The teleoperation system for the experiments consists of two Phantom Premium 1.5 A robots (SensAble Technologies Inc.) see Fig. 2.7 are also conducted. A hard wood box was set at position −0.05 m on the X-direction of the slave manipulator, which represents a stiff wall.

The user pushed the last joint of the master manipulator to move it in the X-direction. The corresponding slave joint followed the movement of the master joint, and then the slave joint contacted the stiff box and could not move any more. Our objective here is: (i) to verify the effectiveness of the proposed controller (7.2) and to check whether the slave joint follows the movement of the master joint and whether they can return eventually to the original positions; (ii) to test how the user feel the feedback force, especially when the slave robot contacts the stiff wall.

With the use of the proposed controller (7.2), the movements of the master and slave joints on the X-direction are shown in Fig. 7.7. In Fig. 7.8, the positions of q_m and q_s are shown. We can see that the slave joint reached −0.05 m and could not move any more. When the user moved the master joint further, force feedback was felt, so he knew that the slave contacted with an obstacle. The felt force was stronger when he moved the master joint farther. Finally, the user moved the master joint

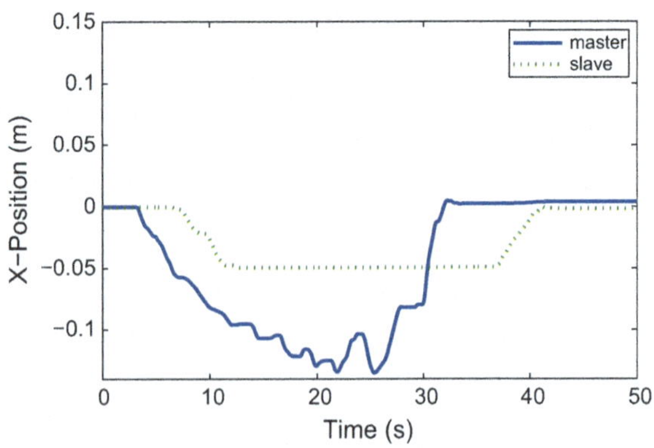

Fig. 7.7 The X-position of the master and the slave with controller (7.2) in the experiment

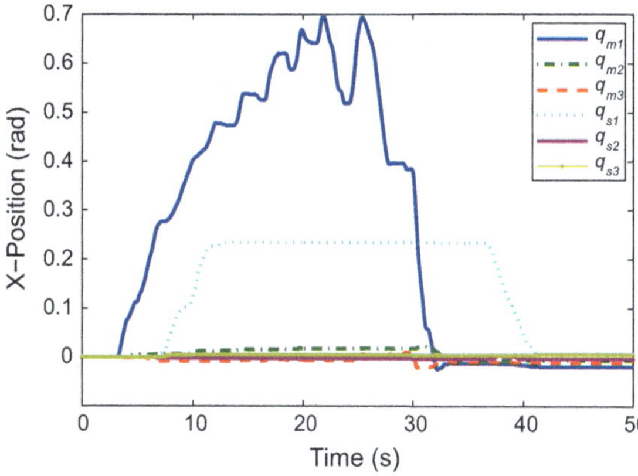

Fig. 7.8 The positions of the master and slave joints with controller (7.2) in the experiment

back to the original position. From the figure, we can see that the slave manipulator followed the movement of the master and reached the origin at last.

7.5 Summary

This chapter focuses on the controller design problem for network-based teleoperation systems with asymmetric and time-varying delays. We propose a master-slave controller, which renders the whole system exponentially stable. The relationships are built among the control design parameters and the allowable maximal time delays. The method is extended to cases that the velocity information is not available. Both the simulations and the experiments performed confirm well the effectiveness of the main results.

References

1. C.C. Hua, X.P. Liu, Delay-dependent stability criteria of teleoperation systems with asymmetric time-varying delays. IEEE Trans. Rob. **26**(5), 925–932 (2010)
2. C.C. Hua, P.X. Liu, Convergence analysis of teleoperation systems with unsymmetric time-varying delays. IEEE Trans. Circuits Syst. II Express Briefs **56**(3), 240–244 (2009)
3. C.C. Hua, X.P. Liu, A new coordinated slave torque feedback control algorithm for network-based teleoperation systems. IEEE/ASME Trans. Mechatron. **18**(2), 764–774 (2013)
4. Z. Chen, Y.J. Pan, J. Gu, Integrated adaptive robust control for multilateral teleoperation systems under arbitrary time delays. Int. J. Robust Nonlinear Control **26**(12), 2708–2728 (2016)
5. H.J. Yang, Y.Q. Xia, P. Shi, Stabilization of networked control systems with nonuniform random sampling periods. Int. J. Robust Nonlinear Control **21**(5), 501–526 (2011)
6. H.K. Khalil, *Nonlinear Systems*, 3rd edn. (Prentice Hall, 2002)
7. M. Vidyasagar, *Nonlinear Systems Analysis*. Society for Industrial and Applied Mathematics (2002)
8. D.B. Zhao, Z.S. Song, J.Q. Yi, X.C. Li, A computed torque controller for uncertain robotic manipulator systems: fuzzy approach. Fuzzy Sets Syst. **154**(2), 208–226 (2005)

Chapter 8
Finite Time Controller Design for Bilateral Teleoperation System

Abstract The master-slave control design problem is considered for networked teleoperation system with friction and external disturbances in this chapter. A new finite-time synchronization control method is proposed with the help of adaptive fuzzy approximation. We develop a new nonsingular fast terminal sliding mode (NFTSM) to provide faster convergence and higher precision than the linear hyperplane sliding mode and the classic terminal sliding mode (TSM). Then, the adaptive fuzzy-logic system is employed to approximate the system uncertainties and the corresponding adaptive fuzzy NFTSM controller is designed. By constructing Lyapunov function, the stability and finite time synchronization performance are proved with the new controller in the presence of system uncertainties and external disturbances. Compared with the traditional teleoperation design method, the new control scheme achieves better transient-state performance and steady-state performance. Finally, the simulations are performed and the comparisons are shown among the proposed method, the P+d method, the PD+d method, the DFF method and the classic TSM, FTSM. The simulation and experiment results further demonstrate the effectiveness of the proposed method.

8.1 Introduction

In recent years, the teleoperation system has been found in many areas from micro to macro scales, for example, space operation, undersea exploration, handling hazardous materials, telesurgery (see [1] and the references cited therein). To provide satisfactory work performance in above working scene, many methods have been proposed to deal with the teleoperation system design problem. The passivity-based method was developed to overcome the instability of teleoperation system caused by constant time delay in [2]. Latter, the wave variables following the former scattering transformation was introduced in [3]. Subsequent schemes, building on the above two approaches, have been suggested in [4, 5]. After that, the passivity-based architecture was extended in [6] to guarantee state synchronization of master/slave without using scattering transformation. Following this line, in [7], a globally stable P+d controller was proposed, and the stability conditions were developed for the constant time delay

C. Hua et al., *Analysis and Design for Networked Teleoperation System*,
https://doi.org/10.1007/978-981-13-7936-9_8

case without the delayed derivative action. In addition, the delay dependent linear matrix inequality (LMI) stability criteria for the closed-loop teleoperation system with time-varying delays was proposed in [8]. In [9], a kinematic control framework for asymmetric teleoperation systems was proposed. Considering the case that the velocity signals can not be obtained directly, a velocity observer-based new controller was proposed in [10]. Moreover the exponential synchronization performance was achieved. However, in the above works, the asymptotic/exponential synchronization performances were achieved. It means that the synchronization error converges to zero when time goes to infinity. It is well known that for an ideal teleoperation system, we expect the slave can follow the master quickly and the synchronization performance could be achieved in finite time.

The finite-time control problem will be considered in this chapter by using sliding mode control (SMC) method. SMC has been widely used because of their relative simplicity of implementation and their robustness against both plant uncertainties and external disturbances. However, the classic linear hyperplane-based sliding mode can only guarantee the asymptotic error convergence [11–13]. Thus the finite-time control theory appeared to get a higher convergence speed. Terminal sliding mode control (TSMC) is an effective finite-time control approach, and it was first proposed in [14]. Similar to the linear sliding mode technique, strong robustness with respect to uncertain dynamics can be obtained with the TSM. Moreover, the tracking error converges to zero in finite time [15, 16]. However, the premier papers about TSMC encountered singularity problem. To deal with the singularity problem, a nonsingular terminal sliding manifold was proposed in [17]. Then a globally continuous nonsingular terminal sliding mode control method was presented in [18]. Afterwards, fuzzy logic system and a continuous nonsingular TSM were combined to control a manipulator system in [19]. Different from the direct method, an indirect method was proposed in [20] to avoid the singularity problem, and the modified terminal sliding manifold could produce a smoother switch. Motivated by [20], a Chebyshev neural network-based nonsingular TSM controller was presented in [21] for spacecraft control. In [22], some new forms of fast TSM strategies were presented, and comparisons about the convergence speed were made between the different fast TSMC.

It is well known that robotic manipulator is complicated, dynamically coupled, highly time-varying and highly nonlinear system. Moreover, with the limitation of modeling method and the complex external task environment, teleoperation system is inevitably subject to structured and unstructured uncertainties. Fuzzy Logical system (FLs) was introduced by Zadeh in 1965, as one of the most popular intelligent computation approaches, which was proved to be an essential tool for solving some various classes of engineering problems. To approximate the system uncertainties and external disturbance, the fuzzy logic system was widely used in robotic control [23, 24]. Recently, the adaptive fuzzy approximation approach was also used in controlling MIMO nonlinear system and nonlinear time-delay system [25–27]. The adaptive fuzzy method was used in controlling teleoperation system under time-varying delay in [28, 29]. In this chapter, we will use the adaptive fuzzy method to deal with the uncertainties in the master-slave model.

This chapter presents a finite time design scheme for networked teleoperation system with uncertainties. A TSM controller is presented at first. Then, by employing the adaptive fuzzy approximation method, we design master-slave controllers. With the new controllers, the finite-time synchronization performance is achieved. The main contents of this chapter are stated as follows: (i) A new FTSM is proposed. Compared with TSM control method, the faster transient-state and higher-precision control performances are both achieved. (ii) The FLs and parameter adaptive method are combined to deal with the system uncertainties and external disturbance, and the system stability is rigorously proved with the designed adaptive fuzzy controllers. (iii) The precise reaching time and the sliding time can be computed with system initial states, sliding mode parameters and controller parameters. Moreover, both simulation and experiment are conducted to show the effectiveness of the proposed control method.

8.2 Problem Formulation

8.2.1 Dynamics of Teleoperator

Consider a master-slave system given by the following model

$$\begin{cases} M_m(q_m)\ddot{q}_m + C_m(q_m, \dot{q}_m)\dot{q}_m + F_m\dot{q}_m \\ +f_{cm}(\dot{q}_m) + \tau_{dm} + G_m(q_m) = \tau_m + F_h \\ M_s(q_s)\ddot{q}_s + C_s(q_s, \dot{q}_s)\dot{q}_s + F_s\dot{q}_s \\ +f_{cs}(\dot{q}_s) + \tau_{ds} + G_s(q_s) = \tau_s - F_e \end{cases} \tag{8.1}$$

where $F_m, F_s \in R$ denote the viscous friction coefficients; $f_{cm}(\dot{q}_m), f_{cs}(\dot{q}_s) \in R^n$ are the Coulomb friction coefficients; $\tau_{dm}, \tau_{ds} \in R^n$ denote the bounded unknown external disturbances.

Remark 8.1 In this chapter, both the viscous friction and the Coulomb friction are considered in the master robot and the slave robot, respectively. Here, the Coulomb friction function $f_{ci}(\dot{q}_i)$ is a bounded and piecewise continuous function.

8.2.2 Fuzzy Logic Systems

During the past years, fuzzy logic systems (FLs) have been extensively used as universal approximators for controllers design of dynamic systems with precise model unknown. A fuzzy system is a collection of fuzzy IF-THEN rules of the form:

$$R^{(j)} : IF\ z_1(t)\ is\ A_1^j\ and\ \cdots\ and\ z_n(t)\ is\ A_n^j$$
$$THEN\ y(t)\ is\ B^j.$$

By using the strategy of singleton fuzzification, product inference and center-average defuzzification, the output of the fuzzy system is

$$y(z(t)) = \frac{\sum_{j=1}^{l} y^j \left(\prod_{i=1}^{n} \mu_{A_i^j}(z_i(t))\right)}{\sum_{j=1}^{l} \prod_{i=1}^{n} \mu_{A_i^j}(z_i(t))}, \tag{8.2}$$

where $\mu_{A_i^j}(z_i(t))$ is the membership function of linguistic variable $z_i(t)$, and y^j is the point in $\mathbb{R}$ at which μ_{B^j} achieves its maximum value (assume that $\mu_{B^i}(y^i) = 1$).

Introducing the concept of the fuzzy basic function vector $\varsigma(z(t))$ gives

$$y(z(t)) = \theta(t)^T \varsigma(z(t)), \tag{8.3}$$

where

$$\theta(t) = \left(y^1(t), y^2(t), \ldots, y^l(t)\right)^T,$$
$$\varsigma(z(t)) = \left(\varsigma_1(z(t)), \varsigma_2(z(t)), \ldots, \varsigma_l(z(t))\right)^T,$$

and $\varsigma_j(z(t))$ is defined as

$$\varsigma_j(z(t)) = \frac{\prod_{i=1}^{n} \mu_{A_i^j}(z_i(t))}{\sum_{j=1}^{l} \prod_{i=1}^{n} \mu_{A_i^j}(z_i(t))}.$$

Based on the universal approximation theorem, there exists the optimal approximation parameter θ^* such that $\theta^{*T}\varsigma(z(t))$ can approximate a nonlinear function $G(z(t))$ to any desired degree over a compact set Ω_z. The parameter θ^* is defined as follows

$$\theta^* = \arg\min_{\theta\in\Omega_\theta}\left(\sup_{z\in\Omega_z}\left|\theta^T(t)\varsigma(z(t)) - G(z(t))\right|\right),$$

where Ω_θ and Ω_z denote the sets of suitable bounds on $\theta(t)$ and $z(t)$, respectively. The minimum approximation error satisfies

$$G(z(t)) = \theta^{*T}\varsigma(z(t)) + \epsilon(z(t)) \tag{8.4}$$

where $\|\epsilon(z(t))\| \le \epsilon^*$ over $z(t) \in \Omega_z$, and ϵ^* is a positive scalar.

Actually, because of the existence of friction and backlash, the dynamic functions of manipulators are piecewise continuous functions possibly. Make a supposition that $G'(z(t))$ is a piecewise function, which can be expressed as $G'(z(t)) = G_1(z(t)) +$

$G_2(z(t))$, where $G_1(z(t))$ is the continuous part and $G_2(z(t))$ is the bounded piecewise term. Therefore, we have

$$\begin{aligned} G'(z(t)) &= \theta^{*T}\varsigma(z(t)) + \epsilon(z(t)) + G_2(z(t)) \\ &= \theta^{*T}\varsigma(z(t)) + \bar{\epsilon}(z(t)) \end{aligned} \tag{8.5}$$

where $\bar{\epsilon}(z(t)) = \epsilon(z(t)) + G_2(z(t))$ with $\|\bar{\epsilon}(z(t))\| \leq \bar{\epsilon}^*$ is the piecewise function approximation error and $\bar{\epsilon}^*$ is an upper bound of the approximation error.

8.3 Main Results

8.3.1 TSM Manifold

In this section, a new nonsingular fast terminal sliding manifold is used for dealing with the problem of position tracking between the master and the slave. To simplify the expression, the next definition and lemmas are needed.

Definition 8.1

$$sig(\xi)^{\alpha} = \left[|\xi_1|^{\alpha_1} sign(\xi_1), |\xi_2|^{\alpha_2} sign(\xi_2) \ldots |\xi_n|^{\alpha_n} sign(\xi_n)\right]^T \tag{8.6}$$

where $\xi = [\xi_1, \xi_2, \ldots, \xi_n]^T \in R^n, \alpha_1, a_2, \ldots, a_n > 0$ and $sign(\cdot)$ being the standard signum function.

Lemma 8.1 (sliding time): *Choose the terminal sliding mode as follows*

$$s = \dot{e} + \alpha sig(e)^{\gamma_1} + \beta sig(e)^{\gamma_2} \tag{8.7}$$

where $\alpha > 0$, $\beta > 0$, $\gamma_1 > 1$ *and* $0 < \gamma_2 < 1$. *if* $s = 0$, *the convergence time* T_1 *of* e *is about*

$$T_1 < \frac{1}{\alpha}\frac{1}{\gamma_1 - 1} + \frac{1}{\beta}\frac{1}{\gamma_2 - 1} \tag{8.8}$$

Lemma 8.2 (reaching time) [18] *Consider the dynamics model* $\dot{x} = f(x)$, $f(0) = 0$ *and* $x \in R^n$. *If there is a positive definite scalar function* $V(x)$ *such that*

$$\dot{V}(x) \leq -\alpha V(x) - \beta V(x)^{\delta} \tag{8.9}$$

where $\alpha, \beta > 0$, $0 < \delta < 1$, *then the system is finite-time stable. Furthermore, the settling time is given by*

$$T \leq \frac{1}{\alpha(1-\delta)} \ln \frac{\alpha V^{1-\delta}(x_0) + \beta}{\beta} \tag{8.10}$$

Lemma 8.3 [18] *Assume $a_1 > 0$, $a_2 > 0$ and $0 < c < 1$, the following inequality holds:*

$$(a_1 + a_2)^c \leq a_1^c + a_2^c \tag{8.11}$$

The NFTSM proposed in this chapter is given as follows

$$\begin{aligned} s_m &= \dot{e}_m + \alpha_{m_1} sig(e_m)^{\gamma_{m_1}} + \alpha_{m_2}\beta_m(e_m) = \dot{e}_m + \lambda_m(e_m) \\ s_s &= \dot{e}_s + \alpha_{s_1} sig(e_s)^{\gamma_{s_1}} + \alpha_{s_2}\beta_s(e_s) = \dot{e}_s + \lambda_s(e_s) \end{aligned} \tag{8.12}$$

where α_{m_1}, α_{s_1}, α_{m_2} and α_{s_2} are positive constants, $\gamma_{m_1} > 1$, $\gamma_{s_1} > 1$; $e_m = q_s(t - d_s) - q_m$ and $e_s = q_m(t - d_m) - q_s$ are the position synchronization errors between the master manipulator and the slave manipulator; $\dot{e}_m = \dot{q}_s(t - d_s) - \dot{q}_m$ and $\dot{e}_s = \dot{q}_m(t - d_m) - \dot{q}_s$ are the velocity tracking errors; d_m represents the signal transmission time delay from the master side to the slave side and d_s stands for the transmission time delay from the slave side to the master side. The time delay d_m and d_s are not known in advance. The $\beta_m(e_m)$ and $\beta_s(e_s)$ are defined as follows

$$\beta_m(e_m) = \begin{cases} sig(e_m)^{\gamma_{m_2}} \ \text{if}\ \bar{s}_m = 0 \ \text{or}\ \bar{s}_m \neq 0,\ |e_m| > \mu_m \\ k_{m_1} e_m + k_{m_2} sign(e_m) e_m^2 \ \text{if}\ \bar{s}_m \neq 0,\ |e_m| \leq \mu_m \end{cases} \tag{8.13}$$

and

$$\beta_s(e_s) = \begin{cases} sig(e_s)^{\gamma_{s_2}} \ \text{if}\ \bar{s}_s = 0 \ \text{or}\ \bar{s}_s \neq 0,\ |e_s| > \mu_s \\ k_{s_1} e_s + k_{s_2} sign(e_s) e_s^2 \ \text{if}\ \bar{s}_s \neq 0,\ |e_s| \leq \mu_s \end{cases} \tag{8.14}$$

where $0 < \gamma_{m_2} < 1$ and $0 < \gamma_{s_2} < 1$; $\bar{s}_m = \dot{e}_m + \alpha_{m_1} sig(e_m)^{\gamma_{m_1}} + \alpha_{m_2} sig(e_m)^{\gamma_{m_2}}$, $\bar{s}_s = \dot{e}_s + \alpha_{s_1} sig(e_s)^{\gamma_{s_1}} + \alpha_{s_2} sig(e_s)^{\gamma_{s_2}}$; $k_{m_1} = (2 - \gamma_{m_2})\mu_m^{\gamma_{m_2}-1}$, $k_{m_2} = (\gamma_{m_2} - 1)\mu_m^{\gamma_{m_2}-2}$, $k_{s_1} = (2 - \gamma_{s_2})\mu_s^{\gamma_{s_2}-1}$, $k_{s_2} = (\gamma_{s_2} - 1)\mu_s^{\gamma_{s_2}-2}$; μ_m, μ_s are small positive constants.

Remark 8.2 Compared with the linear sliding mode (LSM), the system with the TSM can get faster convergence and higher control precision. However, the classic TSM faces with the singularity problem. To deal with the singularity problem, a TSM by using the switching idea was proposed in [20]. However, the convergence speed in [20] was in a relatively slow rate when the system states are far away from the equilibrium points. Considering the speed problem, the comparisons between the different TSM were made in [22]. Motivated by [20, 22], we proposed a modified NFTSM. The new NFTSM not only can avoid the singularity problem, but also can get a higher convergence speed.

Remark 8.3 The new NFTSM is composed of three parts, When the system states stay at a distance from equilibrium points, $\alpha sig(e)^{\gamma_1}$ dominates over $\beta sig(e)^{\gamma_2}$, thus a fast convergence rate can be guaranteed; when the system states are close to the region, the dominant term $\beta sig(e)^{\gamma_2}$ determines finite time convergence.

8.3.2 Finite-Time Controller Design

With the NFTSM, the teleoperation dynamics can be rewritten as the following forms

$$\begin{cases} M_m(q_m)\dot{s}_m + C_m(q_m, \dot{q}_m)s_m \\ = G'_m(Z_m) - \tau_m + F_h + \tau_{dm} \\ M_s(q_s)\dot{s}_s + C_s(q_s, \dot{q}_s)s_s \\ = G'_s(Z_s) - \tau_s - F_e + \tau_{ds} \end{cases} \tag{8.15}$$

where $Z_m = [\ddot{q}_s^T(t-d_s), \dot{q}_s^T(t-d_s), q_s^T(t-d_s), \dot{q}_m^T, q_m^T]^T$, $Z_s = [\ddot{q}_m^T(t-d_m), \dot{q}_m^T(t-d_m), q_m^T(t-d_m), \dot{q}_s^T, q_s^T]^T$, $G'_m(Z_m)$ and $G'_s(Z_s)$ are defined as

$$\begin{aligned} G'_m(Z_m) = {} & M_m(q_m)(\ddot{q}_s(t-d_s) + \lambda_m(\dot{e}_m)) \\ & + C_m(q_m, \dot{q}_m)(\dot{q}_s(t-d_s) + \lambda_m(e_m)) \\ & + F_m\dot{q}_m + f_{cm}(\dot{q}_m) + G_m(q_m) \end{aligned} \tag{8.16}$$

and

$$\begin{aligned} G'_s(Z_s) = {} & M_s(q_s)(\ddot{q}_m(t-d_m) + \lambda_s\dot{e}_s) + C_s(q_s, \dot{q}_s) \\ & \times (\dot{q}_m(t-d_m) + \lambda_s e_s) + F_s\dot{q}_s + f_{cs}(\dot{q}_s) + G_s(q_s) \end{aligned} \tag{8.17}$$

Based on FLs approximation property, we use the functions $\hat{G}'_m(Z_m)$ and $\hat{G}'_s(Z_s)$ to approximate the functions $G'_m(Z_m)$ and $G'_s(Z_s)$ with

$$\hat{G}'_m(Z_m) = \hat{\theta}_m^T\varsigma_m(Z_m), \; \hat{G}'_s(Z_s) = \hat{\theta}_s^T\varsigma_s(Z_s) \tag{8.18}$$

where $i = m, s$; $\hat{\theta}_i$ is a matrix of the fuzzy adaption parameters; $\varsigma_i(Z_i)$ is a vector denoting a known fuzzy basis function.

Additionally, to streamline the presentation, we give the following definitions

$$\begin{aligned} \tilde{G}'_m(Z_m) &= G'_m(Z_m) - \hat{G}'_m(Z_m) \\ &= (\theta_m^{*T} - \hat{\theta}_m^T)\varsigma_m(Z_m) + \bar{\epsilon}_m(Z_m) \\ &= \tilde{\theta}_m^T\varsigma_m(Z_m) + \bar{\epsilon}_m(Z_m) \end{aligned} \tag{8.19}$$

and

$$
\begin{aligned}
\tilde{G}'_s(Z_s) &= G'_s(Z_s) - \hat{G}'_s(Z_s) \\
&= (\theta_s^{*T} - \hat{\theta}_s^T)\varsigma_s(Z_s) + \bar{\epsilon}_s(Z_s) \\
&= \tilde{\theta}_s^T \varsigma_s(Z_s) + \bar{\epsilon}_s(Z_s)
\end{aligned} \tag{8.20}
$$

Because of the existence of piecewise continuous functions $f_{cm}(\dot{q}_m)$ and $f_{cs}(\dot{q}_s)$, $\bar{\epsilon}_i(Z_i) = \epsilon_i(Z_i) + f_{ci}(\dot{q}_i)$, where $i = m, s$.

The adaptive fuzzy control scheme is shown in Fig. 8.1 and the controllers are as follows

$$
\begin{aligned}
\tau_m &= \hat{G}'_m(Z_m) + K_{m_1} s_m + K_{m_2} sig(s_m)^\rho \\
&\quad + \hat{w}_m sign(s_m) + \xi_m sign(s_m) \\
\tau_s &= \hat{G}'_s(Z_s) + K_{s_1} s_s + K_{s_2} sig(s_s)^\rho \\
&\quad + \hat{w}_s sign(s_s) + \xi_s sign(s_s)
\end{aligned} \tag{8.21}
$$

where $\hat{w}_m$ and $\hat{w}_s$ are used to estimate the upper bound of the sum of fuzzy modeling error and the bounded disturbances w_m and w_s, i.e. $\|\bar{\epsilon}_m(Z_m) + \tau_{dm}\| \le w_m$ and $\|\bar{\epsilon}_s(Z_s) + \tau_{ds}\| \le w_s$. In addition, K_{m_1}, K_{m_2}, K_{s_1} and K_{s_2} are positive diagonal matrices; ξ_m and ξ_s are positive scalars, ξ_m and ξ_s will be defined; $0 < \rho < 1$.

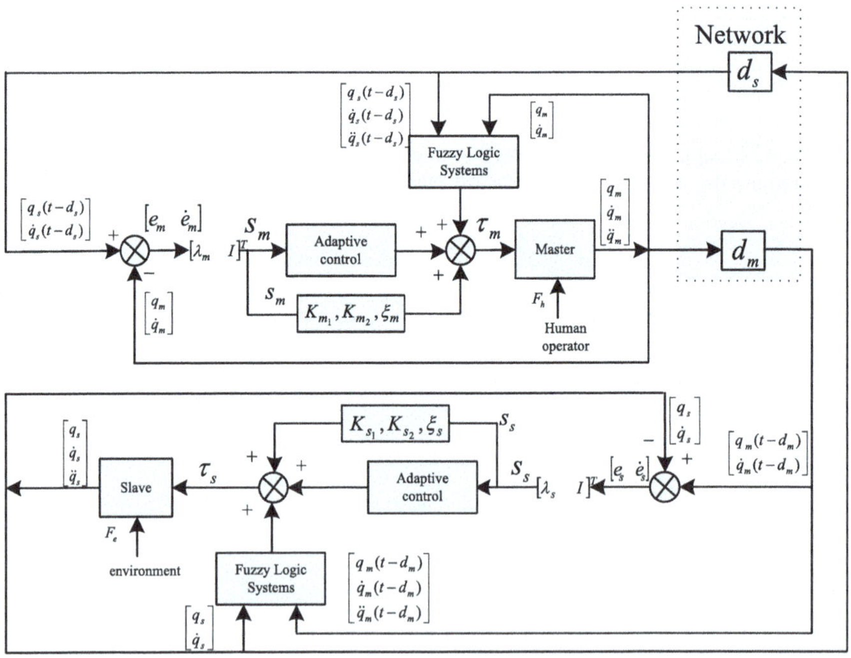

Fig. 8.1 The teleoperation with the new controller

Remark 8.4 For the practical teleoperation design, the synchronization accuracy and synchronization time are very important. For example, the telesurgery can minimize health care cost and make specialist doctors available throughout the world saving people's life and improving health care systems. Meanwhile, the telesurgery requires a higher performance than other applications. Because the work performance influences the patient' health status, even his life directly. However, the existing teleoperation control design methods can guarantee that the synchronization error converges to zero asymptotically. It is well known that the state variables reach zero when $t \to \infty$ for the asymptotical stability. Therefore, the synchronization performance can not suit the demand for telesurgery. The proposed control scheme (8.21) can guarantee the synchronization performance in the finite-time T shown in (8.29) which can be tuned by choosing the design parameters, and then the high-precision and fast synchronization performance are both realized.

8.3.3 Performance Analysis

With the controllers (8.21) for system (8.1), we have the following two theorems. In Theorem 8.1, the boundedness of all the signals is achieved. And the master-slave finite time synchronization performance is obtained in the Theorem 8.2.

Theorem 8.1 *Consider the teleoperator system (8.1) with the controller (8.21) in free motion, and the FLs adaptive laws are given as*

$$\dot{\hat{\theta}}_m = \Lambda_m \varsigma_m(Z_m) s_m^T, \dot{\hat{\theta}}_s = \Lambda_s \varsigma_s(Z_s) s_s^T \tag{8.22}$$

where Λ_m and Λ_s are positive definite matrices, the adaptive tuning laws are

$$\dot{\hat{w}}_m = \|s_m\|, \dot{\hat{w}}_s = \|s_s\| \tag{8.23}$$

where $\hat{w}_m(0) \geq 0$, $\hat{w}_s(0) \geq 0$, then all the signals of the teleoperation system are bounded.

Proof Let us propose the following Lyapunov function candidate

$$\begin{aligned} V =& \frac{1}{2} s_m^T M_m(q_m) s_m + \frac{1}{2} s_s^T M_s(q_s) s_s + \frac{1}{2} trace(\tilde{\theta}_m^T \Lambda_m^{-1} \tilde{\theta}_m) \\ &+ \frac{1}{2} trace(\tilde{\theta}_s^T \Lambda_s^{-1} \tilde{\theta}_s) + \frac{1}{2}(\hat{w}_m - w_m)^2 + \frac{1}{2}(\hat{w}_s - w_s)^2 \end{aligned} \tag{8.24}$$

evaluating V along the system trajectories, yields

$$
\begin{aligned}
\dot{V} = & s_m^T M_m(q_m)\dot{s}_m + \frac{1}{2} s_m^T \dot{M}_m(q_m) s_m + s_s^T M_s(q_s)\dot{s}_s \\
& + \frac{1}{2} s_s^T \dot{M}_s(q_s) s_s - trace(\tilde{\theta}_m^T \Lambda_m^{-1} \dot{\hat{\theta}}_m) - trace(\tilde{\theta}_s^T \Lambda_s^{-1} \dot{\hat{\theta}}_s) \\
& + (\hat{w}_m - w_m)\dot{\hat{w}}_m + (\hat{w}_s - w_s)\dot{\hat{w}}_s
\end{aligned} \tag{8.25}
$$

using the Property 1.2 of the robot manipulators with the controller (8.21) and the FLs adaptive laws (8.22), yields

$$
\begin{aligned}
\dot{V} = & s_m^T (G_m'(Z_m) - \hat{G}_m'(Z_m) - K_{m_1} s_m - K_{m_2} sig(s_m)^\rho \\
& - \hat{w}_m \|s_m\| - \xi_m \|s_m\| + \tau_{dm}) + s_s^T (G_s'(Z_s) - \hat{G}_s'(Z_s) \\
& - K_{s_1} s_s - K_{s_2} sig(s_s)^\rho - \hat{w}_s \|s_s\| - \xi_s \|s_s\| + \tau_{ds}) \\
& - trace(\tilde{\theta}_m^T \varsigma_m(Z_m) s_m^T) - trace(\tilde{\theta}_s^T \varsigma_s(Z_s) s_s^T) \\
& + (\hat{w}_m - w_m)\dot{\hat{w}}_m + (\hat{w}_s - w_s)\dot{\hat{w}}_s
\end{aligned} \tag{8.26}
$$

then with Eqs. (8.19), (8.20) and the adaptive tuning laws (8.23), we can obtain that

$$
\begin{aligned}
\dot{V} = & -s_m^T K_{m_1} s_m - s_m^T K_{m_2} sig(s_m)^\rho - s_s^T K_{s_1} s_s \\
& - s_s^T K_{s_2} sig(s_s)^\rho + s_m^T (\bar{\epsilon}_m(Z_m) + \tau_{dm}) + s_s^T (\bar{\epsilon}_s(Z_s) + \tau_{ds}) \\
& - \hat{w}_m \|s_m\| - \hat{w}_s \|s_s\| + (\hat{w}_m - w_m) \|s_m\| \\
& + (\hat{w}_s - w_s) \|s_s\| - \xi_m \|s_m\| - \xi_s \|s_s\|
\end{aligned} \tag{8.27}
$$

Substituting the upper bound w_m of $\left\|\bar{\epsilon}_m^* + \tau_{dm}\right\|$ and the upper bound w_s of $\left\|\bar{\epsilon}_s^* + \tau_{ds}\right\|$ into (8.27), yields

$$
\begin{aligned}
\dot{V} \leq & - s_m^T K_{m_1} s_m - s_m^T K_{m_2} sig(s_m)^\rho \\
& - s_s^T K_{s_1} s_s - s_s^T K_{s_2} sig(s_s)^\rho
\end{aligned} \tag{8.28}
$$

The above inequality implies that s_m, s_s, $\tilde{\theta}_m$ and $\tilde{\theta}_s$ are bounded. Meanwhile, considering (8.12), we can obtain that the e_m, $\dot{e}_m$, e_s and $\dot{e}_s$ are all bounded. Furthermore, the boundedness of q_m, $\dot{q}_m$, q_s and $\dot{q}_s$ can be achieved. As a result, with the Properties 1.1, 1.4, 1.5, 1.6 the $\dot{s}_m$ and $\dot{s}_s$ are bounded. Therefore, all the signals of the closed-loop system are bounded. This proof is completed.

Next, the finite time convergence performance will be proved.

Theorem 8.2 *Consider the teleoperator system (8.1) with the controller (8.21) in free motion, with the estimation errors* $\tilde{\theta}_m^T \varsigma_m(Z_m)$ *and* $\tilde{\theta}_s^T \varsigma_s(Z_s)$*, and the adaptive approximate error* $w_m - \hat{w}_m$ *and* $w_s - \hat{w}_s$*, if the design parameter* ξ_m *is set as* $\xi_m \geq \left\|\tilde{\theta}_m^T \varsigma_m(X_m)\right\| + \left\|w_m - \hat{w}_m\right\|$ *and* ξ_s *is set as* $\xi_s \geq \left\|\tilde{\theta}_s^T \varsigma_s(X_s)\right\| + \left\|w_s - \hat{w}_s\right\|$*; the system trajectory will converge to* $s_m = 0$ *and* $s_s = 0$ *in finite time, furthermore,* *the synchronization errors e_m and e_s will converge to zero in finite time. And the exact convergence time is*

$$T = T_1 + T_2 \tag{8.29}$$

$$T_1 = \frac{1}{\alpha(1-\delta)} \ln \frac{\alpha U^{1-\delta}(x_0) + \beta}{\beta} \tag{8.30}$$

where $\alpha = \min(\lambda_{\min}(K_{m_1})\frac{2}{m_{m_2}}, \lambda_{\min}(K_{s_1})\frac{2}{m_{s_2}})$,
$\beta = \min(\lambda_{\min}(K_{m_2})(\frac{2}{m_{m_2}})^{(1+\rho)/2}, \lambda_{\min}(K_{s_2})(\frac{2}{m_{s_2}})^{(1+\rho)/2})$ *and* $\delta = (1+\rho)/2$; $\lambda_{\min}$ (K_{i_j}) *denotes the minimum eigenvalue of matrix* K_{i_j}.

$$\begin{aligned} T_2 = \max\{&\frac{1}{\alpha_{m_1}}\frac{1}{\gamma_{m_1}-1} + \frac{1}{\alpha_{m_2}}\frac{1}{\gamma_{m_2}-1}, \\ &\frac{1}{\alpha_{s_1}}\frac{1}{\gamma_{s_1}-1} + \frac{1}{\alpha_{s_2}}\frac{1}{\gamma_{s_2}-1}\} \end{aligned} \tag{8.31}$$

Proof Select another Lyapunov function candidate

$$U = \frac{1}{2}s_m^T M_m s_m + \frac{1}{2}s_s^T M_s s_s \tag{8.32}$$

Differentiating the Lyapunov function U along the system trajectories, yields

$$\dot{U} = s_m^T M_m \dot{s}_m + \frac{1}{2}s_m^T \dot{M}_m s_m + s_s^T M_s \dot{s}_s + \frac{1}{2}s_s^T \dot{M}_s s_s \tag{8.33}$$

with the Property 1.2 of the robot manipulators with the controller (8.21), we have

$$\begin{aligned} \dot{U} \leq &-s_m^T K_{m_1} s_m - s_m^T K_{m_2} sig(s_m)^\rho - s_s^T K_{s_1} s_s \\ &- s_s^T K_{s_2} sig(s_s)^\rho + s_m^T \tilde{\theta}_m^T \varsigma_m(Z_m) + s_s^T \tilde{\theta}_s^T \varsigma_s(Z_s) \\ &+ \|s_m\| (w_m - \hat{w}_m) + \|s_s\| (w_s - \hat{w}_s) \\ &- \xi_m \|s_m\| - \xi_s \|s_s\| \end{aligned} \tag{8.34}$$

Then with $\left\| s_m^T \tilde{\theta}_m^T \varsigma_m(z_m) \right\| \leq \|s_m\| \left\| \tilde{\theta}_m^T \varsigma_m(z_m) \right\|$ and $\left\| s_s^T \tilde{\theta}_s^T \varsigma_s(Z_s) \right\| \leq \|s_s\|$ $\left\| \tilde{\theta}_s^T \varsigma_s(Z_s) \right\|$, we can obtain that

$$
\begin{aligned}
\dot{U} \leq & - s_m^T K_{m_1} s_m - s_m^T K_{m_2} sig(s_m)^\rho - s_s^T K_{s_1} s_s \\
& - s_s^T K_{s_2} sig(s_s)^\rho + \|s_m\| \left\| \tilde{\theta}_m^T \varsigma_m(z_m) \right\| \\
& + \|s_s\| \left\| \tilde{\theta}_s^T \varsigma_s(Z_s) \right\| + \|s_m\| (w_m - \hat{w}_m) \\
& + \|s_s\| (w_s - \hat{w}_s) - \xi_m \|s_m\| - \xi_s \|s_s\|
\end{aligned} \tag{8.35}
$$

With the definitions of ξ_m and ξ_s, one has

$$
\begin{aligned}
\dot{U} \leq & - s_m^T K_{m_1} s_m + s_m^T K_{m_2} sig(s_m)^\rho \\
& - s_s^T K_{s_1} s_s - s_s^T K_{s_2} sig(s_s)^\rho
\end{aligned} \tag{8.36}
$$

With the manipulator Property 1.1, we have

$$
\begin{aligned}
\dot{U} \leq & - \lambda_{\min}(K_{m_1}) \frac{2}{m_{m_2}} (\frac{1}{2} s_m^T M_m s_m) \\
& - \lambda_{\min}(K_{m_2}) (\frac{2}{m_{m_2}})^{(1+\rho)/2} (\frac{1}{2} s_m^T M_m s_m)^{(1+\rho)/2} \\
& - \lambda_{\min}(K_{s_1}) \frac{2}{m_{s_2}} (\frac{1}{2} s_s^T M_s s_s) \\
& - \lambda_{\min}(K_{s_2}) (\frac{2}{m_{s_2}})^{(1+\rho)/2} (\frac{1}{2} s_s^T M_s s_s)^{(1+\rho)/2}
\end{aligned} \tag{8.37}
$$

With Lemma 8.3, we have

$$
\begin{aligned}
\dot{U} \leq & - \min\{\lambda_{\min}(K_{m_1}) \frac{2}{m_{m_2}}, \lambda_{\min}(K_{s_1}) \frac{2}{m_{s_2}}\} U \\
& - \min\{\lambda_{\min}(K_{m_2}) (\frac{2}{m_{m_2}})^{(1+\rho)/2}, \\
& \lambda_{\min}(K_{s_2}) (\frac{2}{m_{s_2}})^{(1+\rho)/2}\} U^\delta \\
= & - \alpha U - \beta U^\delta
\end{aligned} \tag{8.38}
$$

By using Lemma 8.2, the reaching time can be computed accurately as showed in Eq. (8.30). Moreover, with Lemma 8.1, the accurate sliding time also can be computed as shown in Eq. (8.31). This completes the proof.

Remark 8.5 Because switching idea is used to avoid the singularity problem, when the terminal sliding mode s_{m_i} and s_{s_i} reach zero, two possible cases for synchronization errors e_{m_i} and e_{s_i} approaching zero will occur. Take the s_{m_i} as example, at the first case, $\left|e_{m_i}\right| \geq \mu_m$ as the terminal sliding mode reaches zero, the convergence time can be computed with Lemma 8.2. For the second case $\left|e_{m_i}\right| \leq \mu_m$, tracking error will approach zero along the general sliding manifold. Because of the enough small μ_m, the convergence time can be neglected.

Remark 8.6 The NFTSM methods have been used in [19, 22], and other papers. This paper presents a new nonsingular fast finite-time sliding mode surface, which is different from the existing ones. The faster transient-state and higher-precision control performances are both achieved. Moreover, when sliding mode surfaces presented in [19, 22] are used to design controller. The term of $diag(|\dot{e}|)$ will be composed in the reaching time, so the finite reaching time could not be computed exactly. In [22], the system uncertainties and disturbance were not considered.

Remark 8.7 The control design parameters are chosen such that the system is stable and has good transient performance. With the sliding mode surface designed in (8.12), we choose the positive parameters K_{m_1}, K_{m_2}, K_{s_1}, K_{s_2}, then the synchronization error system is stable. Next, let's analyze the transient performance. The settling time is $T = T_1 + T_2$ shown in (8.29). Based on (8.30), one can choose big parameters K_{m_2}, K_{s_2} and small parameters K_{m_1}, K_{s_1} to achieve small T_1. Based on (8.31), the sliding mode surface parameters α_{m_1}, α_{m_2}, α_{s_1} and α_{s_2} are chosen to be big to obtain small T.

8.4 Numerical Simulations and Experiments

8.4.1 Simulation Results

In this chapter, in order to verify the effectiveness of our main results, the simulations are performed on 2-DOF manipulators:

$$\begin{array}{c} M_m(q_m)\ddot{q}_m + C_m(q_m, \dot{q}_m)\dot{q}_m + F_m\dot{q}_m \\ +f_{cm}(\dot{q}_m) + G_m(q_m) + \tau_{dm} = \tau_m + F_h \\ M_s(q_s)\ddot{q}_s + C_s(q_s, \dot{q}_s)\dot{q}_s + F_s\dot{q}_s \\ +f_{cs}(\dot{q}_s) + G_s(q_s) + \tau_{ds} = \tau_s - F_e \end{array}, \tag{8.39}$$

the friction functions are as follows

$$F_i\dot{q}_i + f_{ci}(\dot{q}_i) = F\dot{q}_i + f_c(\dot{q}_i),\ \text{for } i = m, s \tag{8.40}$$

with

$$F\dot{q} + f_c(\dot{q}) = [\begin{array}{c} f_{d1}\dot{q}_1 + k_1 sgn(\dot{q}_1) \\ f_{d2}\dot{q}_2 + k_2 sgn(\dot{q}_2) \end{array}] \tag{8.41}$$

For simulation, we choose the parameters $m_1 = 0.5\,\text{kg}$, $m_2 = 1\,\text{kg}$, $L_1 = 1\,\text{m}$, $L_2 = 0.8\,\text{m}$, $g = 9.81\,\text{m/s}^2$, $f_{d1} = 3$, $f_{d2} = 4$, $k_1 = 5$, $k_2 = 4$, $\tau_{di} = [\begin{array}{c} 0.3q_1\dot{q}_1 \sin t \\ 0.3q_2\dot{q}_2 \sin t \end{array}]$.

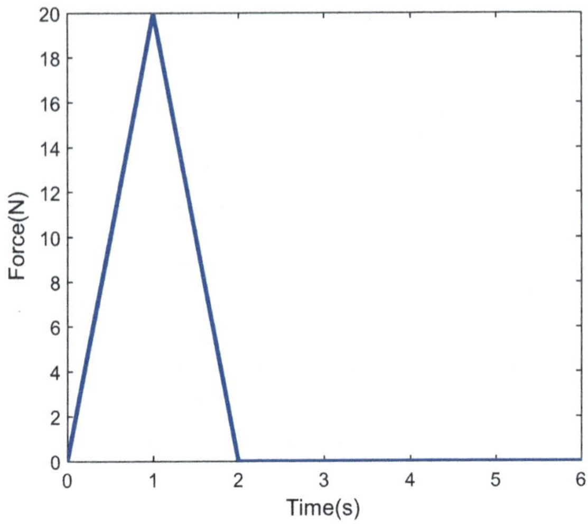

Fig. 8.2 The human insert force

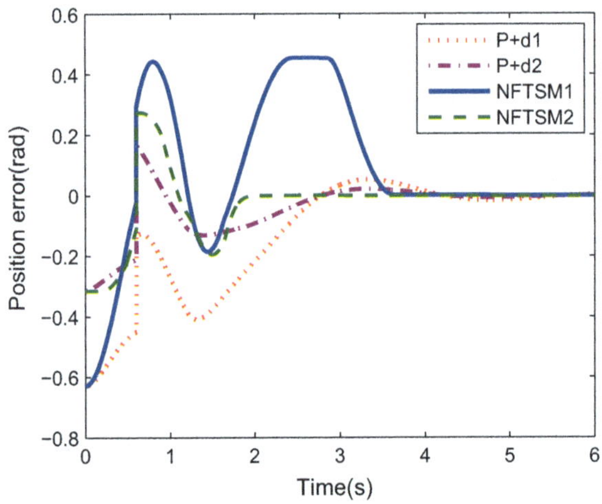

Fig. 8.3 The position tracking errors at master side

The controller (8.21) with the parameters as $K_{m_1} = K_{s_1} = diag(1, 0.8)$, $K_{m_2} = K_{s_2} = diag(0.4, 0.6)$, $\Lambda_m = \Lambda_s = diag(1, 1)$, $\gamma_{m_1} = \gamma_{s_1} = 4$, $\gamma_{m_2} = \gamma_{s_2} = 5/7$, $\rho = 9/11$, $\alpha_{m_1} = \alpha_{s_1} = 10$, $\alpha_{m_2} = \alpha_{s_2} = 5$, $\mu_m = \mu_s = 0.0001$ and $d_m = d_s = 0.6$ is adopted. The initial states are chosen as $q_m(0) = \left[0.2\pi\ 0.1\pi\right]^T$ $\dot{q}_m(0) = \left[0\ 0\right]^T$, $q_s(0) = \left[0.1\pi\ 0.12\pi\right]^T$, $\dot{q}_s(0) = \left[0\ 0\right]^T$. The initial values of the fuzzy system is chosen as $[0.01, 0.01, 0.02, 0.01, 0.02, 0.001, 0.01, 0.02, 0.03, 0.01]$ and the membership function is $\mu_{A^1}(z) = e^{-\frac{(z-4)^2}{2}}$, $\mu_{A^2}(z) = e^{-\frac{(z-2)^2}{2}}$, $\mu_{A^1}(z) = e^{-\frac{z^2}{2}}$, $\mu_{A^1}(z) = e^{-\frac{(z+2)^2}{2}}$ and $\mu_{A^1}(z) = e^{-\frac{(z+4)^2}{2}}$. The human input force is showed in Fig. 8.2.

To illustrate the effectiveness of the Fuzzy-NFTSM controller, some comparisons with the P+d, PD+d, DFF will be made. And it should be noticed that the controller

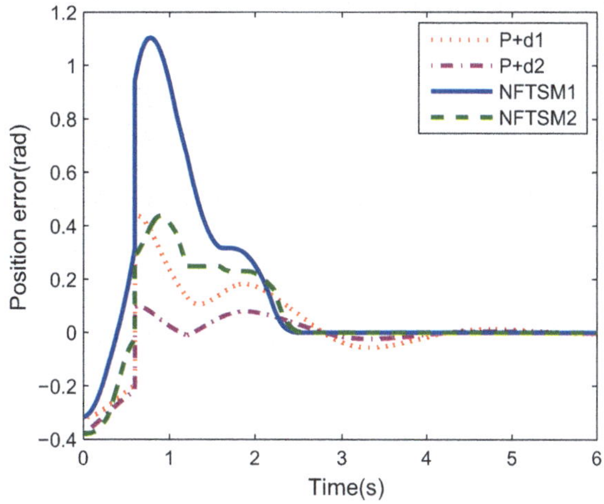

Fig. 8.4 The position tracking errors at slave side

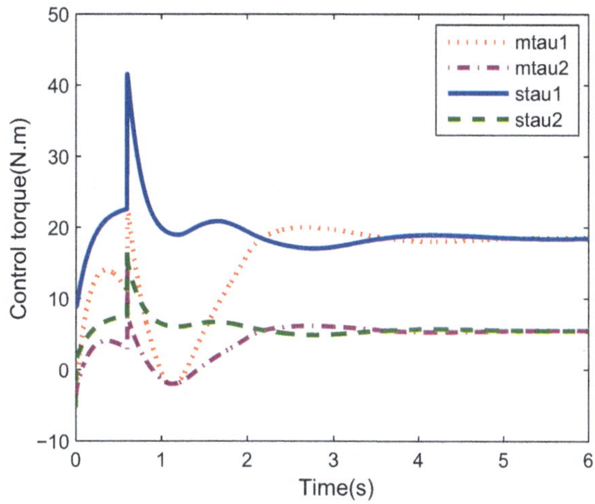

Fig. 8.5 The master/slave control torques

parameters of P+d, PD+d and DFF are chosen to get the same maximum control torques with the NFTSM.

First we carried out the comparison with the commonly used P+d control approach proposed in paper [4]. The P+d control laws are given by

$$\begin{aligned} \tau_m &= -L_m e_m - N_m \dot{q}_m + G_m \\ \tau_s &= -L_s e_s - N_s \dot{q}_s + G_s \end{aligned} \tag{8.42}$$

where $L_m = L_s = 30$ and $N_m = N_s = 30$. Figures 8.3 and 8.4 illustrate the position tracking errors at the master side and the slave side, respectively. The master and slave control torques are showed in Fig. 8.5. It can be seen that the slave completed

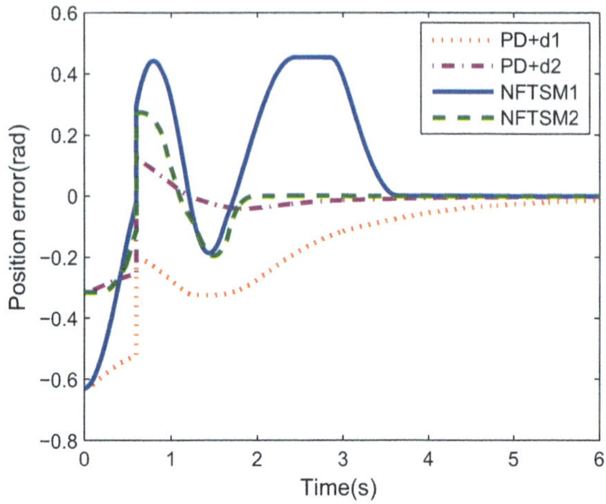

Fig. 8.6 The position tracking error at master side

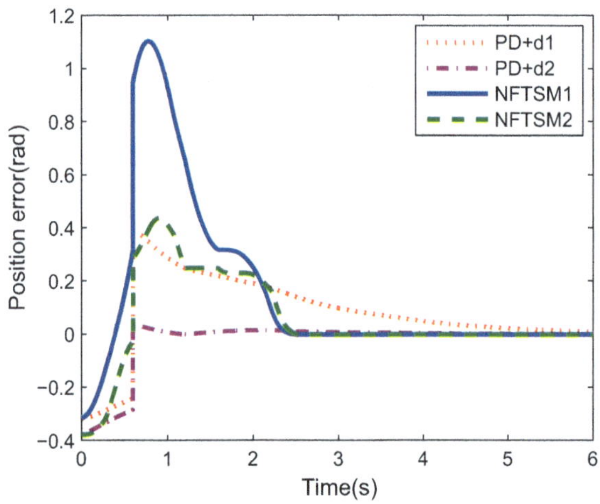

Fig. 8.7 The position tracking error at slave side

the tracking successfully, and after a transient due to initial errors condition, the position tracking errors tend to zero. Furthermore, a faster response is achieved in comparison with the P+d control approach.

After the comparison with P+d controller, the comparison with the PD+d control laws proposed in paper [5] is also conducted. The PD+d control laws are given as follows:

$$\begin{aligned}\tau_m &= -Z_m e_m - V_m \dot{e}_m - I_m \dot{q}_m + G_m \\ \tau_s &= -Z_s e_s - V_s \dot{e}_s - I_s \dot{q}_s + G_s\end{aligned} \tag{8.43}$$

where $Z_m = Z_s = 30$, $V_m = V_s = 30$ and $I_m = I_s = 30$.

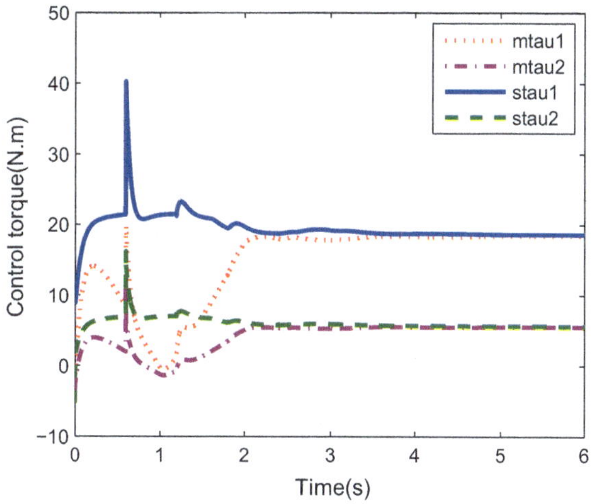

Fig. 8.8 The master/slave control torques

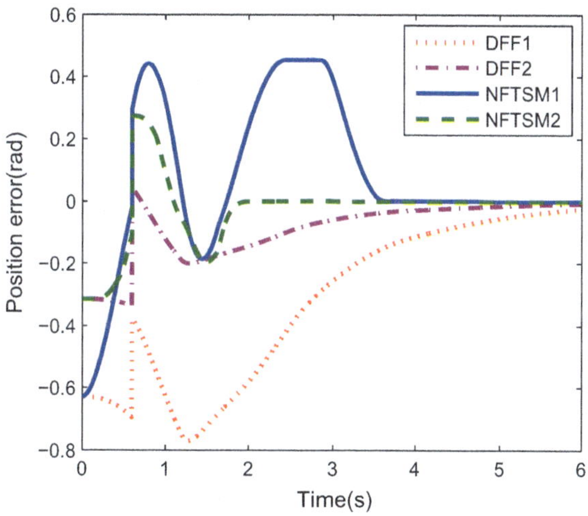

Fig. 8.9 The position tracking error at master side

Simulation results are showed in Figs. 8.6, 8.7 and 8.8. It can be clearly seen that the proposed finite-time nonsingular fast sliding mode controller obtains a much faster transient performance over the PD+d controller.

Moreover, the comparison with the direct force feedback (DFF) control laws proposed in paper [8] is also carried out. The DFF control laws are showed as follows:

$$\begin{aligned}\tau_m &= -\tau_s(t-d_s) - B_m\dot{q}_m \\ \tau_s &= -D_s e_s - F_s\dot{q}_s\end{aligned} \tag{8.44}$$

where $B_m = 30$, $D_s = 30$ and $F_s = 30$.

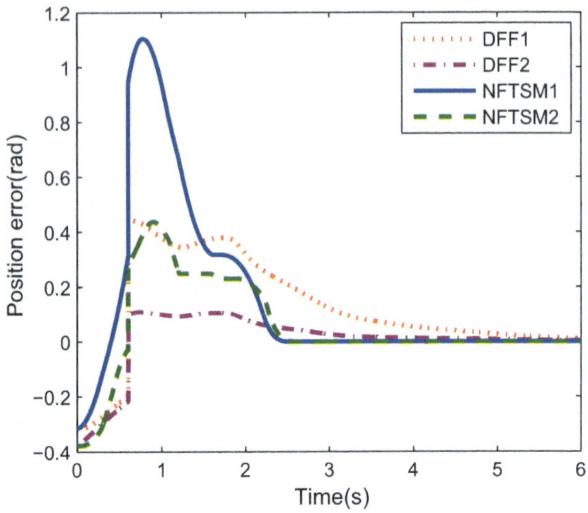

Fig. 8.10 The position tracking error at slave side

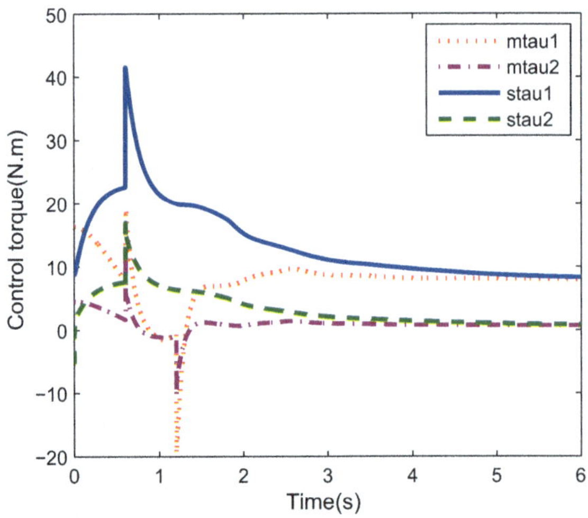

Fig. 8.11 The master/slave control torques

The simulation results are illustrated in Figs. 8.9, 8.10 and 8.11. It can be seen that the proposed NFTSM controller also produces a much higher synchronization speed over the DFF controller.

As a result, we can conclude that the proposed finite-time nonsingular fast terminal sliding control scheme for teleoperation system will produce a faster transient performance and higher precision synchronization performance over P+d, PD+d and DFF control schemes.

To further illustrate the effectiveness of the control schemes proposed in this chapter, the comparisons between the following two sliding mode surfaces with our NFTSM are also implemented.

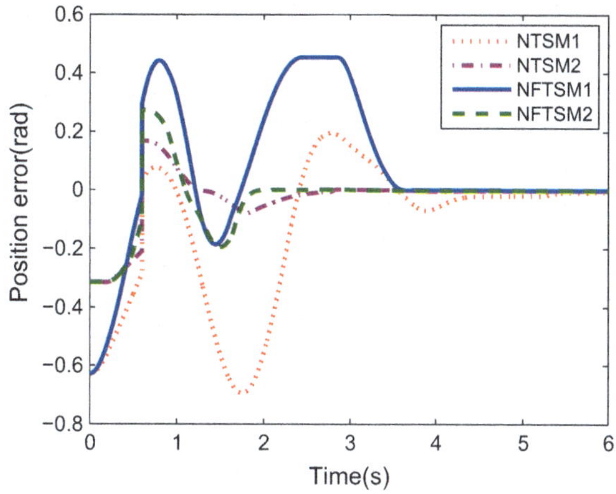

Fig. 8.12 The position tracking errors at master side

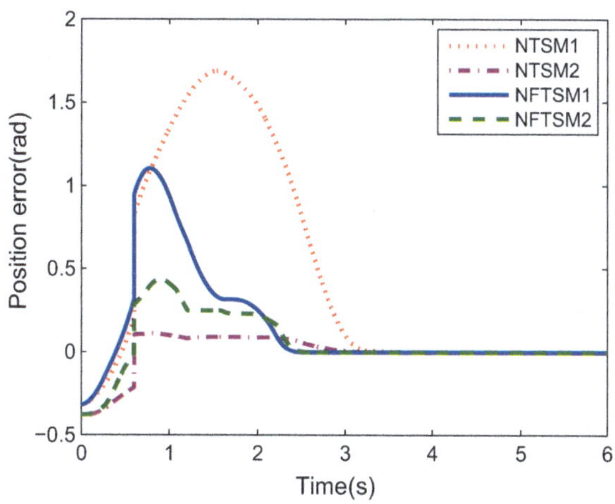

Fig. 8.13 The position tracking errors at slave side

$$s = \dot{e} + \alpha_2 sig(e)^{\gamma_2} \tag{8.45}$$

$$s = \dot{e} + \alpha_1 e + \alpha_2 sig(e)^{\gamma_2} \tag{8.46}$$

The sliding mode surfaces (8.45) and (8.46) were proposed in [16, 30]. We know that the finite-time sliding mode surfaces (8.45) and (8.46) will result in the singular problem. To avoid the singular problem, the switch idea used in this chapter is also employed for the cases of (8.45) and (8.46). First, the comparison between the sliding mode surface (8.45) and our NFTSM is presented. The simulation results are showed in Figs. 8.12 and 8.13. As we can see, the teleoperation system with NFTSM has a faster convergence speed. Next the comparison between the sliding mode surface (8.46) and our NFTSM is also presented. The simulation results are given in Figs. 8.14 and 8.15. The faster convergence performance also can be achieved with the NFTSM.

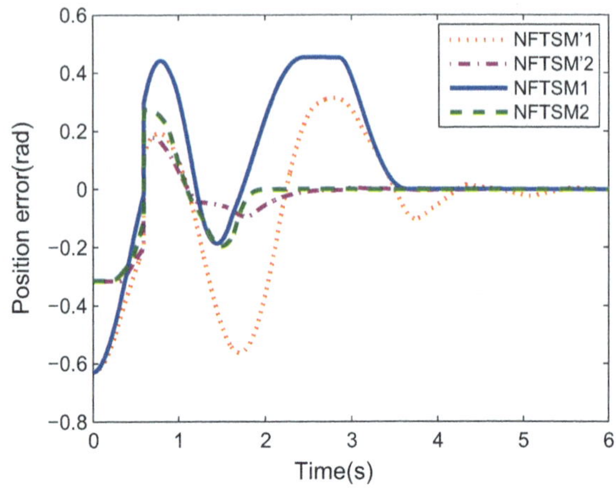

Fig. 8.14 The position tracking errors at master side

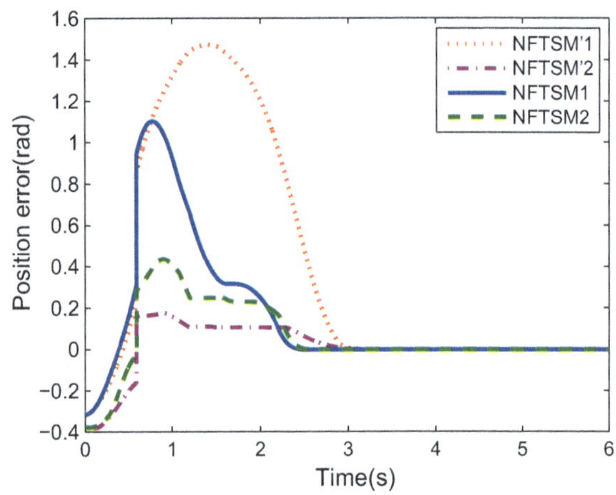

Fig. 8.15 The position tracking errors at slave side

From the simulation results, we can see the proposed method is effective and can achieve faster convergence performance.

The values of nonsingular fast sliding mode surface are showed in Fig. 8.16. The values of sliding mode can reach zero in finite-time. And we also can see that the reaching time is smaller than the system total convergence time. It shows the correctness of Theorem 8.2. In Figs. 8.17 and 8.18, the control torques of master and slave are given. The control torques for the master and the slave are bounded.

The Fuzzy tuning parameters are showed in Figs. 8.19 and 8.20 for the master and slave, respectively. From these two figures, we can see that the tuning parameters converge to constants. The adaptive approximation upper bounds are showed in Figs. 8.21 and 8.22 for the master and slave, respectively. Figures 8.19, 8.20, 8.21 and 8.22 have verified the correctness of the Theorem 8.1. Overall, from the simulation results obtained in this section, one can see that the networked bilateral teleoperation

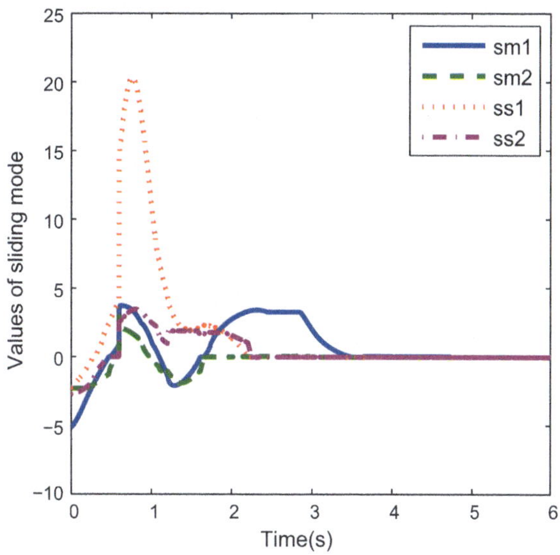

Fig. 8.16 The values of sliding mode surface

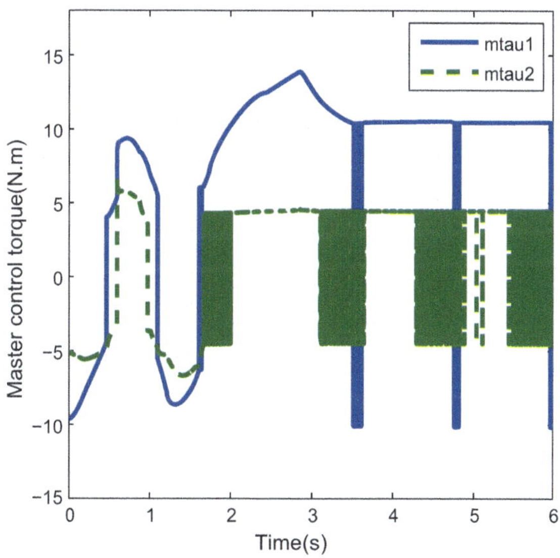

Fig. 8.17 The control torque of master

system with the new adaptive fuzzy finite-time controller can achieve good steady-state performance and good transient-state performance in the presence of system uncertainties and external disturbance.

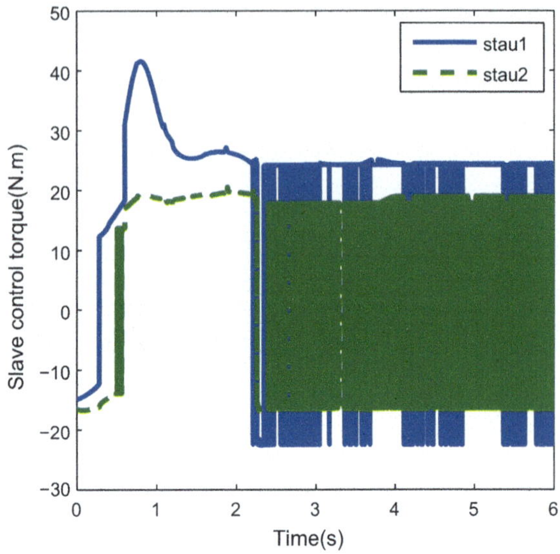

Fig. 8.18 The control torque of slave

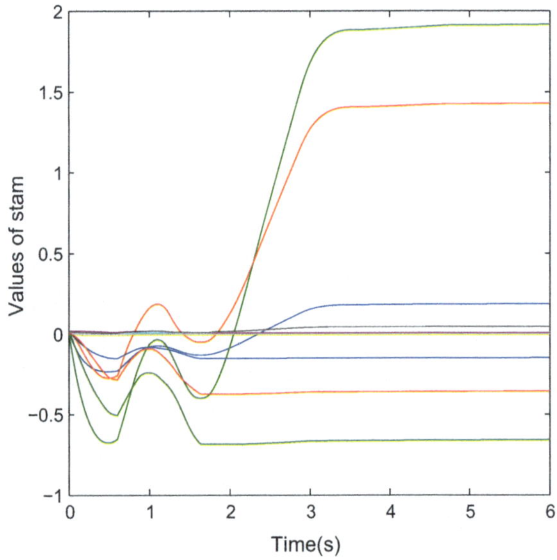

Fig. 8.19 The values of fuzzy tuning parameters at master side

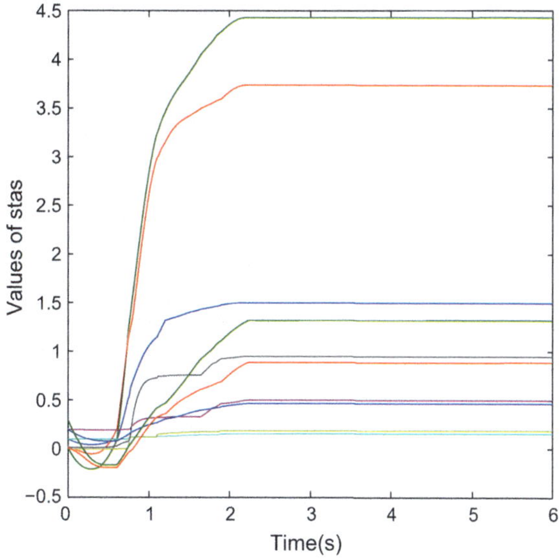

Fig. 8.20 The values of fuzzy tuning parameters at slave side

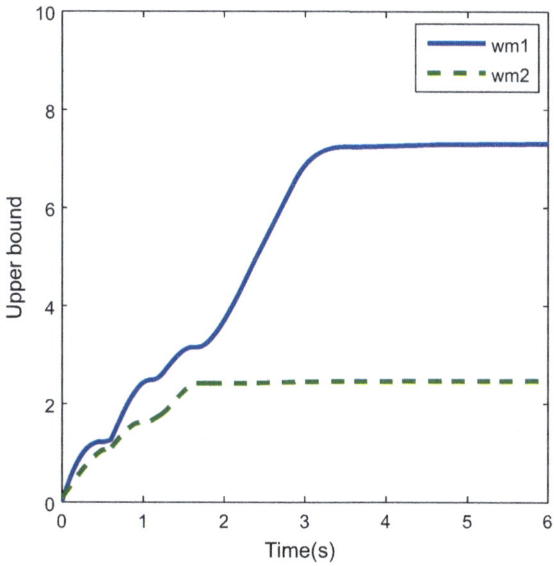

Fig. 8.21 The upper bound at master side

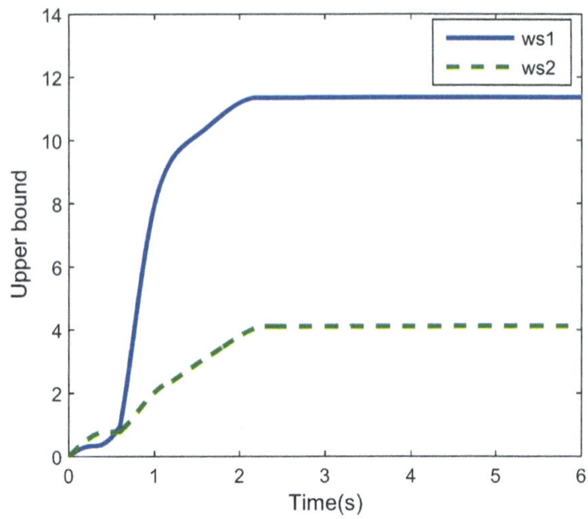

Fig. 8.22 The upper bound at slave side

8.4.2 Experiment Results

To further prove the superior performances of the closed-loop teleoperation system with the proposed finite-time controller, the experiment results on the experiment platform are given. In experiment, the parameters of the controller are set as $K_{m1} = K_{s1} = diag(0.6, 0.8, 0.8)$, $K_{m2} = K_{s2} = diag(0.4, 0.6, 0.6)$ and $\gamma = 9/11$. The sliding mode parameters are given as $\gamma_{m1} = \gamma_{s1} = 9/11$, $\gamma_{m2} = \gamma_{s2} = 4$, $\alpha_{m1} = \alpha_{s1} = 5$, $\alpha_{m2} = \alpha_{s2} = 5$ and $\mu_m = \mu_s = 0.0001$. The parameters for the fuzzy system tuning law are chosen as $\Lambda_m = \Lambda_s = diag(1, 1, 1)$.

Firstly, the communication delays between the master and the slave are set as $d_m = d_s = 200$ ms. The human operator will move the master manipulator, then the trajectories of the master joints $q_m = [q_{m1}; q_{m2}; q_{m3}]$ are shown as Fig. 8.23. Then with the designed finite time control scheme, the joints of the slave will track the trajectories of the master manipulator. Finally, after $t = 70$ s, the synchronization errors will converge to zero in finite time when the human operator insert force converge to zero. Furthermore, the joint synchronization error trajectories are shown as Fig. 8.24. From Fig. 8.24, it is clear that the synchronization errors are bounded when the master and slave are moving. And when the human insert force converges to zero, the synchronization errors will converge to zero with high speed.

In addition, the communication delays are increased as $d_m = d_s = 800$ ms, the trajectories of the joints and the synchronization errors are presented as Figs. 8.25 and 8.26. Compared with the experiment results with small time delays, the convergence time will increase, relatively.

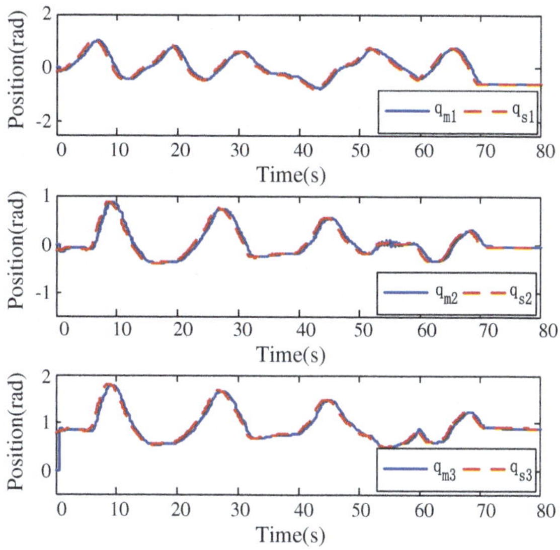

Fig. 8.23 The joint trajectories of the master and the slave

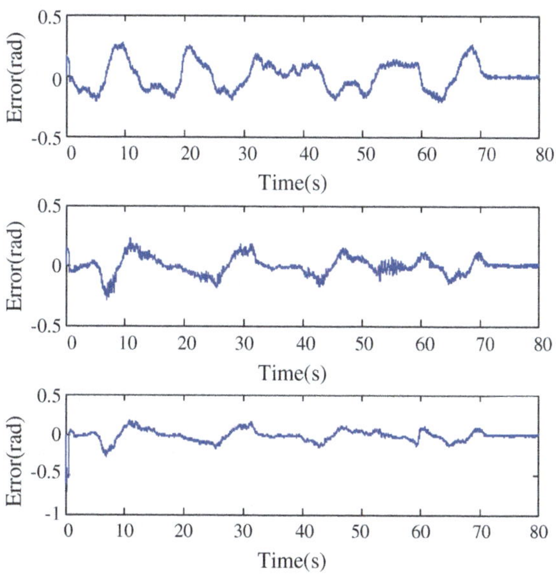

Fig. 8.24 The joint position synchronization errors between the master and the slave

Fig. 8.25 The joint trajectories of the master and the slave

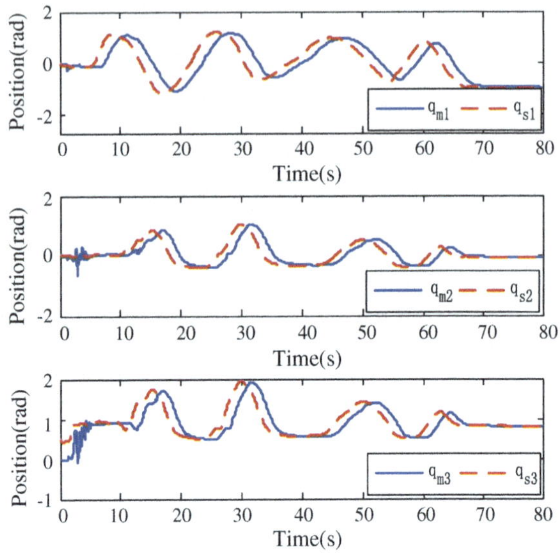

Fig. 8.26 The joint position synchronization errors between the master and the slave

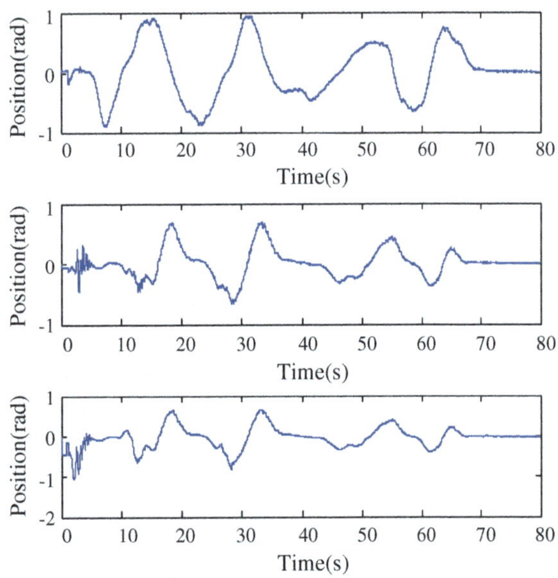

8.5 Summary

This chapter has studied the finite-time control problem for networked bilateral teleoperation system. The terminal sliding mode controller is designed and the tracking error converges to zero in finite time. Compared with the classic teleoperation design method, the proposed control design achieves the accurate synchronization in the finite time. With the control design parameters and initial values of the state variables, one can compute the converging time. With the given required converging time, we can design the controller parameters. The proposed method will have more wide applications in the practical teleoperation systems.

References

1. P.F. Hokayem, M.W. Spong, Bilateral teleoperation: an history survey. Automatica **42**(12), 2035–2057 (2006)
2. R.J. Anderson, M.W. Spong, Bilateral control of teleoperators with time delay. IEEE Trans. Autom. Control **34**(5), 494–501 (1989)
3. G. Niemeyer, J.E. Slotine, Stable adaptive teleoperation. IEEE J. Ocean. Eng. **16**(1), 152–162 (1991)
4. N. Chopra, M.W. Spong, R. Ortega, N.E. Barabanov, On tracking performance in bilateral teleoperation. IEEE Trans. Rob. **22**(4), 861–866 (2006)
5. D. Lee, M.W. Spong, Passive bilateral teleoperation with constant time delay. IEEE Trans. Rob. Autom. **22**(2), 269–281 (2006)
6. N. Chopra, M.W. Spong, Synchronization of bilateral teleoperators with time delay. Automatica **44**, 2142–2148 (2008)
7. E. Nuño, R. Ortega, N. Barabanov, L. Basañez. A globally stable pd controller for bilateral teleoperators. IEEE Trans. Rob. **24**(3), 753–758 (2008)
8. C.C. Hua, X.P. Liu, Delay-dependent stability criteria of teleoperation systems with asymmetric time-varying delays. IEEE Trans. Rob. **26**(5), 925–932 (2010)
9. P. Malysz, S. Sirouspour, A kinematic control framework for single-slave asymmetric teleoperation systems. IEEE Trans. Rob. **44**(5), 901–917 (2011)
10. C.C. Hua, X.P. Liu, Teleoperation over the internet with/without velocity signal. IEEE Trans. Instrum. Meas. **60**, 4–13 (2011)
11. J.H. Park, H.C. Cho, Sliding mode control of bilateral teleoperation systems with force-refection on the internet, in *IEEE International Conference on Intelligent Robots and Systems* (2000), pp. 1187–1192
12. J.H. Park, H.C. Cho, Sliding-mode controller for bilateral teleoperation with varying time delay, in *IEEE/ASME International Conference on Advanced Intelligent Mechatronics* (1999), pp. 311–316
13. M. Guiatni, A. Kheddar, H. Melouah, Sliding mode bilateral control and four channels schemes control of a force reflecting master/slave teleoperator, in *IEEE International Conference on Mechatronics and Automation* (2005), pp. 1660–1665
14. S.T. Venkataraman, S. Gulati, Control of nonlinear systems using terminal sliding modes. J. Dyn. Syst. Meas. Control **115**(3), 554–560 (1993)
15. A. Levant, Universal single-input-single-output (siso) sliding-mode controllers with finite-time convergence. IEEE Trans. Autom. Control **46**(9), 1447–1451 (2001)
16. X.H. Yu, Z.H. Man, Fast terminal sliding-mode control design for nonlinear dynamical systems. IEEE Trans. Circuits Syst. I Fundam. Theory Appl. **49**(2), 261–264 (2002)

17. Y. Feng, X.H. Yu, Z.H. Man, Non-singular terminal sliding mode control of rigid manipulators. Automatica **38**(12), 2159–2167 (2002)
18. S.H. Yu, X.H. Yu, B. Shirinzadeh, Z.H. Man, Continuous finite-time control for robotic manipulators with terminal sliding mode. Automatica **41**(11), 1957–1964 (2005)
19. V. Nekoukar, A. Erfanian, Adaptive fuzzy terminal sliding mode control for a class of mimo uncertain nonlinear systems. Fuzzy Sets Syst. **179**(1), 34–49 (2011)
20. T.Y. Chai, L.Y. Wang, L.F. Zhai, Neural-network-based terminal sliding-mode control of robotic manipulators including actuator dynamics. IEEE Trans. Ind. Electron. **56**(9), 3296–3304 (2009)
21. A.M. Zou, K.D. Kumar, Z.G. Hou, X. Liu, Finite-time attitude tracking control for spacecraft using terminal sliding mode and chebyshev neural network. IEEE Trans. Syst. Man Cybern. Part B Cybern. **41**(4), 950–963 (2011)
22. L. Yang, J.Y. Yang, Nonsingular fast terminal sliding-mode control for nonlinear dynamical systems. Int. J. Robust Nonlinear Control **21**(16), 1865–1879 (2011)
23. B.K. Yoo, W.C. Ham, Adaptive control of robot manipulator using fuzzy compensator. IEEE Trans. Fuzzy Syst. **8**(2), 186–199 (2000)
24. D.B. Zhao, Z.S. Song, J.Q. Yi, X.C. Li, A computed torque controller for uncertain robotic manipulator systems: fuzzy approach. Fuzzy Sets Syst. **154**(2), 208–226 (2005)
25. B. Chen, X.P. Liu, K.F. Liu, C. Lin, Fuzzy approximation-based adaptive control of strict-feedback nonlinear systems with time delays. IEEE Trans. Fuzzy Syst. **18**(5), 883–892 (2010)
26. S.C. Tong, Y.M. Li, P. Shi, Observer-based adaptive fuzzy backstepping output feedback control of uncertain mimo pure-feedback nonlinear systems. IEEE Trans. Fuzzy Syst. **20**(4), 771–785 (2012)
27. C.C. Hua, S.X. Ding, Decentralized networked control system design using t-s fuzzy approach. IEEE Trans. Fuzzy Syst. **20**(1), 9–21 (2012)
28. Z.J. Li, Trilateral tele-operation of adaptive fuzzy force/motion control for nonlinear teleoperators with communication random delays. IEEE Trans. Fuzzy Syst. **21**(4), 610–624 (2013)
29. Z.J. Li, X.Q. Cao, N. Ding, Adaptive fuzzy control for synchronization of nonlinear teleoperators with stochastic time-varying communication delays. IEEE Trans. Fuzzy Syst. **19**(4), 745–757 (2011)
30. X.H. Yu, Z.H. Man, Multi-input uncertain linear systems with terminal sliding-mode control. Automatica **34**(3), 389–392 (1998)

Chapter 9
Constrained Control Design for Bilateral Teleoperation System

Abstract Due to the cognitive limitations of the human operator and lack of complete information about the remote environment, the work performance of such teleoperation systems can not be guaranteed in most cases. However, some practical tasks conducted by the teleoperation system require high performances, such as telesurgery needs satisfactory high speed and more precision control results to guarantee patient health status. To obtain some satisfactory performances, the error constrained control is employed by applying the Barrier Lyapunov function (BLF). With the constrained synchronization errors, some high performances: high convergence speed, small overshoot and an arbitrarily predefined small residual constrained synchronization error can be achieved simultaneously. Nevertheless, like many classical control schemes only the asymptotic/exponential convergence, i.e., the synchronization errors converge to zero as time goes infinity can be achieved with the error constrained control. It is clear that finite time convergence is more desirable. To obtain a finite-time synchronization performance, the terminal sliding mode (TSM) based finite time control method is developed for teleoperation system with position error constrained in this chapter. Firstly, a nonsingularity fast TSM surface with transformed synchronization errors is proposed. Secondly, adaptive neural network (ANN) system is applied for dealing with the system uncertainties and the external disturbances. Thirdly, the BLF is applied to prove the stability and the non-violation of the synchronization errors constraints. Finally, some comparisons are conducted in simulation and experiment results are also presented to show the effectiveness of the proposed method.

9.1 Introduction

In this chapter, we consider the constrained control design problem for a class of bilateral nonlinear teleoperation system. Due to the cognitive limitations of the human operator and lack of complete information about the remote environment, the work performance of such teleoperation systems can not be guaranteed in most cases.

C. Hua et al., *Analysis and Design for Networked Teleoperation System*,
https://doi.org/10.1007/978-981-13-7936-9_9

However, some practical tasks conducted by the teleoperation system require high performances, such as telesurgery needs satisfactory high speed and more precision control results to guarantee patient' health status. To obtain some satisfactory performances, the error constrained control is employed by applying the Barrier Lyapunov function (BLF). With the constrained synchronization errors, then not only the steady-state performance, but also the transient-state performance can be guaranteed. It has been early recognized that transient-state control performance is very important and deserves further research [1]. In most cases, the transient-state performance can be guaranteed by setting some constraints on the system errors [2–7]. By setting constraints on the system errors, the transient-state performances: high convergence speed, small overshoot and an arbitrarily predefined small errors can be provided [2]. The Barrier Lyapunov Function (BLF) is a concept used for dealing with the control problems with constraints [3, 4]. In [3], the output constraint problem was addressed by using a BLF for the single-input single-output (SISO) nonlinear systems in strict feedback form. The time-varying output constraint satisfaction was guaranteed by employing the asymmetric time-varying BLF in [4] for strict feedback nonlinear systems. The output constrained problem had been considered for some practical systems [5–7]. However, in above literatures, only the exponential convergence can be obtained which means that the errors converge to zero as time goes infinity. In most cases, the asymptotic/exponential convergence can not satisfy the requirement of high convergence speed from some practical applications. It is clear that finite time convergence is more desirable. To obtain a finite-time synchronization performance, the terminal sliding mode (TSM) based finite time control method is developed for teleoperation system with position error constrained in this chapter. Firstly, a non-singularity fast TSM surface with transformed synchronization errors is proposed. Secondly, adaptive neural network (ANN) system is applied for dealing with the system uncertainties and the external disturbances. Thirdly, the BLF is applied to prove the stability and the non-violation of the synchronization errors constraints. Finally, some comparisons are conducted in simulation and experiment results are also presented to show the effectiveness of the proposed method.

In this chapter, a TSM control approach with defined error transformed variables is designed to provide some superior performances: higher synchronization speed, more accurate convergence, finite time synchronization, safety operating and so on. The main contents of this chapter can be summarized as follows: Firstly, a fast non-singularity terminal sliding mode based finite time control scheme is designed for nonlinear teleoperation system with time delay. Compared with the existing control works: P+d control, PD+d control, directed force feedback control and adaptive control for teleoperation, finite time control can provide some superior control performances: higher convergence speed, more precision convergence, finite time convergence, stronger robustness. In most cases, the finite time control can meet the requirements of speed and precision for many real applications of teleoperation. Secondly, the synchronization error constrained is considered in this chapter. The synchronization error constrained is significant for real applications. On one hand it provides a good steady-state performance: the errors will be bounded by an arbitrarily predefined small constraint. On the other hand it provides a good transient-state performance: the errors change in a bounded area and the overshoot can be limited. Finally, the system model parameters and external disturbances are not necessary to be known. The neural networks are applied to estimate the system uncertainties. Moreover, the parameter adaptive method is used to eliminate the effect from the neural network estimation error and the bounded external disturbances.

9.1.1 Dynamics Models of Master and Slave

The Euler-Lagrange equations for n-link master and slave manipulators are given as

$$\begin{cases} M_m(q_m)\ddot{q}_m + C_m(q_m, \dot{q}_m)\dot{q}_m + G_m(q_m) \\ \quad +B_m(\dot{q}_m) = \tau_m + F_h \\ M_s(q_s)\ddot{q}_s + C_s(q_s, \dot{q}_s)\dot{q}_s + G_s(q_s) \\ \quad +B_s(\dot{q}_s) = \tau_s - F_e \end{cases} \tag{9.1}$$

$B_i(\dot{q}_i) \in R^n$ is the unknown bounded external disturbances, i.e., $\|B_i(\dot{q}_i)\| \le \bar{B}_i$, $\bar{B}_i$ is an unknown positive constant.

In reality, similar to many engineering applications, it is impossible or very difficult to obtain an exact dynamics model of the master or the slave manipulators, due to the presence of large flexibility, coulomb friction, wear and so on. So we have $M_i(q_i) = M_{oi}(q_i) + \Delta M_i(q_i)$; $C_i(q_i, \dot{q}_i) = C_{oi}(q_i, \dot{q}_i) + \Delta C_i(q_i, \dot{q}_i)$; $G_i(q_i) = G_{oi}(q_i) + \Delta G_i(q_i)$; here $M_{oi}(q_i)$, $C_{oi}(q_i, \dot{q}_i)$ and $G_{oi}(q_i)$ are the nominal parts, whereas $\Delta M_i(q_i)$, $\Delta C_i(q_i, \dot{q}_i)$ and $\Delta G_i(q_i)$ represent the uncertain parts in the system model. Thus, the dynamics models can be rewritten as

$$\begin{cases} M_{om}(q_m)\ddot{q}_m + C_{om}(q_m, \dot{q}_m)\dot{q}_m + G_{om}(q_m) \\ \quad = \tau_m + P_m(q_m, \dot{q}_m, \ddot{q}_m) + F_h \\ M_{os}(q_s)\ddot{q}_s + C_{os}(q_s, \dot{q}_s)\dot{q}_s + G_{os}(q_s) \\ \quad = \tau_s + P_s(q_s, \dot{q}_s, \ddot{q}_s) - F_e \end{cases} \tag{9.2}$$

where $P_m(q_m, \dot{q}_m, \ddot{q}_m) = -\Delta M_m(q_m)\ddot{q}_m - \Delta C_m(q_m, \dot{q}_m)\dot{q}_m - \Delta G_m(q_m) - B_m(\dot{q}_m) \in R^n$ and $P_s(q_s, \dot{q}_s, \ddot{q}_s) = -\Delta M_s(q_s)\ddot{q}_s - \Delta C_s(q_s, \dot{q}_s)\dot{q}_s - \Delta G_s(q_s) - B_s(\dot{q}_s) \in R^n$.

9.1.2 RBF Neural Networks

The motivation of using NNs is to take advantages of NNs global approximation capabilities to handle nonlinearies [8–11]. Based on the universal approximation theorem a wide range of nonlinear functions can be estimated by a NN with sufficient neurons. RBF NN is an artificial NN that uses radial basis functions as activation functions. RBF networks are used widely for function approximation and system control for its simpleness and effectiveness. The RBF NN approximates a continuous function $f(X) : R^q \rightarrow R^p$ can be expressed as follows:

$$f(X) = W^T \varphi(X) + \epsilon(X) \tag{9.3}$$

where $X \in R^{n \times p}$ is the weight matrix; $n > 1$ is the number of the neurons; and $\varphi(X) = [\varphi_1(X), \varphi_2(X), ..., \varphi_n(X)]^T$, with $\varphi_i(X)$ being the RBF functions, where the Gaussian RBF function is

$$\varphi_i(X) = \exp(-\frac{\|X - C_i\|^2}{2b_i^2}), i = 1, 2, ..., n \tag{9.4}$$

where C_i, b_i are the center and the width of the ith neuron.

With the universal approximation property of NNs, for any continuous function $f(X)$, there exists a NN such that

$$f(X) = W^{*T} \varphi(X) + \epsilon(X), \|\epsilon(X)\| \leq \epsilon_N \tag{9.5}$$

where W^* is the ideal weight in the approximation; $\epsilon(X)$ is the approximation error; ϵ_N is an upper bound of the approximation error $\epsilon(X)$.

9.1.3 Terminal Sliding Mode

In this section, a nonsingular fast terminal sliding mode surface will be defined with defined error transformed variables. To streamline the presentation, below, the definition needed in the subsequent analysis is presented as follows.

Definition 9.1

$$sig(\zeta)^a = \left[|\zeta_1|^{a_1} sign(\zeta_1), ..., |\zeta_n|^{a_n} sign(\zeta_n)\right]^T \tag{9.6}$$

where $\zeta = [\zeta_1, \zeta_2, ..., \zeta_n]^T \in R^n$; $a_1, a_2, ..., a_n > 0$ and $sign(\cdot)$ being the standard signum function.

Define the position synchronization errors between the master and the slave as follows

$$\begin{aligned} e_m &= q_m - q_s(t - d_s) \\ e_s &= q_s - q_m(t - d_m) \end{aligned} \tag{9.7}$$

where d_m represents the signal transmission time delay from the master side to the slave side and d_s stands for the transmission time delay in the inverse direction. Then, the velocity errors are given as follows

$$\dot{e}_m = \dot{q}_m - \dot{q}_s(t - d_s)$$

$$\dot{e}_s = \dot{q}_s - \dot{q}_m(t - d_m)$$

It should be noticed that in this chapter $i = m/s$ and $j = 1, 2, ..., n$.

To achieve some superior performances, some constraints are set for the joint synchronization errors as

$$\begin{aligned} -k_{amj}(t) < e_{mj}(t) < k_{bmj}(t) \\ -k_{asj}(t) < e_{sj}(t) < k_{bsj}(t) \end{aligned} \tag{9.8}$$

where $k_{amj}(t), k_{bmj}(t), k_{asj}(t), k_{bsj}(t) : R_+ \to R$ are positive time-varying boundaries, additionally, $-k_{amj}(t) < k_{bmj}(t)$, $-k_{asj}(t) < k_{bsj}(t)$. Then the error transformed variables are defined as

$$\begin{aligned} \xi_{amj}(t) = \frac{e_{mj}(t)}{k_{amj}(t)}, \xi_{bmj}(t) = \frac{e_{mj}(t)}{k_{bmj}(t)} \\ \xi_{asj}(t) = \frac{e_{sj}(t)}{k_{asj}(t)}, \xi_{bsj}(t) = \frac{e_{sj}(t)}{k_{bsj}(t)} \end{aligned} \tag{9.9}$$

In the following, the time argument is dropped from above defined variables for presentation compactness. With the defined error transformed variables, we can further have that

$$\begin{aligned} \xi_{mj} = p_m(e_{mj})\xi_{bmj} + (1 - p_m(e_{mj}))\xi_{amj} \\ \xi_{sj} = p_s(e_{sj})\xi_{bsj} + (1 - p_s(e_{sj}))\xi_{asj} \end{aligned} \tag{9.10}$$

where $p(x) = \begin{cases} 1 \text{ if } x > 0 \\ 0 \text{ if } x \le 0 \end{cases}$.

Differentiating (9.10) with respect to the time t, yields

$$\begin{aligned} \dot{\xi}_{mj} = \frac{p_{mj}}{k_{bmj}}(\dot{e}_{mj} - \frac{\dot{k}_{bmj}}{k_{bmj}}e_{mj}) + \frac{(1 - p_{mj})}{k_{amj}}(\dot{e}_{mj} - \frac{\dot{k}_{amj}}{k_{amj}}e_{mj}) \\ \dot{\xi}_{sj} = \frac{p_{sj}}{k_{bsj}}(\dot{e}_{sj} - \frac{\dot{k}_{bsj}}{k_{bsj}}e_{sj}) + \frac{(1 - p_{sj})}{k_{asj}}(\dot{e}_{sj} - \frac{\dot{k}_{asj}}{k_{asj}}e_{sj}) \end{aligned} \tag{9.11}$$

where $p_{mj} = p_m(e_{mj})$ and $p_{sj} = p_s(e_{sj})$.

To streamline the presentation, the following definitions are given as $\xi_{bi} = [\xi_{bi1}, \xi_{bi2}, ..., \xi_{bin}]^T$, $\xi_{ai} = [\xi_{ai1}, \xi_{ai2}, ..., \xi_{ain}]^T$, $\xi_i = [\xi_{i1}, \xi_{i2}, ..., \xi_{in}]^T$, $p_i(e_i) = diag(p_{i1}(e_{i1}), p_{i2}(e_{i2}), ..., p_{in}(e_{in}))$, $s_i = [s_{i1}, s_{i2}, ..., s_{in}]^T$. With above defined error transformed variables, the nonsingular fast TSM surfaces are designed as

$$\begin{aligned} s_m &= \dot{\xi}_m + \alpha_{m_1} sig(\xi_m)^{\gamma_{m_1}} + \alpha_{m_2}\beta_m(\xi_m) \\ &= \dot{\xi}_m + \lambda_m(\xi_m) \end{aligned}$$

$$
\begin{aligned}
s_s &= \dot{\xi}_s + \alpha_{s_1} sig(\xi_s)^{\gamma_{s_1}} + \alpha_{s_2}\beta_s(\xi_s) \\
&= \dot{\xi}_s + \lambda_s(\xi_s)
\end{aligned} \tag{9.12}
$$

where $\alpha_{m_1}, \alpha_{s_1}, \alpha_{m_2}$ and α_{s_2} are positive constants, $\gamma_{m_1} > 1, \gamma_{s_1} > 1$; With $\beta_m(\xi_m) = [\beta_m(\xi_{m1}), ..., \beta_m(\xi_{mn})]^T$ and $\beta_s(\xi_s) = [\beta_s(\xi_{s1}), ..., \beta_s(\xi_{sn})]^T$, the $\beta_m(\xi_{mj})$ and $\beta_s(\xi_{sj})$ are defined as follows

$$
\beta_m(\xi_{mj}) = \begin{cases} sig(\xi_{mj})^{\gamma_{m_2}} \text{ if } \bar{s}_{mj} = 0 \\ \text{or } \bar{s}_{mj} \neq 0, \ |\xi_{mj}| > \mu_m \\ k_{m_1}\xi_{mj} + k_{m_2} sign(\xi_{mj})\xi_{mj}^2 \\ \text{if } \bar{s}_{mj} \neq 0, \ |\xi_m| \leq \mu_m \end{cases} \tag{9.13}
$$

and

$$
\beta_s(\xi_{sj}) = \begin{cases} sig(\xi_{sj})^{\gamma_{s_2}} \text{ if } \bar{s}_{sj} = 0 \\ \text{or } \bar{s}_{sj} \neq 0, \ |\xi_{sj}| > \mu_s \\ k_{s_1}\xi_{sj} + k_{s_2} sign(\xi_{sj})\xi_{sj}^2 \\ \text{if } \bar{s}_{sj} \neq 0, \ |\xi_{sj}| \leq \mu_s \end{cases} \tag{9.14}
$$

where $0 < \gamma_{m_2} < 1$ and $0 < \gamma_{s_2} < 1$; $\bar{s}_{mj} = \dot{\xi}_{mj} + \alpha_{m_1} sig(\xi_{mj})^{\gamma_{m_1}} + \alpha_{m_2} sig(\xi_{mj})^{\gamma_{m_2}}$, $\bar{s}_{sj} = \dot{\xi}_{sj} + \alpha_{s_1} sig(\xi_{sj})^{\gamma_{s_1}} + \alpha_{s_2} sig(\xi_{sj})^{\gamma_{s_2}}$; $k_{m_1} = (2 - \gamma_{m_2})\mu_m^{\gamma_{m_2}-1}$, $k_{m_2} = (\gamma_{m_2} - 1)\mu_m^{\gamma_{m_2}-2}$, $k_{s_1} = (2 - \gamma_{s_2})\mu_s^{\gamma_{s_2}-1}$, $k_{s_2} = (\gamma_{s_2} - 1)\mu_s^{\gamma_{s_2}-2}$; μ_m, μ_s are small positive constants.

The derivative of the sliding mode surface is

$$
\begin{aligned}
\dot{s}_{ij} &= \ddot{\xi}_{ij} + \lambda'_{ij}(\xi_{ij}) \\
&= \frac{p_{ij}}{k_{bij}}(\ddot{e}_{ij} - \frac{\ddot{k}_{bij}e_{ij}}{k_{bij}} - \frac{2\dot{k}_{bij}\dot{e}_{ij}}{k_{bij}} \\
&\quad + \frac{2\dot{k}_{bij}^2}{k_{bij}^2}e_{ij}) + \frac{1 - p_{ij}}{k_{aij}}(\ddot{e}_{ij} - \frac{\ddot{k}_{aij}e_{ij}}{k_{aij}} \\
&\quad - \frac{2\dot{k}_{aij}\dot{e}_{ij}}{k_{aij}} + \frac{2\dot{k}_{aij}^2}{k_{aij}^2}e_{ij}) + \lambda'_{ij}(\xi_{ij})
\end{aligned} \tag{9.15}
$$

where $\lambda'_{ij}(\xi_{ij})$ denotes the derivative of $\lambda_{ij}(\xi_{ij})$.

Remark 9.1 The choice for synchronization error constraints has great significance for the control performances. The values of the constraints can determine the minimum convergence speed, maximum overshoot and the maximum convergence errors. In this chapter, to guarantee idea synchronization performances, the $k_{amj}(t)$, $k_{bmj}(t)$, $k_{asj}(t)$, $k_{bsj}(t)$ are defined as decaying functions of time like exponential functions.

Remark 9.2 It should be noticed that, in teleoperation system, the slave manipulator completes tasks in remote side while the master set in local side. Due to the cognitive limitations of the human operator and lack of complete information about the remote environment, the control performance can not be guaranteed, and worse still it will lead to collision between the slave manipulator and the object. By guarantee the synchronization performance, the absolute safety can be obtained, which is significant for the teleoperation system.

Remark 9.3 Compared with the LSM, the TSM offers some superior properties such as higher precision, more robust, higher speed and finite-time convergence. However, the classic TSM faces the singularity problem. To deal with the singularity problem, the switching method is used in this chapter. The sliding mode of the system is switched between the TSM and the LSM, i.e., when a singularity appears, the sliding mode is switched from TSM to LSM. Then the sliding mode is switched back from LSM to TSM as soon as the system trajectory passes the singularity area. The choice of the k_{i_1} and k_{i_2} make the switching is smooth.

9.2 Main Results

In this section, finite time controllers will be designed for the master manipulator and the slave manipulator with the sliding mode surfaces defined in above section. The proposed finite time controllers not only can provide the finite time synchronization performance, but also can guarantee the synchronization errors of the system moving in the constraints.

9.2.1 Controller Design

In this chapter, a TSM surface is proposed with the defined error transformed variables. Then with the nonsingular fast TSM designed in above section, adaptive neural network based finite time controllers are designed for the master and the slave as follows

$$\begin{aligned}\tau_m &= M_{om}(q_m)\ddot{q}_s(t-d_s) + C_{om}(q_m,\dot{q}_m)\dot{q}_m + G_{om}(q_m) + \bar{\tau}_{m1} + \bar{\tau}_{m2}\\ \tau_s &= M_{os}(q_s)\ddot{q}_m(t-d_m) + C_{os}(q_s,\dot{q}_s)\dot{q}_s + G_{os}(q_s) + \bar{\tau}_{s1} + \bar{\tau}_{s2}\end{aligned} \tag{9.16}$$

The specific definitions of $\bar{\tau}_{m1}$ and $\bar{\tau}_{s1}$ are given as

$$\begin{aligned}\bar{\tau}_{m1} = M_{om}(q_m)(&-\frac{s_m\dot{\xi}_m^T}{1-\xi_m^T\xi_m}e_m - \lambda_m'(\xi_m)\frac{e_m}{\xi_m}\\ &- p_m(-\frac{2\dot{k}_{bm}\dot{e}_m}{k_{bm}} - \frac{\ddot{k}_{bm}e_m}{k_{bm}} + \frac{2\dot{k}_{bm}^2}{k_{bm}^2}e_m)\\ &- (1-p_m)(-\frac{2\dot{k}_{am}\dot{e}_m}{k_{am}} - \frac{\ddot{k}_{am}e_m}{k_{am}} + \frac{2\dot{k}_{am}^2}{k_{am}^2}e_m)\end{aligned}$$

$$- K_{m1}s_m\frac{e_m}{\xi_m} - K_{m2}\frac{sig(s_m)^\rho}{\left(1-\xi_m^T\xi_m\right)^{(\rho-1)/2}}\frac{e_m}{\xi_m}) \tag{9.17}$$

and

$$\begin{aligned}\bar{\tau}_{s1} = & M_{os}(q_s)(-\frac{s_s\dot{\xi}_s^T}{1-\xi_s^T\xi_s}e_s - \lambda_s'(\xi_s)\frac{e_s}{\xi_s} \\ & - p_s(-\frac{2\dot{k}_{bs}\dot{e}_s}{k_{bs}} - \frac{\ddot{k}_{bs}e_s}{k_{bs}} + \frac{2\dot{k}_{bs}^2}{k_{bs}^2}e_s) \\ & - (1-p_s)(-\frac{2\dot{k}_{as}\dot{e}_s}{k_{as}} - \frac{\ddot{k}_{as}e_s}{k_{as}} + \frac{2\dot{k}_{as}^2}{k_{as}^2}e_s) \\ & - K_{s1}s_s\frac{e_s}{\xi_s} - K_{s2}\frac{sig(s_s)^\rho}{\left(1-\xi_s^T\xi_s\right)^{(\rho-1)/2}}\frac{e_s}{\xi_s})\end{aligned} \tag{9.18}$$

where $K_{m1} = diag(K_{m1_1}, K_{m1_2}, \ldots, K_{m1_n})$, $K_{s1} = diag(K_{s1_1}, K_{s1_2}, \ldots, K_{s1_n})$, $K_{m2} = diag(K_{m2_1}, K_{m2_2}, \ldots, K_{m2_n})$, $K_{s2} = diag(K_{s2_1}, K_{s2_2}, \ldots, K_{s2_n})$ are positive definite diagonal matrices.

Considering the system uncertainties and the external disturbances, the $\bar{\tau}_{m2}$ and $\bar{\tau}_{s2}$ are designed as follows

$$\begin{aligned}\bar{\tau}_{m2} &= -\hat{W}_m^T\varphi_m(X_m) - \hat{\phi}_m(\frac{s_m^T}{1-\xi_m^T\xi_m}\times(\frac{p_m}{k_{bm}} + \frac{1-p_m}{k_{am}})M_{om}^{-1}(q_m))^T \\ \bar{\tau}_{s2} &= -\hat{W}_s^T\varphi_s(X_s) - \hat{\phi}_s(\frac{s_s^T}{1-\xi_s^T\xi_s}\times(\frac{p_s}{k_{bs}} + \frac{1-p_s}{k_{as}})M_{os}^{-1}(q_s))^T\end{aligned} \tag{9.19}$$

where $\hat{\phi}_m = \frac{1}{2}\delta_{m1}^{-1}\hat{\psi}_m$ and $\hat{\phi}_s = \frac{1}{2}\delta_{s1}^{-1}\hat{\psi}_s$; δ_{m1} and δ_{s1} are positive constants; $\psi_m = \epsilon_{mN}^2$ and $\psi_s = \epsilon_{sN}^2$; ϵ_{mN} and ϵ_{sN} are the upper bounds of the NN approximation errors i.e., $\|\epsilon_m(X_m)\| \le \epsilon_{mN}$ and $\|\epsilon_s(X_s)\| \le \epsilon_{sN}$; $X_m = [q_m^T, \dot{q}_m^T, \ddot{q}_m^T]^T$ and $X_s = [q_s^T, \dot{q}_s^T, \ddot{q}_s^T]^T$ The teleoperation system with the designed control algorithm is shown in Fig. 9.1. The NT denotes nonlinear transformation.

Remark 9.4 In this chapter, the finite time control design is investigated for the nonlinear teleoperation system with synchronization error constrained. Different from the error constrained control proposed for high-order SISO system, the two-order MIMO nonlinear systems are considered as the master and the slave, respectively. With MIMO control systems, the controller design becomes more complex. Moreover, the expressions for derivation and analysis are more complicated because there are so many derivations for vectors of variables such as the term $-p_m\frac{2\dot{k}_{bm}\dot{e}_m}{k_{bm}}$, which is a vector likes $[-p_m(e_{m1})\frac{2\dot{k}_{bm1}\dot{e}_{m1}}{k_{bm1}}, \ldots, -p_m(e_{mn})\frac{2\dot{k}_{bmn}\dot{e}_{mn}}{k_{bmn}}]^T$. To avoid repetition explanations and save space, this expression will not be explained any more.

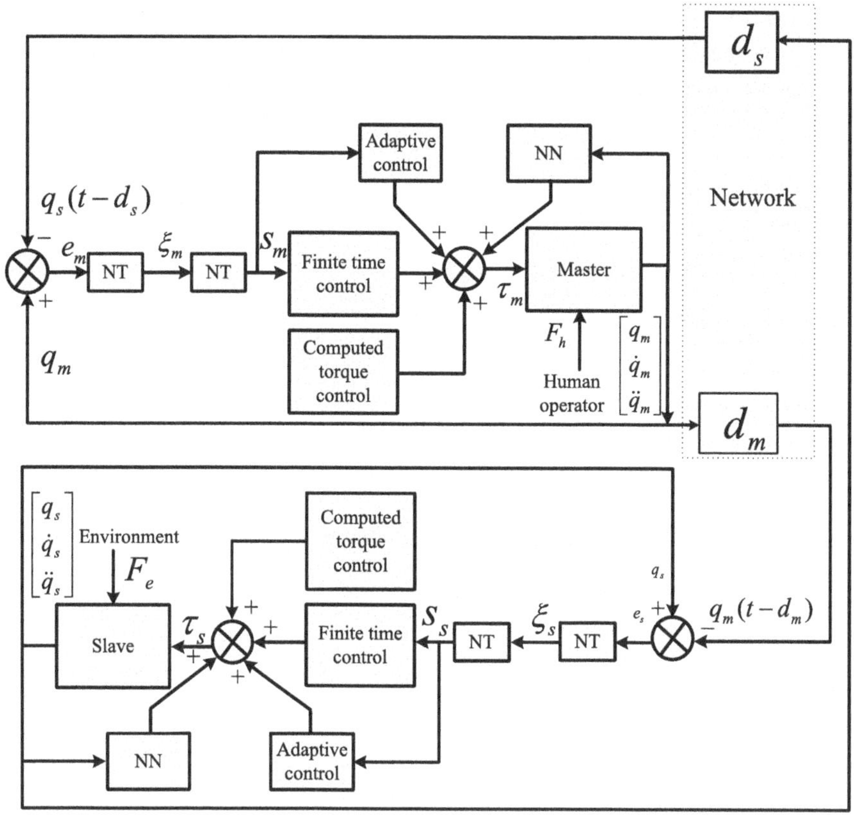

Fig. 9.1 Teleoperation system with designed controller

Remark 9.5 In this chapter, the TSM based finite time control method and the BLF based position synchronization error constrained control method are combined to provide some super control performances. The control terms (9.17) and (9.18) are used to counteract the effect from the known nonlinear terms. Moreover, the (9.19) is designed to deal with the system uncertainties and the unknown external disturbances. Different from the classical TSM finite time control, the term $-K_{i1}s_i\frac{e_i}{\xi_i} - K_{i2}\frac{sig(s_i)^{\rho}}{\left(1-\xi_i^T\xi_i\right)^{(\rho-1)/2}}\frac{e_i}{\xi_i}$ are designed to provide the finite time synchronization performance.

9.2.2 *Performance Analysis*

In this section, the rigorous prove process will be presented.

Theorem 9.1 *For the teleoperation system (9.2) with the human operator and remote environment insert forces are zero, i.e., $F_h = 0$, $F_e = 0$ in the presence of system uncertainties and external disturbances, if the NFTSM manifold is chosen as (9.12), the NN tuning law and adaptive tuning law are designed as follows*

$$\dot{\hat{W}}_i = \Lambda_i \varphi_i(X_i)(\frac{s_i^T}{1-\xi_i^T\xi_i}(\frac{p_i}{k_{bi}} + \frac{1-p_i}{k_{ai}})M_{oi}^{-1}(q_i)) - \delta_{i2}\hat{W}_i \tag{9.20}$$

$$\dot{\hat{\psi}}_i = \frac{\Delta_i}{2\delta_{i1}} \left\| \begin{matrix} \frac{s_i^T}{1-\xi_i^T\xi_i}(\frac{p_i}{k_{bi}} \\ +\frac{1-p_i}{k_{ai}})M_{oi}^{-1}(q_i)) \end{matrix} \right\|^2 - \delta_{i3}\hat{\psi}_i$$

where Λ_i, Δ_i, δ_{i1}, δ_{i2} and δ_{i3} are positive constants. With the continuous NFTSM controllers designed as (9.16), then

(i) All signals of the closed-loop system are bounded.

(ii) The time-varying position synchronization error constraints are never violated.

(iii) The NFTSM manifold s_{ij} converges to region Φ_i in finite time, the ξ_{ij} finally converge into region $\bar{\Phi}_i$ in finite time.

$$\Phi_i = \min(\Phi_{i1}, \Phi_{i2}),$$

$$\bar{\Phi}_i = \max\left(\mu_i, \sqrt[\gamma_{i1}]{\frac{\Phi_i}{2\alpha_{i_1}}}, \sqrt[\gamma_{i2}]{\frac{\Phi_i}{2\alpha_{i_2}}}\right).$$

where $\Phi_{i1} = \frac{H_i}{2\bar{K}_{i1}}$, $\Phi_{i2} = \sqrt[\rho]{\frac{H_i(1-\xi_i^T\xi_i)^{(\rho-1)/2}}{2\bar{K}_{i2}}}$; $\bar{K}_{i1} = \lambda_{\min}(K_{i1})$ *and* $\bar{K}_{i2} = \lambda_{\min}(K_{i2})$; $\lambda_{\min}(A)$ *denotes the minimum eigenvalue of A; H_i is a positive constant, which will be defined later.*

Proof (i) Let us consider the Lyapunov function candidates

$$V_1 = \sum_{i=\{m,s\}} \frac{1}{2}\frac{s_i^T s_i}{1-\xi_i^T\xi_i} \tag{9.21}$$

$$V_2 = \sum_{i=\{m,s\}} \frac{1}{2}tr(\tilde{W}_i^T \Lambda_i^{-1}\tilde{W}_i) + \frac{1}{2}\Delta_i^{-1}\tilde{\psi}_i^2 \tag{9.22}$$

where $\tilde{W}_i = W_i - \hat{W}_i$, $\tilde{\psi}_i = \psi_i - \hat{\psi}_i$,

The derivation of the above function V_1 is given as follows

$$\dot{V}_1 = \sum_{i=\{m,s\}} \frac{s_i^T\dot{s}_i}{1-\xi_i^T\xi_i} + \frac{\xi_i^T\dot{\xi}_i s_i^T s_i}{(1-\xi_i^T\xi_i)^2} \tag{9.23}$$

With the differentiation of the sliding mode surface, we have

$$
\begin{aligned}
\dot{V}_1 &= \sum_{i=\{m,s\}} \frac{s_i^T(\ddot{\xi}_i + \lambda_i'(\xi_i))}{1-\xi_i^T\xi_i} + \frac{\xi_i^T\dot{\xi}_i s_i^T s_i}{(1-\xi_i^T\xi_i)^2} \\
&= \sum_{i=\{m,s\}} \frac{s_i^T}{1-\xi_i^T\xi_i}(\frac{p_i}{k_{bi}}(\ddot{e}_i - \frac{2\dot{k}_{bi}\dot{e}_i}{k_{bi}} - \frac{\ddot{k}_{bi}e_i}{k_{bi}} \\
&\quad + \frac{2\dot{k}_{bi}^2}{k_{bi}^2}e_i)) + \frac{s_i^T}{1-\xi_i^T\xi_i}(\frac{1-p_i}{k_{ai}}(\ddot{e}_i - \frac{2\dot{k}_{ai}\dot{e}_i}{k_{ai}} \\
&\quad - \frac{\ddot{k}_{ai}e_i}{k_{ai}} + \frac{2\dot{k}_{ai}^2}{k_{ai}^2}e_i)) + \frac{s_i^T}{1-\xi_i^T\xi_i}\lambda_i'(\xi_i) + \frac{\xi_i^T\dot{\xi}_i s_i^T s_i}{(1-\xi_i^T\xi_i)^2}
\end{aligned} \tag{9.24}
$$

Applying the system model (9.2) with $F_h = 0$, $F_e = 0$, yields
$\ddot{e}_m = \ddot{q}_m - \ddot{q}_s(t-d_s) = M_{om}^{-1}(q_m)\ (\tau_m + P_m(q_m, \dot{q}_m, \ddot{q}_m) - C_{om}(q_m, \dot{q}_m)\dot{q}_m - G_{om}(q_m)) - \ddot{q}_s(t-d_s)$ and $\ddot{e}_s = \ddot{q}_s - \ddot{q}_m(t-d_m) = M_{os}^{-1}(q_s)(\tau_s + P_s(q_s, \dot{q}_s, \ddot{q}_s) - C_{os}(q_s, \dot{q}_s)\dot{q}_s - G_{os}(q_s)) - \ddot{q}_m(t-d_m)$.

Employing the controller (9.16) and the NN approximation $P_i(q_i, \dot{q}_i, \ddot{q}_i) = W_i^{*T}\varphi_i(X_i) + \epsilon_i(X_i)$, yields

$$
\begin{aligned}
\dot{V}_1 &= \sum_{i=\{m,s\}} \frac{s_i^T}{1-\xi_i^T\xi_i}(\frac{p_i}{k_{bi}}(M_{oi}^{-1}(q_i)\bar{\tau}_{i1} - \frac{2\dot{k}_{bi}\dot{e}_i}{k_{bi}} \\
&\quad - \frac{\ddot{k}_{bi}e_i}{k_{bi}} + \frac{2\dot{k}_{bi}^2}{k_{bi}^2}e_i)) + \frac{s_i^T}{1-\xi_i^T\xi_i}(\frac{1-p_i}{k_{ai}} \\
&\quad \times (M_{oi}^{-1}(q_i)\bar{\tau}_{i1} - \frac{2\dot{k}_{ai}\dot{e}_i}{k_{ai}} - \frac{\ddot{k}_{ai}e_i}{k_{ai}} + \frac{2\dot{k}_{ai}^2}{k_{ai}^2}e_i)) \\
&\quad + \frac{s_i^T}{1-\xi_i^T\xi_i}\lambda_i'(\xi_i) + \frac{\xi_i^T\dot{\xi}_i s_i^T s_i}{(1-\xi_i^T\xi_i)^2} \\
&\quad + \frac{s_i^T}{1-\xi_i^T\xi_i}(\frac{p_i}{k_{bi}} + \frac{1-p_i}{k_{ai}})M_{oi}^{-1}(q_i)(W_i^{*T}\varphi_i(X_i) \\
&\quad + \epsilon_i(X_i)) + \frac{s_i^T}{1-\xi_i^T\xi_i}(\frac{p_i}{k_{bi}} + \frac{1-p_i}{k_{ai}})M_{oi}^{-1}(q_i)\bar{\tau}_{i2}
\end{aligned} \tag{9.25}
$$

Further, with the designed $\bar{\tau}_{m1}$ (9.17), $\bar{\tau}_{s1}$ (9.18) and $\bar{\tau}_{m2}$, $\bar{\tau}_{s2}$ (9.19), we have

$$
\begin{aligned}
\dot{V}_1 &= \sum_{i=\{m,s\}} -\frac{s_i^T K_{i1}s_i}{1-\xi_i^T\xi_i} - \frac{s_i^T K_{i2}sig(s_i)^\rho}{\left(1-\xi_i^T\xi_i\right)^{(\rho+1)/2}} \\
&\quad + \frac{s_i^T}{1-\xi_i^T\xi_i}(\frac{p_i}{k_{bi}} + \frac{1-p_i}{k_{ai}})M_{oi}^{-1}(q_i)W_i^{*T}\varphi_i(X_i)
\end{aligned}
$$

$$
\begin{aligned}
&- \frac{s_i^T}{1-\xi_i^T\xi_i}(\frac{p_i}{k_{bi}} + \frac{1-p_i}{k_{ai}})M_{oi}^{-1}(q_i)\,\hat{W}_i^T\varphi_i(X_i) \\
&+ \frac{s_i^T}{1-\xi_i^T\xi_i}(\frac{p_i}{k_{bi}} + \frac{1-p_i}{k_{ai}})M_{oi}^{-1}(q_i)\,(\epsilon_i(X_i) - \hat{\phi}_i \\
&\times (\frac{s_i^T}{1-\xi_i^T\xi_i}(\frac{p_i}{k_{bi}} + \frac{1-p_i}{k_{ai}})M_{oi}^{-1}(q_i))^T \\
= &\sum_{i=\{m,s\}} -\frac{s_i^T K_{i1} s_i}{1-\xi_i^T\xi_i} - \frac{s_i^T K_{i2} sig(s_i)^\rho}{\left(1-\xi_i^T\xi_i\right)^{(\rho+1)/2}} \\
&+ \frac{s_i^T}{1-\xi_i^T\xi_i}(\frac{p_i}{k_{bi}} + \frac{1-p_i}{k_{ai}})M_{oi}^{-1}(q_i)\,\tilde{W}_i^T\varphi_i(X_i) \\
&+ \frac{s_i^T}{1-\xi_i^T\xi_i}(\frac{p_i}{k_{bi}} + \frac{1-p_i}{k_{ai}})M_{oi}^{-1}(q_i)\,\epsilon_i(X_i) \\
&- \left\| \frac{s_i^T}{1-\xi_i^T\xi_i}(\frac{p_i}{k_{bi}} + \frac{1-p_i}{k_{ai}})M_{oi}^{-1}(q_i) \right\|^2 \hat{\phi}_i
\end{aligned} \tag{9.26}
$$

Since $\|\epsilon_i(X_i)\| \le \epsilon_{iN}$ and for any constants x and y have $xy \le \frac{\delta}{2}x^2 + \frac{1}{2\delta}y^2$, where δ is a positive constant. Then we obtain that

$$
\begin{aligned}
\dot{V}_1 \le &\sum_{i=\{m,s\}} -\frac{s_i^T K_{i1} s_i}{1-\xi_i^T\xi_i} - \frac{s_i^T K_{i2} sig(s_i)^\rho}{\left(1-\xi_i^T\xi_i\right)^{(\rho+1)/2}} \\
&+ \frac{s_i^T}{1-\xi_i^T\xi_i}(\frac{p_i}{k_{bi}} + \frac{1-p_i}{k_{ai}})M_{oi}^{-1}(q_i)\,\tilde{W}_i^T\varphi_i(X_i) \\
&- \left\| \frac{s_i^T}{1-\xi_i^T\xi_i}(\frac{p_i}{k_{bi}} + \frac{1-p_i}{k_{ai}})M_{oi}^{-1}(q_i) \right\|^2 \hat{\phi}_i \\
&+ \frac{1}{2}\delta_{i1} + \frac{1}{2\delta_{i1}}\epsilon_{iN}^2 \left\| \frac{s_i^T}{1-\xi_i^T\xi_i}(\frac{p_i}{k_{bi}} + \frac{1-p_i}{k_{ai}})M_{oi}^{-1}(q_i) \right\|^2
\end{aligned} \tag{9.27}
$$

With the definitions for $\hat{\phi}_i = \frac{1}{2\delta_{i1}}\hat{\psi}_i$ and $\hat{\psi}_i = \epsilon_{iN}^2$, we have

$$
\begin{aligned}
\dot{V}_1 \le &\sum_{i=\{m,s\}} -\frac{s_i^T K_{i1} s_i}{1-\xi_i^T\xi_i} - \frac{s_i^T K_{i2} sig(s_i)^\rho}{\left(1-\xi_i^T\xi_i\right)^{(\rho+1)/2}} + \frac{1}{2}\delta_{i1} \\
&+ \frac{s_i^T}{1-\xi_i^T\xi_i}(\frac{p_i}{k_{bi}} + \frac{1-p_i}{k_{ai}})M_{oi}^{-1}(q_i)\,\tilde{W}_i^T\varphi_i(X_i) \\
&+ \frac{1}{2\delta_{i1}} \left\| \frac{s_i^T}{1-\xi_i^T\xi_i}(\frac{p_i}{k_{bi}} + \frac{1-p_i}{k_{ai}})M_{oi}^{-1}(q_i) \right\|^2 \tilde{\psi}_i
\end{aligned} \tag{9.28}
$$

The derivation of V_2 is given as follows

$$\dot{V}_2 = \sum_{i=\{m,s\}} -tr(\tilde{W}_i^T \Lambda_i^{-1} \dot{\hat{W}}_i) - \Delta_i^{-1} \tilde{\psi}_i \dot{\hat{\psi}}_i \tag{9.29}$$

With the NN tuning law and the adaptive tuning law (9.20), we have

$$\begin{aligned}\dot{V}_2 = & \sum_{i=\{m,s\}} -tr(\tilde{W}_i^T \varphi_i(X_i)(\frac{s_i^T}{1-\xi_i^T \xi_i} \\ & \times (\frac{p_i}{k_{bi}} + \frac{1-p_i}{k_{ai}}) M_{oi}^{-1}(q_i))) \\ & - \frac{1}{2\delta_{i1}} \left\| \frac{s_i^T}{1-\xi_i^T \xi_i} (\frac{p_i}{k_{bi}} + \frac{1-p_i}{k_{ai}}) M_{oi}^{-1}(q_i)) \right\|^2 \tilde{\psi}_i \\ & + tr(\tilde{W}_i^T \frac{\delta_{i2}}{\Lambda_i} \hat{W}_i) + \tilde{\psi}_i \frac{\delta_{i3}}{\Delta_i} \hat{\psi}_i\end{aligned} \tag{9.30}$$

Thus, we can obtain that

$$\begin{aligned}\dot{V} &= \dot{V}_1 + \dot{V}_2 \\ &= \sum_{i=\{m,s\}} -\frac{s_i^T K_{i1} s_i}{1-\xi_i^T \xi_i} - \frac{s_i^T K_{i2} sig(s_i)^\rho}{(1-\xi_i^T \xi_i)^{(\rho+1)/2}} \\ &\quad + tr(\tilde{W}_i^T \frac{\delta_{i2}}{\Lambda_i} \hat{W}_i) + \tilde{\psi}_i \frac{\delta_{i3}}{\Delta_i} \hat{\psi}_i + \frac{\delta_{i1}}{2}\end{aligned} \tag{9.31}$$

further, with $\hat{W}_i = W_i - \tilde{W}_i$ and $\hat{\psi}_i = \psi_i - \tilde{\psi}_i$, we have $tr(\tilde{W}_i^T \frac{\delta_{i2}}{\Lambda_i}(W_i - \tilde{W}_i)) = tr(\tilde{W}_i^T \frac{\delta_{i2}}{\Lambda_i} W_i - \tilde{W}_i^T \frac{\delta_{i2}}{\Lambda_i} \tilde{W}_i)$ and $tr(\tilde{W}_i^T \frac{\delta_{i2}}{\Lambda_i} W_i) \le tr(\tilde{W}_i^T \frac{\delta_{i2}\delta_{i4}}{2\Lambda_i} \tilde{W}_i + W_i^T \frac{\delta_{i2}}{2\delta_{i4}\Lambda_i} W_i)$. Similarly, $\tilde{\psi}_i \frac{\delta_{i3}}{\Delta_i} \hat{\psi}_i \le \frac{\delta_{i3}\delta_{i5}}{2\Delta_i} \tilde{\psi}_i^2 + \frac{\delta_{i3}}{2\delta_{i5}\Delta_i} \hat{\psi}_i^2$ also can be obtained.

Therefore, we have

$$\begin{aligned}\dot{V} \le & \sum_{i=\{m,s\}} -\frac{s_i^T K_{i1} s_i}{1-\xi_i^T \xi_i} - \frac{s_i^T K_{i2} sig(s_i)^\rho}{(1-\xi_i^T \xi_i)^{(\rho+1)/2}} \\ & + \frac{\delta_{i1}}{2} + \tilde{W}_i^T \frac{\delta_{i2}}{2\Lambda_i \delta_{i4}} \tilde{W}_i + W_i^T \frac{\delta_{i2}\delta_{i4}}{2\Lambda_i} W_i \\ & - \tilde{W}_i^T \frac{\delta_{i2}}{\Lambda_i} \tilde{W}_i + \frac{\delta_{i3}}{2\Delta_i \delta_{i5}} \tilde{\psi}_i^2 + \frac{\delta_{i3}\delta_{i5}}{2\Delta_i} \hat{\psi}_i^2 - \frac{\delta_{i3}}{\Delta_i} \tilde{\psi}_i^2 \\ = & \sum_{i=\{m,s\}} -\frac{s_i^T K_{i1} s_i}{1-\xi_i^T \xi_i} - \frac{s_i^T K_{i2} sig(s_i)^\rho}{(1-\xi_i^T \xi_i)^{(\rho+1)/2}}\end{aligned}$$

$$+\frac{\delta_{i1}}{2}-tr(\tilde{W}_i^T\eta_{i1}\tilde{W}_i)+tr(W_i^T\eta_{i2}W_i)$$
$$-\eta_{i3}\tilde{\psi}_i^2+\eta_{i4}\psi_i^2 \tag{9.32}$$

where $\eta_{i1}=\dfrac{\delta_{i2}(2\delta_{i4}-1)}{2\Lambda_{ii}\delta_{i4}}$, $\eta_{i2}=\dfrac{\delta_{i2}\delta_{i4}}{2\Lambda_i}$, $\eta_{i3}=\dfrac{\delta_{i3}(2\delta_{i5}-1)}{2\Delta_i\delta_{i5}}$ and $\eta_{i4}=\dfrac{\delta_{i3}\delta_{i5}}{2\Delta_i}$; δ_{i4} and δ_{i5} are two positive constants and chosen satisfying $\delta_{i4}>\dfrac{1}{2}$, $\delta_{i5}>\dfrac{1}{2}$.

Finally, we can obtain that

$$\dot{V}\le-\Psi V+\Omega \tag{9.33}$$

where $\Psi=\min(2\bar{K}_{m1},2\bar{K}_{s1},R)$, $R=\min(2\eta_{m1}\Lambda_m,2\eta_{m3}\Delta_m,2\eta_{s1}\Lambda_s,2\eta_{s3}\Delta_s)$, $\Omega=\dfrac{\delta_{m1}}{2}+\dfrac{\delta_{s1}}{2}+tr(W_m^T\eta_{m2}W_m)+tr(W_s^T\eta_{s2}W_s)+\eta_{m4}\psi_m^2+\eta_{s4}\psi_s^2$.

Then, based on the boundedness Theorem in Khalil (2002) and the Properties 1.1, 1.4, 1.6, all signals in the closed-loop system are bounded. Furthermore, we have $|\xi_{mj}|<1$ and $|\xi_{sj}|<1$, which implies that the time-varying error constraint is never violated. Next the finite time synchronization performance will be proved.

Define $\Xi_i=(\dfrac{p_i}{k_{bi}}+\dfrac{1-p_i}{k_{ai}})M_{oi}^{-1}(q_i)\,P_i(q_i,\dot{q}_i,\ddot{q}_i)+(\dfrac{p_i}{k_{bi}}+\dfrac{1-p_i}{k_{ai}})M_{oi}^{-1}(q_i)$ $\bar{\tau}_{i2}$. The boundedness of system signals shows that Ξ_i is bounded and $\|\Xi_i\|\le H_i$ with H_i a positive constant.

Next the finite time synchronization performance is analyzed. Consider the Lyapunov candidate function V_1

$$V_1=\sum_{i=\{m,s\}}\frac{1}{2}\frac{s_i^Ts_i}{1-\xi_i^T\xi_i} \tag{9.34}$$

The derivative of V_1 is

$$\begin{aligned}\dot{V}_1&=\sum_{i=\{m,s\}}-\frac{s_i^TK_{i1}s_i}{1-\xi_i^T\xi_i}-\frac{s_i^TK_{i2}sig(s_i)^\rho}{(1-\xi_i^T\xi_i)^{(\rho+1)/2}}+\frac{s_i^T}{1-\xi_i^T\xi_i}\Xi_i\\&\le\sum_{i=\{m,s\}}-(\bar{K}_{i1}-\frac{H_i}{2\|s_i\|})\frac{\|s_i\|^2}{1-\xi_i^T\xi_i}\\&\quad-(\bar{K}_{i2}-\frac{H_i(1-\xi_i^T\xi_i)^{(\rho-1)/2}}{2\|s_i\|^\rho})\frac{\|s_i\|^{\rho+1}}{(1-\xi_i^T\xi_i)^{(\rho+1)/2}}\end{aligned} \tag{9.35}$$

It follows that if $\|s_i\|>\Phi_i$, it obtains that $\dot{V}_1+\sigma_1V_1+\sigma_2V_1^{(1+\rho)/2}\le0$, $\sigma_1=\min(2\tilde{K}_{m1},2\tilde{K}_{s1})$ and $\sigma_2=\min(2\tilde{K}_{m2}2^{(\rho+1)/2},2\tilde{K}_{s2}2^{(\rho+1)/2})$, where $\tilde{K}_{i1}=\bar{K}_{i1}-$

Table 9.1 System parameters and controller parameters

L_1	1 m	L_2	0.8 m	m_1	0.5 kg	m_2	0.6 kg
$\tilde{L}_1$	0.8 m	$\tilde{L}_2$	0.9 m	$\tilde{m}_1$	0.55 kg	$\tilde{m}_2$	0.56 kg
g	9.81 m/s^2	K_{i11}	2	K_{i12}	1.2	K_{i21}	0.4
K_{i21}	0.6	α_1	5	α_2	5	ϱ_1	9/11
ϱ_2	9/11	μ	0.001	γ_1	4	γ_2	5/7

$\dfrac{H_i}{2\left\|s_i\right\|}$ and $\tilde{K}_{i2} = \bar{K}_{i2} - \dfrac{H_i\left(1-\xi_i^T\xi_i\right)^{(\rho-1)/2}}{2\left\|s_i\right\|^{\rho}}$. In addition, Φ_i is a small region containing the origin with large enough K_{i1} and K_{i2}.

Next the convergence region of ξ_{ij} will be derived. Because of $\|s_i\| \leq \Phi_i$ for the case $\xi_{ij} \geq \mu_i, i = m, s, j = 1, 2, ..., n$, then we get

$$\begin{aligned} &\dot{\xi}_{ij} + \alpha_{i_1} sig(\xi_{ij})^{\gamma_{i_1}} + \alpha_{i_2} sig(\xi_{ij})^{\gamma_{i_2}} \\ &= \varpi_{ij}, \left|\varpi_{ij}\right| \leq \Phi_i \end{aligned} \tag{9.36}$$

Then we have

$$\begin{aligned} &\dot{\xi}_{ij} + (\alpha_{i_1} - \frac{\varpi_{ij}}{2sig(\xi_{ij})^{\gamma_{i_1}}}) sig(\xi_{ij})^{\gamma_{i_1}} \\ &+ (\alpha_{i_2} - \frac{\varpi_{ij}}{2sig(\xi_{ij})^{\gamma_{i_2}}}) sig(\xi_{ij})^{\gamma_{i_2}} = 0 \end{aligned} \tag{9.37}$$

Therefore as long as $\alpha_{i_1} - \dfrac{\varpi_{ij}}{2sig(\xi_{ij})^{\gamma_{i_1}}} > 0$ and $\alpha_{i_2} - \dfrac{\varpi_{ij}}{2sig(\xi_{ij})^{\gamma_{i_2}}} > 0$, the ξ_{ij} will converge to the region $\left|\xi_{ij}\right| \leq \bar{\Phi}_i$ in finite time. This completes the proof.

9.3 Simulation and Experiment

9.3.1 Simulation Results

In order to show the effectiveness of the proposed control scheme, some simulations are conducted in this section, in which two identical 2-DOF serial links manipulators are chosen as the local and remote manipulators. Thus the same controller parameters are set for the master and the slave. The specific parameters of the teleoperation dynamics model and the controller parameters are shown in Table 9.1.

In Table 9.1 the nominal mass and length are expressed as m_1, m_2, L_1 and L_2. The actual values are expressed as $\tilde{m}_1$, $\tilde{m}_2$, $\tilde{L}_1$ and $\tilde{L}_2$. The initial joint position of the master and slave are set as $q_m(0) = [0.2\pi\ 0.12\pi]^T$ and $q_s(0) = [0.1\pi\ 0.12\pi]^T$ and the velocity of master and the slave manipulators are set as $\dot{q}_m = [0\ 0]^T$, $\dot{q}_s =$

$[0\ 0]^T$. The time delays are set as $d_m = 600$ ms and $d_s = 600$ ms. To obtain some superior work performances, the synchronization error constraints are set as $k_{aij} = (1.5 - 0.05)\exp(-1.6t) + 0.05$, $k_{bij} = (1.5 - 0.05)\exp(-1.6t) + 0.05$, $j = 1, 2$, $i = m, s$. To simplify the presentation, the same constraints are set on the different joint synchronization errors.

To illustrate the control performance of the proposed controller, comparisons will be conducted. We carry out the comparison with the general nonsingularity terminal sliding mode controller, which is designed as follows

$$
\begin{aligned}
\tau_m = {} & M_{om}(q_m)(\ddot{q}_s(t - d_s) - \lambda'_m(e_m) \\
& - K_{m1}s_m - K_{m2}sig(s_m)^{\rho}) + C_{om}(q_m, \dot{q}_m)\dot{q}_m \\
& + G_{om}(q_m) - \hat{W}_m^T \varphi_m(X_m) - \frac{1}{2\sigma_{m1}}\hat{\phi}_m s_m \\
\tau_s = {} & M_{os}(q_s)(\ddot{q}_m(t - d_m) - \lambda'_s(e_s) \\
& - K_s s_s - K_s sig(s_s)^{\rho}) + C_{os}(q_s, \dot{q}_s)\dot{q}_s \\
& + G_{os}(q_s) - \hat{W}_s^T \varphi_s(X_s) - \frac{1}{2\sigma_{s1}}\hat{\phi}_s s_s
\end{aligned}
\tag{9.38}
$$

The human insert force is shown in Fig. 9.2. Figures 9.3 and 9.4 illustrate the position tracking errors at the master side and the slave side with controllers (9.38), respectively. The joint positions of the slave are shown in Fig. 9.5. With the controller (9.16), the position synchronization errors with the constraints at the master side and the slave side are shown in Figs. 9.6 and 9.7. Then in Fig. 9.8, the joint positions of the slave are given. Compared Figs. 9.3, 9.4 with Figs. 9.6, 9.7, it is obvious that with the controller (9.16) proposed in this chapter, the synchronization errors are always bounded by the decaying boundaries. Moreover, by comparing Fig. 9.5 and Fig. 9.8, it can be seen that the positions of the slave links is always restricted in a boundary. This can help the slave avoid having collision with the remote object.

9.3.2 Experiment Results

The experiment on the 3-DOF PHANTOM manipulator (see Fig. 8.23) is also conducted to demonstrate the effectiveness of the proposed control scheme. The constraints are set as $k_{aij} = (2.5 - 0.05)\exp(-0.2t) + 0.05$, $k_{bij} = (2.5 - 0.05)\exp(-0.2t) + 0.05$, $i = m, s$, $j = 1, 2, 3$. In the experiment, we set the master follows a desired trajectory as $q_{md} = 0.1 + 0.2\sin(0.2t)$, and the slave tracks the trajectory of the master. First, we set the time delay as $d_m = 200$ ms and $d_s = 200$ ms. The joint positions are shown in Fig. 9.9. The joint position error with the constraints are shown in Fig. 9.10. The control torques for the three joints are shown in Fig. 9.11.

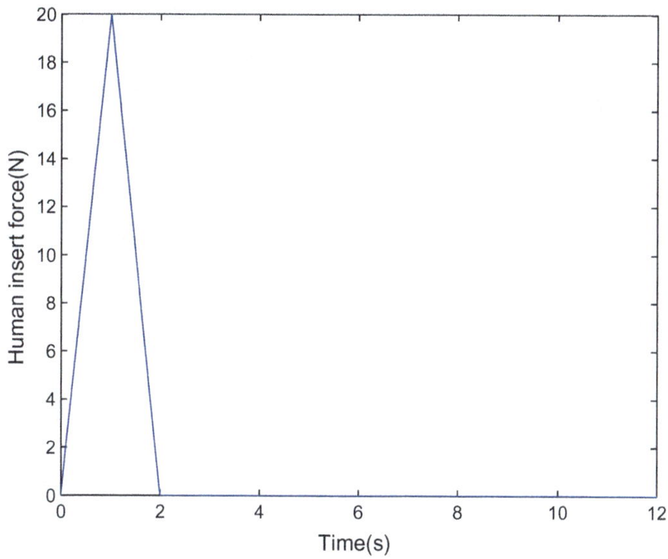

Fig. 9.2 Human-force input F

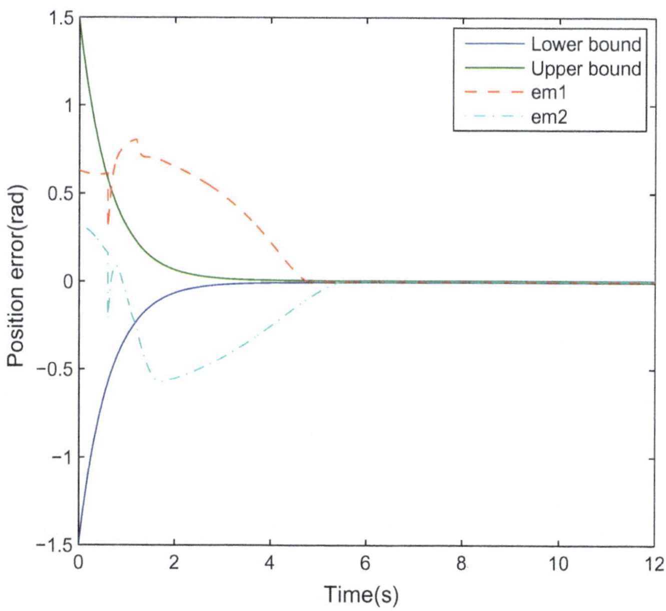

Fig. 9.3 The joint position errors at the master side

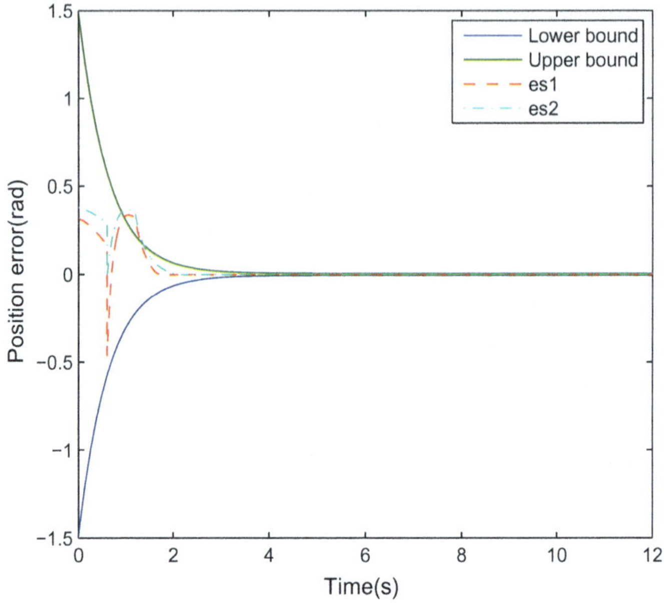

Fig. 9.4 The joint position errors at the slave side

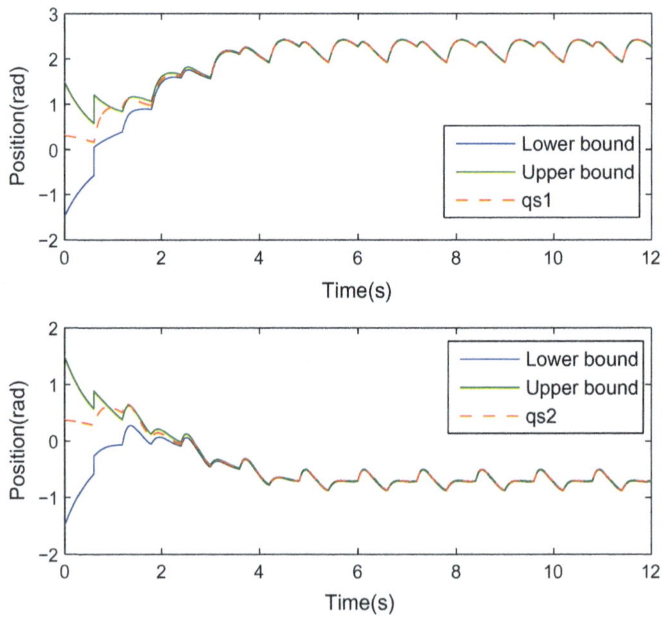

Fig. 9.5 The joint positions of the slave

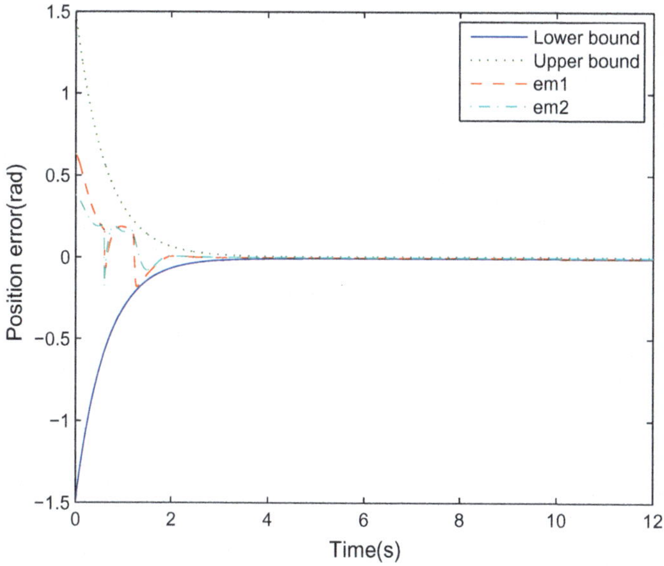

Fig. 9.6 The joint position errors at the master side

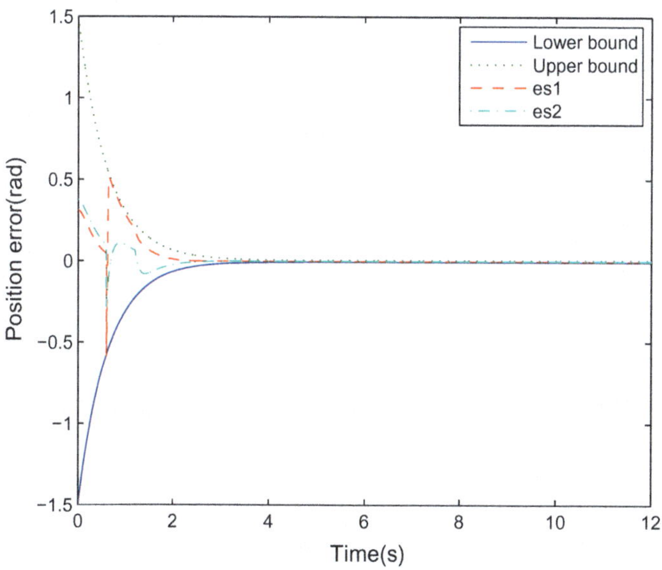

Fig. 9.7 The joint position errors at the slave side

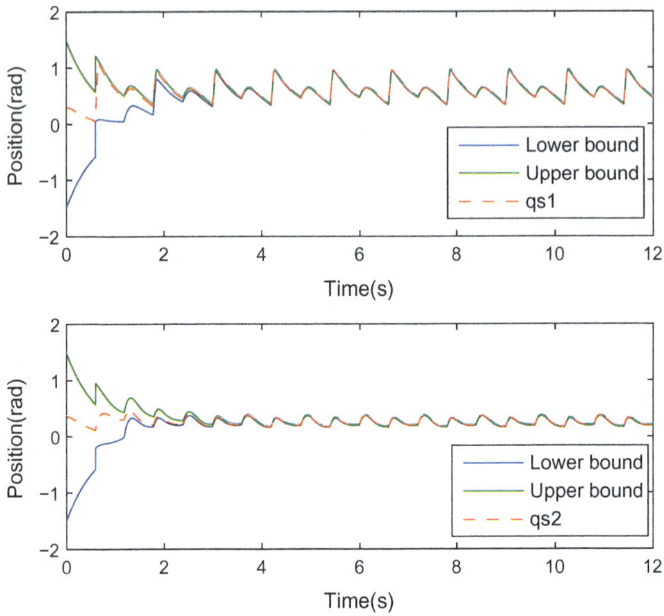

Fig. 9.8 The joint positions of the slave

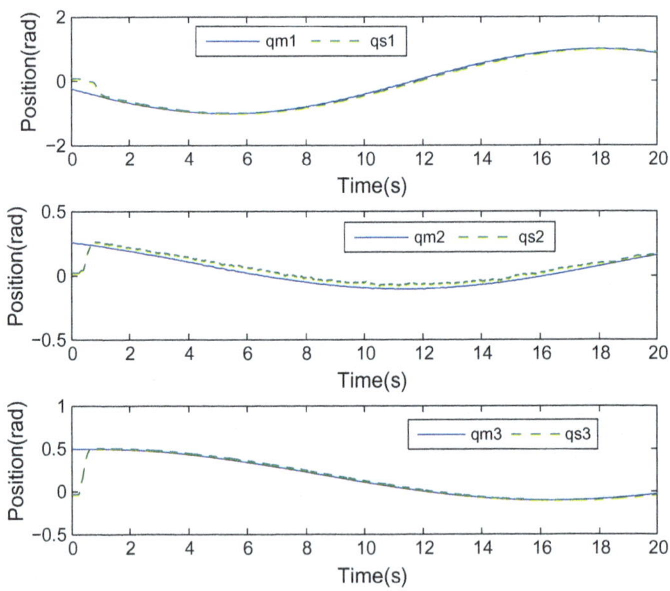

Fig. 9.9 The joint positions of the master and the slave

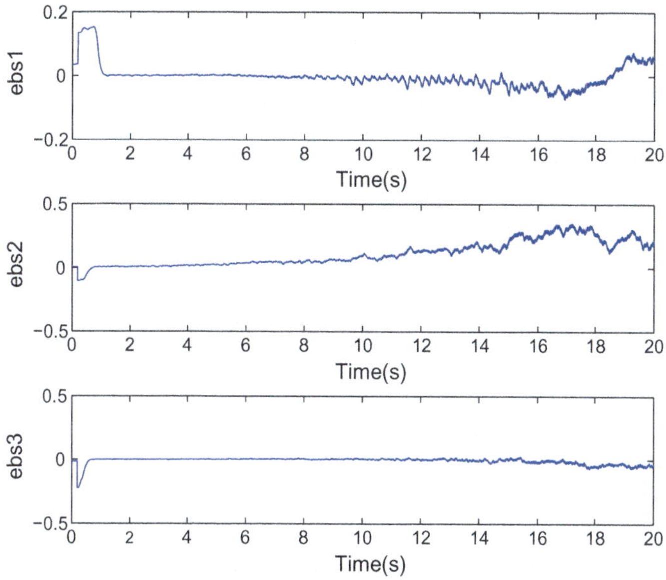

Fig. 9.10 The position errors with the constraint

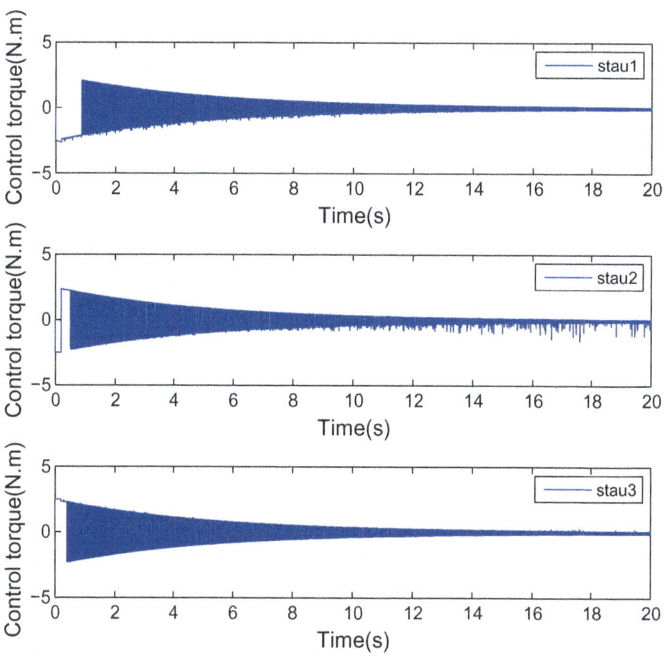

Fig. 9.11 The control torque at the slave side

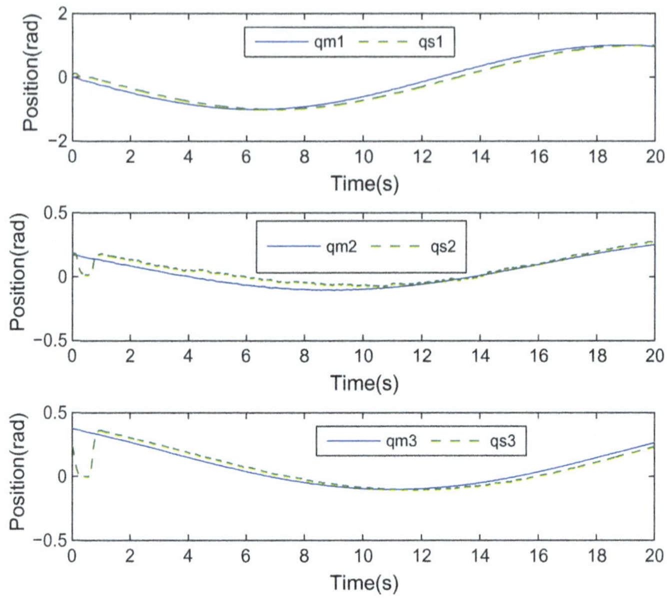

Fig. 9.12 The joint positions of the master and the slave

Then the time delays are set as $d_m = 600\,\text{ms}$ and $d_s = 600\,\text{ms}$. The joint positions are shown in Fig. 9.12. The joint position errors with the constraints are shown in Fig. 9.13. The control torques for the three joints are shown in Fig. 9.14.

As we can see from Fig. 9.9 and Fig. 9.12, the slave tracks the trajectory of the master in a high speed with initial tracking errors. Then the slave will always track the trajectory of the master. In Fig. 9.10 and Fig. 9.13, the $ebs1$, $ebs2$, $ebs3$ stand for the ξ_{s1}, ξ_{s2} and ξ_{s3}. As we can see in these two figures, the values of ξ_{s1}, ξ_{s2} and ξ_{s3} are always smaller than 1, which means that the synchronization errors satisfy $-k_{amj}(t) < e_{mj}(t) < k_{bmj}(t)$ and $-k_{asj}(t) < e_{sj}(t) < k_{bsj}(t)$. Therefore, the position synchronization error constraints are not violated.

9.4 Summary

This chapter has studied the issues associated with the finite time synchronization control of bilateral teleoperation system with position synchronization errors constrained. The proposed control algorithm not only imbibes advantages of both methods: finite time control method and BLF-based control method but also provides additional advantage. Both the transient-state performance and the steady-state performance can be guaranteed in this chapter. In simulation, by comparing the designed controller with the NFTSM controller, we show that the proposed BLF-based NFTSM

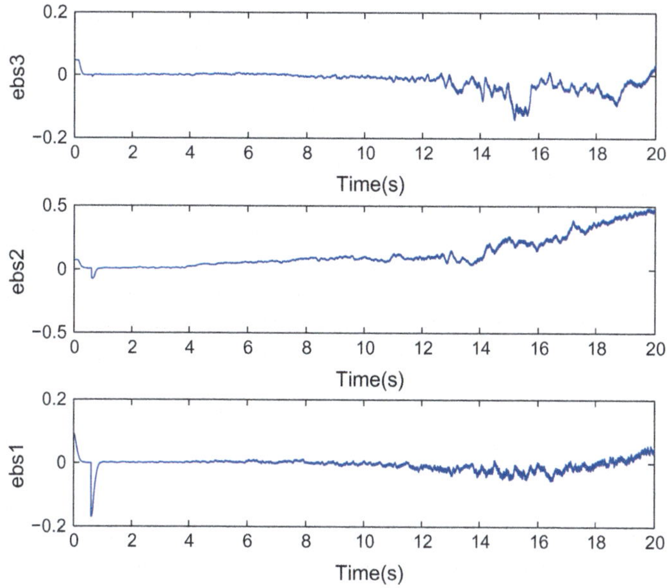

Fig. 9.13 The position errors with the constraint

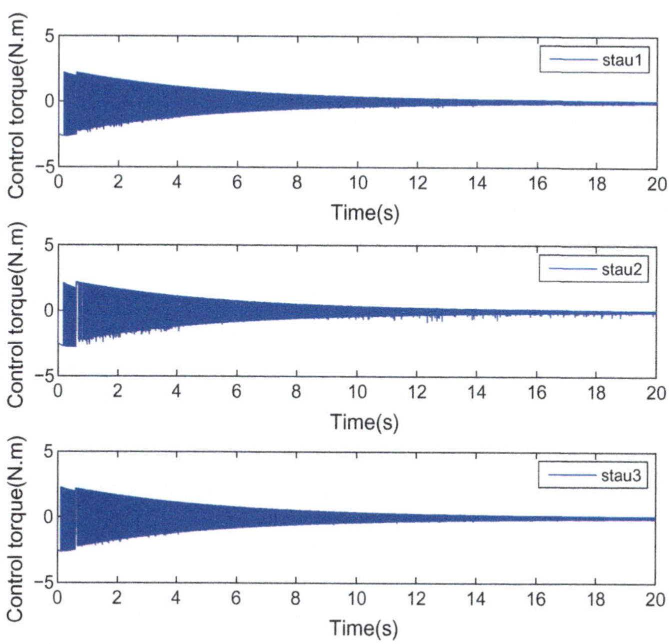

Fig. 9.14 The control torque at the slave side

controller can achieve super transient-state performances. Finally, the experiments are conducted with different time delays to show the effectiveness of the proposed methods.

References

1. R. Kelly, A tuning procedure for stable pid control of robot manipulators. Robotica **13**, 141–148 (1995)
2. C.P. Bechlioulis, G.A. Rovithakis, Adaptive control with guaranteed transient and steady state tracking error bounds for strict feedback systems. Automatica **45**(2), 532–538 (2009)
3. K.B. Ngo, R. Mahony, Z.P. Jiang, Integrator backstepping using barrier functions for systems with multiple state constraints, in *Proceeding of 44th IEEE Conference on Decision and Control*, pp. 8306–8312 (2005)
4. K.P. Tee, B.B. Ren, S.S. Ge, Control of nonlinear systems with time-varying output constraints. Automatica **47**(11), 2511–2616 (2011)
5. Z. Zhao, S.S. Ge, Adaptive neural network control of a fully actuated marine surface vessel with multiple output constraint. IEEE Trans. Control Syst. Technol. **22**(4), 1536–1543 (2013)
6. X. Jin, J.X. Xu, Iterative learning control for output-constrained systems with both parametric and nonparametric uncertainties. Automatica **49**(8), 2508–2516 (2013)
7. Y.M. Li, S.C. Tong, T.S. Li, Adaptive fuzzy output-feedback control for output constrained nonlinear systems in the presence of input saturation. Fuzzy Sets Syst. **248**, 138–155 (2014)
8. D. Wang, J. Huang, Neural network-based adaptive dynamic surface control for a class of uncertain nonlinear systems in strict-feedback form. IEEE Trans. Neural Netw. **16**(1), 195–202 (2005)
9. J.R. Noriega, H. Wang, A direct adaptive neural-network control for unknown nonlinear systems and its application. Fuzzy Sets Syst. **9**(1), 27–34 (1998)
10. W. He, Y.H. Chen, Z. Yin, Adaptive neural network control of an uncertain robot with full-state constraints. IEEE Trans. Cybern. **46**(3), 620–629 (2016)
11. Y.J. Liu, J. Li, S.C. Tong, C.L. Philip Chen, Neural network control-based adaptive learning design for nonlinear systems with full-state constraints. IEEE Trans. Neural Netw. Learn. Syst. **27**(7), 1562–1571 (2016)

Part III
Velocity Observer Design for Teleoperation System

Chapter 10
Output Feedback Controller Design for Teleoperation System via High-Gain Observer

Abstract In this chapter, the state-feedback control result achieved in Chap. 7 will be extended to the case that the velocity information is not available. A high gain observer is designed to estimate the velocities of the maser and slave joints online and the controller is constructed on the basis of the estimated velocities. The stability of the closed-loop system is proved. Both simulations and experiments are performed to verify the correctness and effectiveness of the proposed method.

10.1 Introduction

In above chapters, many control results have been achieved by designing P+d controller, PD+d controller, directed force feedback controller and so on. However, the controller designs shown in the above chapters and the quoted literature require the velocities of the master and slave robots. The joint velocities are usually obtained by differentiating the position signals, which leads to severe noise, or otherwise are very expensive to measure. Moreover, due to the complexity of the communication network, the transmission delays of data packets are not only time-varying, but also asymmetric. It is an interesting problem for design a stabilizing controller for the Internet-based teleoperation systems for which the velocities of the joints are unknown and communication delays are asymmetric and time-varying.

Motivated by the above analysis, in this chapter, we will consider the problem of output feedback stability control for a class of nonlinear teleoperation systems. A simple but effective high gain observer [1–5] is designed to estimate the velocities of the master and slave joints online. Then the controller is designed with the estimated velocities. With the proposed output feedback control scheme, we prove the stability of the closed-loop system and show that the synchronization error converges exponentially to a bounded region. Finally, simulations and experiments are performed to show the effectiveness of the proposed main results.

C. Hua et al., *Analysis and Design for Networked Teleoperation System*,
https://doi.org/10.1007/978-981-13-7936-9_10

10.2 Preliminaries and Problem Formulation

Similar with Chap. 7, in this chapter we consider a master-slave system described by

$$\begin{cases} M_m(q_m)\ddot{q}_m + C_m(q_m, \dot{q}_m)\dot{q}_m = \tau_m + F_h \\ M_s(q_s)\ddot{q}_s + C_s(q_s, \dot{q}_s)\dot{q}_s = \tau_s - F_e \end{cases} \tag{10.1}$$

Likewise, the same properties and assumptions on the inserted forces and the time-delays are needed.

In this chapter, we will extend the control result achieved in Chap. 7 to the case that velocity information is not available. Then the corresponding high-gain observer based control is developed.

10.3 Main Results

We first propose the high gain observer to estimate the velocities, and then the controller is constructed. Let $x_{m1} = q_m, x_{s1} = q_s, x_{m2} = \dot{q}_m$ and $x_{s2} = \dot{q}_s$, we have

$$\begin{cases} \dot{x}_{m1} = x_{m2} \\ \dot{x}_{m2} = f_m(x_{m1}, x_{m2}, \tau_m + F_h) \\ \dot{x}_{s1} = x_{s2} \\ \dot{x}_{s2} = f_s(x_{s1}, x_{s2}, \tau_s - F_e) \end{cases} \tag{10.2}$$

where $f_m(\cdot) = M_m^{-1}(-C_m x_{m2} + \tau_m + F_h)$ and $f_s(\cdot) = M_s^{-1}(-C_s x_{s2} + \tau_s - F_e)$.

Without the mathematical model, we design the following observer form

$$\begin{cases} \dot{z}_{m1} = z_{m2} + \frac{1}{\varepsilon}K_{m1}(x_{m1} - z_{m1}) \\ \dot{z}_{m2} = \frac{1}{\varepsilon^2}K_{m2}(x_{m1} - z_{m1}) \\ \dot{z}_{s1} = z_{s2} + \frac{1}{\varepsilon}K_{s1}(x_{m1} - z_{m1}) \\ \dot{z}_{s2} = \frac{1}{\varepsilon^2}K_{s2}(x_{s1} - z_{s1}) \end{cases} \tag{10.3}$$

where z_{m1}, z_{m2}, z_{s1} and z_{s2} are the state variables of the observer, ε is a small positive parameter, matrices K_{m1}, K_{m2}, K_{s1} and K_{s2} are chosen such that matrices $A_m = \begin{bmatrix} -K_{m1} & I \\ -K_{m2} & 0 \end{bmatrix}$ and $A_s = \begin{bmatrix} -K_{s1} & I \\ -K_{s2} & 0 \end{bmatrix}$ are Hurwitz stable. We have the following lemma:

Lemma 10.1 *For system (10.2) with the high-gain observer (10.3), the error $e_m = x_m - z_m$ and $e_s = x_s - z_s$ can be rendered sufficiently small by adjusting the control design parameter ε.*

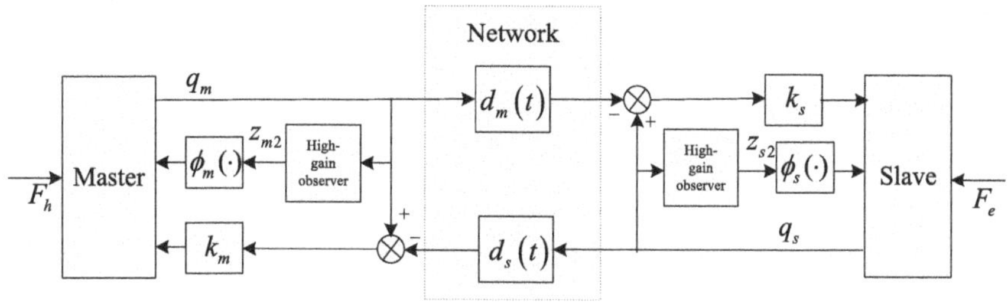

Fig. 10.1 The control scheme for teleoperation system with high-gain observer

Now, we use the estimated velocities for the teleoperation controller design. The master and slave controllers are

$$\begin{cases} \tau_m = -k_m\left(q_m - q_s\left(t - d_s\left(t\right)\right)\right) - \alpha_m z_{m2} - \beta_m \left\|z_{m2}\right\|^2 z_{m2}, \\ \tau_s = k_s\left(q_m\left(t - d_m\left(t\right)\right) - q_s\right) - \alpha_s z_{s2} - \beta_s \left\|z_{s2}\right\|^2 z_{s2} \end{cases} \tag{10.4}$$

where $d_m(t)$ and $d_s(t)$ are forward time delay (from the master site to slave site) and backward time delay (from the slave site to master site), respectively; k_m and k_s are proportional coefficients; α_m, α_s, β_m and β_s are positive parameters. The control scheme is shown in Fig. 10.1, in which the function $\phi_i(z_{i2}) = -\alpha_i z_{i2} - \beta_i \|z_{i2}\|^2 z_{i2}$. We will show that the controller (10.4) can realize the whole system exponentially stable.

Theorem 10.1 *For the teleoperation system with the velocity estimator (10.3), if the matrix $\Phi < 0$, then the velocities $\dot{q}_m, \dot{q}_s$ and position tracking error $q_m - q_s$ converge to the bounded regions, and their bounded regions are determined by the estimation error E_m, and the bounds $\overline{F}_h$ and $\overline{F}_e$.*

Proof Choose the Lyapunov functional $V = V_1 + V_2$ with

$$V_1 = a_m \dot{q}_m^T M_m(q_m)\dot{q}_m + b_m\left(\dot{q}_m^T M_m(q_m)\dot{q}_m\right)^2 - 2\epsilon_s\left(q_m - q_s\right)^T M_s(q_s)\dot{q}_s \tag{10.4}$$

$$+ a_s\left(\dot{q}_s^T M_s(q_s)\dot{q}_s\right) + b_s\left(\dot{q}_s^T M_s(q_s)\dot{q}_s\right)^2 + 2\epsilon_m\left(q_m - q_s\right)^T M_m(q_m)\dot{q}_m$$
$$+ k\left(q_m - q_s\right)^T\left(q_m - q_s\right),$$

$$V_2 = \int_{-\bar{d}_m}^{0}\int_{t+\theta}^{t} e^{-\gamma(t-\xi)}\dot{q}_m(\xi)^T Z\dot{q}_m(\xi)\,d\xi d\theta$$
$$+ \int_{-\bar{d}_s}^{0}\int_{t+\theta}^{t} e^{-\gamma(t-\xi)}\dot{q}_s(\xi)^T S\dot{q}_s(\xi)\,d\xi d\theta,$$

where a_i, b_i and ϵ_i are positive scalars for $i = m, s$, Z and S are positive definite matrices, k is a positive scalar satisfying $k > a_m^{-1}\mu_{m1}^{-1}(\mu_{m2}\epsilon_m)^2 + a_s^{-1}\mu_{s1}^{-1}(\mu_{s2}\epsilon_s)^2$.

Moreover, with the consideration of the following inequalities

$$\begin{aligned}
&\rho_i\alpha_i\dot{q}_i^T(\dot{q}_i - z_{i2}) + \rho_i\dot{q}_i^T(\beta_i\|\dot{q}_i\|^2\dot{q}_i - \beta_i\|z_{i2}\|^2 z_{i2})\\
&= \rho_i\alpha_i\dot{q}_i^T e_i + \rho_i\beta_i\dot{q}_i^T e_i(\|e_i\|^2 - 3\dot{q}_i^T(\dot{q}_i - e_i))\\
&\leq \rho_i\alpha_i\|\dot{q}_i\|\|e_i\| + \rho_i\beta_i\|\dot{q}_i\|\|e_i\| * (\|e_i\|^2 + 3\|\dot{q}_i\|^2 + 3\|\dot{q}_i\|\|e_i\|)\\
&\leq \sigma_{i1}(\|\dot{q}_i\|^2 + \|\dot{q}_i\|^6) + \phi_{i1}(\sigma_{i1}^{-1}, \alpha_i, \beta_i)(\|e_i\|^2 + \|e_i\|^6)
\end{aligned} \tag{10.5}$$

and

$$\begin{aligned}
&\pm 2\epsilon_i(q_m - q_s)^T(\alpha_i(\dot{q}_i - z_{i2}) + \beta_i(\|\dot{q}_i\|^2\dot{q}_i - \|z_{i2}\|^2 z_{i2}))\\
&= \pm 2\epsilon_i(q_m - q_s)^T(\alpha_i\dot{q}_i^T e_i + \rho_i\beta_i\dot{q}_i^T e_i(\|e_i\|^2 - 3\dot{q}_i^T(\dot{q}_i - e_i))\\
&\quad \leq \sigma_{i2}\|q_m - q_s\|^2 + \sigma_{i3}\left(\|\dot{q}_i\|^2 + \|\dot{q}_i\|^6\right)\\
&\quad + \phi_{i2}\left(\sigma_{i2}^{-1}, \alpha_i, \beta_i, \epsilon_i, \sigma_{i3}^{-1}\right)\left(\|e_i\|^2 + \|e_i\|^6\right)
\end{aligned} \tag{10.6}$$

where σ_{i1}, σ_{i2} and σ_{i3} are positive scalars, and $\phi_{i1}(\cdot)$ and $\phi_{i2}(\cdot)$ are positive parameters.

Then we have

$$\begin{aligned}
\dot{V} &\leq -\gamma V + (\sigma_{m1} + \sigma_{m3})(\|\dot{q}_m\|^2 + \|\dot{q}_m\|^6) + \Theta^T\Phi\Theta\\
&\quad + (\sigma_{s1} + \sigma_{s3})(\|\dot{q}_s\|^2 + \|\dot{q}_s\|^6) + (\sigma_{m2} + \sigma_{s2})\|q_m - q_s\|^2 + \Psi(\|e_i\|) + \nu
\end{aligned} \tag{10.7}$$

where $\Psi(\|e_i\|) = (\phi_{m1} + \phi_{m2} + \phi_{s1} + \phi_{s2})(\|e_i\|^2 + \|e_i\|^6)$.
$\Theta^T = \left[q_m^T - q_s^T\ \dot{q}_m^T\ \dot{q}_s^T\ \|\dot{q}_m\|^2\dot{q}_m^T\ \|\dot{q}_s\|^2\dot{q}_s^T\right]$ and

$$\Phi = \begin{bmatrix}
\Phi_{(1,1)} & \Phi_{(1,2)} & \Phi_{(1,3)} & \Phi_{(1,4)} & \Phi_{(1,5)}\\
* & \Phi_{(2,2)} & 0 & \Phi_{(2,4)} & 0\\
* & * & \Phi_{(3,3)} & 0 & \Phi_{(3,5)}\\
* & * & * & \Phi_{(4,4)} & 0\\
* & * & * & * & \Phi_{(5,5)}
\end{bmatrix}$$

where $*$ is the transpose of the corresponding matrix, and the elements are shown below:

$\Phi_{(1,1)} = \eta_1^{-1}I + \eta_2^{-1}I - 2k_m\epsilon_m I - 2k_s\epsilon_s I + 2h_m^{-1}\epsilon_m^2 I + 2h_s^{-1}\epsilon_s^2 I + \overline{d}_s k_m^2\epsilon_m^2\overline{S}^{-1} + \overline{d}_m k_s^2\epsilon_s^2\overline{Z}^{-1} + \eta_3^{-1}(\overline{d}_s k_m\epsilon_m)^2\overline{S}^{-2} + \eta_4^{-1}(\overline{d}_m k_s\epsilon_s)^2\overline{Z}^{-2} + \gamma l_3 I + \vartheta_3\epsilon_m^2 I + \vartheta_4\epsilon_s^2 I$,

$\Phi_{(1,2)} = -k_m a_m I + \overline{d}_s k_m^2 a_m\epsilon_m\overline{S}^{-1} + kI$,

$\Phi_{(1,3)} = k_s a_s I - \overline{d}_m k_s^2\epsilon_s a_s\overline{Z}^{-1} - kI$,

$\Phi_{(1,4)} = -(\mu_{m1} + \mu_{m2})k_m b_m I + \overline{d}_s b_m k_m^2\epsilon_m(\mu_{m1} + \mu_{m2})\overline{S}^{-1}$,

$\Phi_{(1,5)} = (\mu_{s1} + \mu_{s2})b_s k_s I - \overline{d}_m b_s k_s^2\epsilon_s(\mu_{s1} + \mu_{s2})\overline{Z}^{-1}$,

$\Phi_{(2,2)} = -2a_m\alpha_m I + 2\mu_{m2}\epsilon_m I + \omega_1^{-1} I + \omega_2^{-1} I + \overline{d}_m Z + \overline{d}_s k_m^2 a_m^2 \overline{S}^{-1} + \gamma l_{m1} I + \vartheta_1 a_m^2$,

$\Phi_{(2,4)} = 2.5h_m K_{cm}^2 I + \overline{d}_s k_m^2 a_m b_m (\mu_{m1} + \mu_{m2}) \overline{S}^{-1} + \overline{d}_s k_m^2 a_m b_m \overline{\delta}_{\mu_m} \overline{S}^{-1} - a_m \beta_m I - 2\alpha_m \mu_{m1} b_m I + c_m I + \gamma l_{m2} I + 2\vartheta_1 \mu_m a_m b_m I$,

$\Phi_{(3,3)} = -2\alpha_s a_s I + 2\mu_{s2}\epsilon_s I + \omega_1 \epsilon_m^2 \mu_{m2}^2 I + \omega_2 \epsilon_s^2 \mu_{s2}^2 I + \overline{d}_s S + \overline{d}_m k_s^2 a_s^2 \overline{Z}^{-1} + \gamma l_{s1} I + \vartheta_2 a_s^2$,

$\Phi_{(3,5)} = 2.5h_s K_{cs}^2 I + \overline{d}_m (\mu_{s1} + \mu_{s2}) k_s^2 a_s b_s \overline{Z}^{-1} + \overline{d}_m k_s^2 a_s b_s \overline{\delta}_{\mu_s} \overline{Z}^{-1} - a_s \beta_s I - 2\alpha_s \mu_{s1} b_s I + c_s I + \gamma l_{s2} I + 2\vartheta_2 \mu_s a_s b_s I$,

$\Phi_{(4,4)} = \eta_1 \overline{\delta}_{\mu_m}^2 b_m^2 k_m^2 I + \overline{d}_s k_m^2 (\mu_{m1} + \mu_{m2})^2 b_m^2 \overline{S}^{-1} + \overline{d}_s \left(k_m b_m \overline{\delta}_{\mu_m}\right)^2 \overline{S}^{-1} + 2\overline{d}_s k_m^2 (\mu_{m1} + \mu_{m2}) b_m^2 \overline{\delta}_{\mu_m} \overline{S}^{-1} + \eta_3 \left(k_m b_m \overline{\delta}_{\mu_m}\right)^2 I - 4b_m \mu_{m1} \beta_m I + 4\vartheta_1 b_m^2 \mu_m^2 I$,

$\Phi_{(5,5)} = \eta_2 \overline{\delta}_{\mu_s}^2 b_s^2 k_s^2 I + \overline{d}_m k_s^2 (\mu_{s1} + \mu_{s2})^2 b_s^2 \overline{Z}^{-1} + \overline{d}_m k_s^2 \left(b_s \overline{\delta}_{\mu_s}\right)^2 \overline{Z}^{-1} + 2\overline{d}_m k_s^2 (\mu_{s1} + \mu_{s2}) b_s^2 \overline{\delta} -_{\mu_s} \overline{Z}^{-1} + \eta_4 \left(k_s b_s \overline{\delta}_{\mu_s}\right)^2 I - 4b_s \mu_{s1} \beta_s I + 4\vartheta_2 b_s^2 \mu_s^2 I$.

We can choose small parameters σ_{i1}, σ_{i2} and σ_{i3} such that

$$\Theta^T \Phi \Theta + (\sigma_{m1} + \sigma_{m3}) \left(\|\dot{q}_m\|^2 + \|\dot{q}_m\|^6\right) + (\sigma_{s1} + \sigma_{s3}) \left(\|\dot{q}_s\|^2 + \|\dot{q}_s\|^6\right) + (\sigma_{m2} + \sigma_{s2}) \|q_m - q_s\|^2 \leq 0. \tag{10.8}$$

Finally, one can obtain that

$$\dot{V} \leq -\gamma V + \Psi(\|e_i\|) + \nu \tag{10.9}$$

so the closed-loop system is exponentially stable and the converging regions are determined by the estimation error e_i, $\overline{F}_h$ and $\overline{F}_e$. The proof is completed.

10.4 Simulations and Experiments

Similar with above chapters, in this section, simulations are first presented to verify the effectiveness of the main result. Then, experiments performed on a teleoperation system consisting of two Phantom Premium 1.5A robot manipulators are shown.

The user exerted force F to the master robot on the Y-direction, which is shown in Fig. 10.2. The force that the user applied was zero at 0 s, and then was increased to 5 N at 10 s. It was kept constant from time 10 s to 40 s, and decreased to zero from time 40 s to 50 s. At the slave side, there was a wall at $y = 0.5m$. When the slave joint reached the wall and moved further, the feedback force was $10000 * (y - 0.5)\, N$. We can see that the wall is quite stiff.

With the human force input, we know the torque $F_h = -J_m^T * \begin{bmatrix} 0 & 1 \end{bmatrix}^T * F$, where J_m is the Jacobian matrix of the master manipulator. The torque F_e is $F_e = -J_s^T * \begin{bmatrix} 0 & 1 \end{bmatrix}^T * 10000N * (y - 0.5)$. The simulation were used to verify the following: (i) When the user moves the master robot, does the slave follow the movement of the master? (ii) When the slave contacts the wall, is the torque feedback F_e from the wall

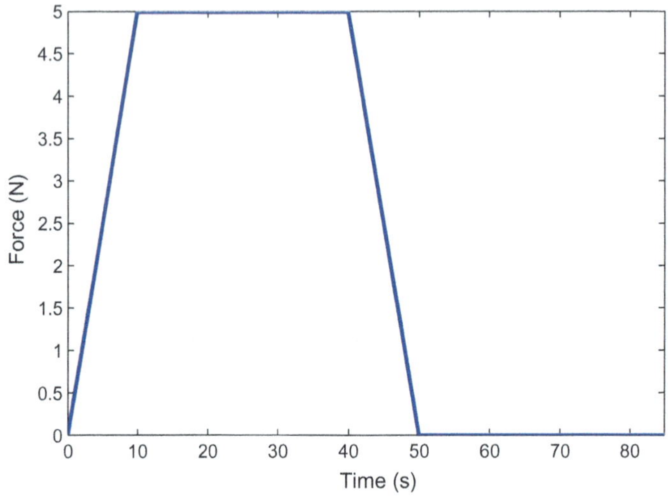

Fig. 10.2 The human force input

close to the human input F_h? (iii) When the human input force is changed to zero, does the error between the master and slave positions disappear? For (i) and (iii), our aim is to show the stability of the closed-loop system. For (ii), it is used to show whether the slave torque changes with the changes of the master torque. If we see F_e as the environment force, the change of F_e can be used to show the transparency performance. To show the effectiveness of the proposed delay-dependent controller, all the above questions will be checked with the computed maximum allowable time delays.

Based on the properties of the teleoperation system, it is easy to have $\mu_{i1} = 0.1250$, $\mu_{i2} = 0.2500$, $K_{ci} = 0.3$. Based on the values of the physical parameters, we chose $\epsilon_i = \epsilon_s = 0.1, k = 1, l_{m1} = l_{m2} = l_{s1} = l_{s2} = 0.5, l_3 = 1$.

With theorem, the control design parameters are given as $k_i = 5, \alpha_i = \beta_i = 20, \gamma = 0.1$. Choose the forward delay parameter $\overline{d}_m = 1$, solving the LMI gives the maximum allowable backward time delay $\overline{d}_s = 2.7084$. The time delays chosen for simulations were $d_m(t) = 0.8 + 0.2\sin(t)$ and $d_s(t) = 2 + 0.7084\sin(t)$. When the velocity information is unavailable, the parameters for the high gain observer are chosen as follows: $K_{i1} = 5$, $K_{i2} = 6$ and $\varepsilon = 0.01$. The simulation results are shown in Figs. 10.3, 10.4, 10.5 and 10.6. Figures 10.3 and 10.6 show the Y-positions and the joint positions, while the torque responses are shown in Figs. 10.4 and 10.5. From the figures, we can see that the high-gain observer based controller (10.4) is effective and renders the closed-loop system stable.

The experiments consists of two Phantom Premium 1.5A robots (SensAble Technologies Inc.) are performed. The experimental results are shown as follows (Figs. 10.7 and 10.8). The user pushed the last joint of the master manipulator to move it in the X-direction. The corresponding slave joint followed the movement

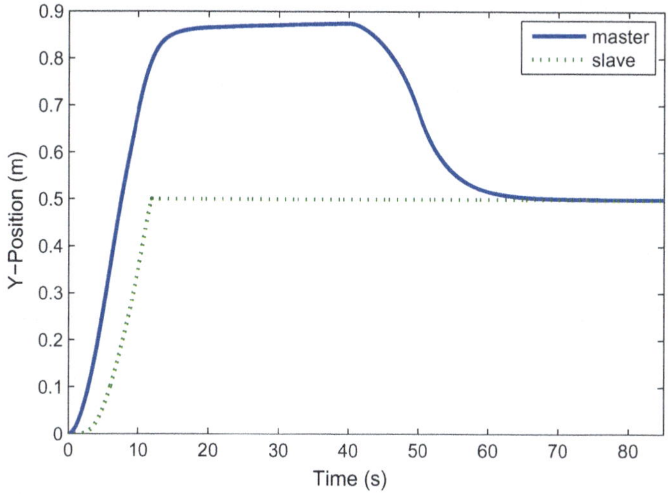

Fig. 10.3 The Y-position of the master and the slave with controller (10.4)

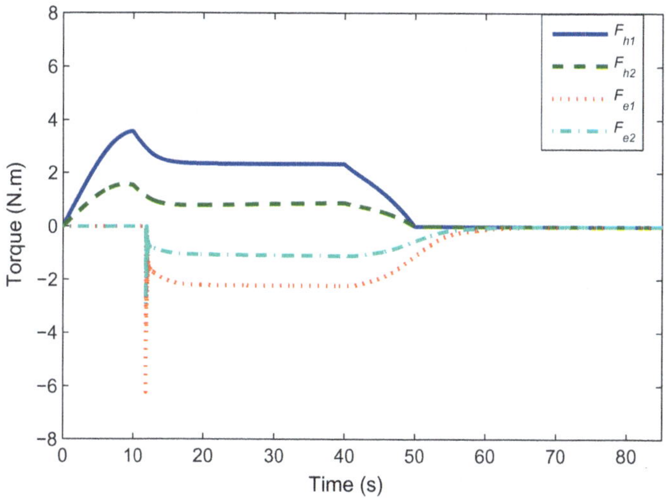

Fig. 10.4 The torques of F_h and F_e with controller (10.4)

of the master joint, and then the slave joint contacted the stiff box and could not move any more. Our objective here is: (i) to verify the effectiveness of the proposed controller (10.4) and to check whether the slave joint follows the movement of the master joint and whether they can return eventually to the original positions; (ii) to test how the user feel the feedback force, especially when the slave robot contacts the stiff wall.

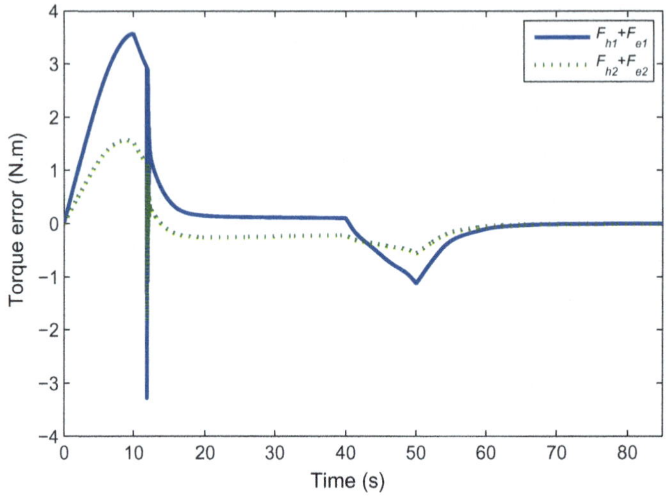

Fig. 10.5 The error between F_h and F_e with controller (10.4)

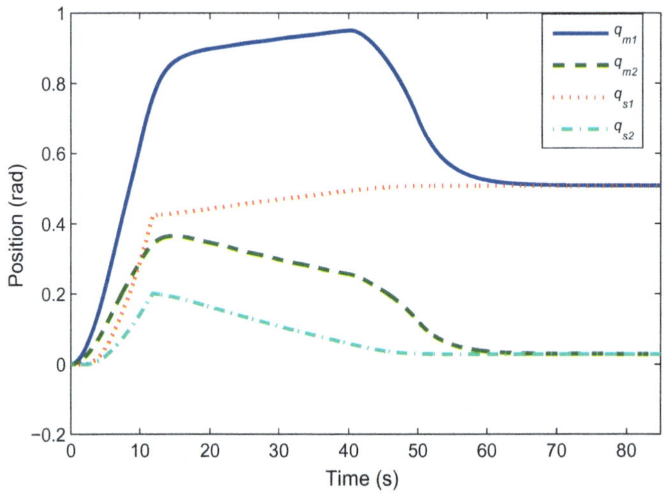

Fig. 10.6 The positions of the master and slave joints with controller (10.4)

With the controller (10.4), We can see that the slave joint reached −0.05 m and could not move any more. When the user moved the master joint further, force feedback was felt, so he knew that the slave contacted with an obstacle. The felt force was stronger when he moved the master joint farther. Finally, the user moved the master joint back to the original position. From these figures, we can see that the slave manipulator followed the movement of the master and reached the origin at last even without the velocity information.

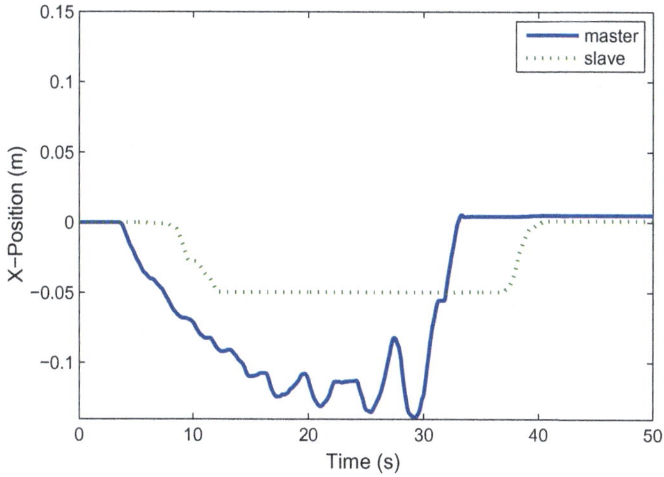

Fig. 10.7 The X-position of the master and the slave with controller (10.4) in the experiment

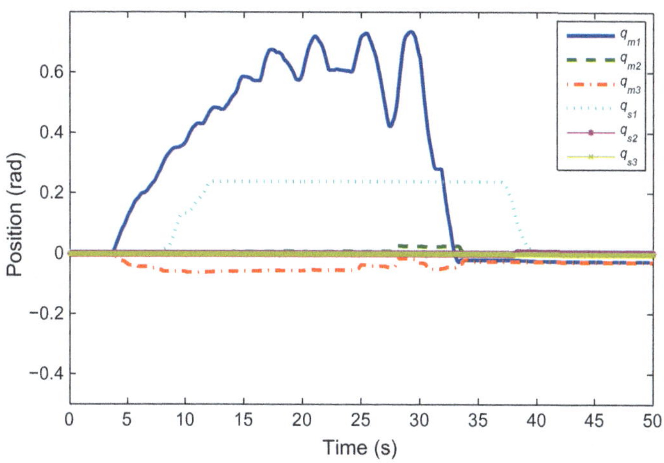

Fig. 10.8 The positions of the master and slave joints with controller (10.4) in the experiment

10.5 Summary

In this chapter the problem of output feedback control design for a class of nonlinear teleoperation system is considered. First, we construct a high-gain observer to estimate the velocity online. Then the controller is designed based on the estimated velocities. The stability of the closed-loop system is proved by employing proper Lyapunov functional. Finally, numerical simulations and experiments are conducted to verify the effectiveness of the main theoretic results obtained.

References

1. G. Bornard, H. Hammouri, A high gain observer for a class of uniformly observable systems, in *Proceedings of the 30th IEEE Conference on Decision and Control*, pp. 1494–1496 (1991)
2. H. Hammouri, B. Targui, F. Armanet, High gain observer based on a triangular structure. Int. J. Robust Nonlinear Control **12**(6), 497–518 (2002)
3. H. Hammouri, G. Bornard, K. Busawon, High gain observer for structured multi-output nonlinear systems. IEEE Trans. Autom. Control **55**(4), 987–992 (2010)
4. M. Farza, M.M. Saad, M. Triki et al., High gain observer for a class of non-triangular systems. Syst. Control Lett. **60**(1), 27–35 (2011)
5. D. Won, W. Kim, M. Tomizuka, High gain observer based integral sliding mode control for position tracking of electro-hydraulic servo systems. IEEE/ASME Trans. Mechatron. **pp**(99), 1–1 (2017)

Chapter 11
Finite Time Observer Design for Teleoperation System

Abstract The finite-time output feedback synchronization control problem is considered for a bilateral teleoperation system in the presence of the modeling error and disturbance. A new observer is designed for the velocity estimation by using neural network approximation and fast terminal sliding mode method. It is shown that the resulting velocity error system is semi-globally finite-time stable. The observer based output feedback finite-time controller is developed by employing a novel nonsingular fast integral terminal sliding mode. The closed-loop system is proved to be semi-globally stable and the master-slave synchronization error converges to zero in finite time. Compared with the existing controllers, the designed controller of this paper only uses the position information and renders the master-slave synchronization error reaching zero in the given finite time. Simulation and experiment are performed and the results demonstrate the effectiveness of the proposed method.

11.1 Introduction

As referred in above two chapters, the result reported on the output-feedback control for teleoperation system with high-gain velocity. Even though good velocity estimation results have been achieved in above chapter and some existing results [1, 2], the velocity estimation errors will converge to zero when time goes to infinity. To provide faster velocity estimation rate and higher estimation precision, in recent years, the finite time velocity observer has been widely used in [3, 4] and the references cited therein. A second-order sliding mode observer was proposed for mechanical systems in [5]. Then the result was extended to MIMO time-delayed teleoperation system to estimate the external forces and velocity signals in [6]. An output feedback terminal sliding mode control approach was designed in [7]. The terminal sliding mode observer was designed for complex nonlinear spacecraft in [8]. In [9] the reduced-order finite time velocity observer was proposed for PWM-based DC-DC buck power converters. However, similar with the general TSM control, for the above finite-time velocity observer, the observation ability will be limited when the initial conditions of estimation errors grow, which will seriously prohibits their practical applications. What's worse, the system uncertainties are not adequately considered when design-

C. Hua et al., *Analysis and Design for Networked Teleoperation System*,
https://doi.org/10.1007/978-981-13-7936-9_11

ing the velocity observer in above literatures. Consider these problems, this chapter designs a new NNs-based FTSM finite-time velocity observer for the uncertain nonlinear bilateral teleoperation system. To the best knowledge of the authors, the problem of finite-time synchronization control has not been well explored for networked teleoperation system with finite-time velocity observer, which finally motivates this study.

In this chapter, the finite-time output-feedback synchronization control problem is addressed for a nonlinear bilateral teleoperation system. The main contents of this chapter are shown as follows. (i) A new NNs based FTSM finite-time velocity observer is designed in the presence of system uncertainties. Compared with the existing velocity observers, faster estimation rate and higher estimation precision can be achieved by employing both high-degree and low-degree terms. (ii) A novel nonsingular fast integral terminal sliding mode (NFITSM) surface is developed without any switching operation to guarantee the synchronization errors converging to zero in finite time when the sliding motion occurs [10–13]. Moreover, the system convergence time will be reduced with the designed NFITSM, because of the neglectable reaching phase by choosing proper initial state of the integral function. Then based on the NFITSM the corresponding finite-time control algorithm is proposed to guarantee the finite-time synchronization between the master and the slave.

11.2 Problem Formulation

Consider a n-link master-slave manipulator system given by the following model

$$\begin{aligned} M_m(q_m)\ddot{q}_m + C_m(q_m, \dot{q}_m)\dot{q}_m + G_m(q_m) &= \tau_m + F_h + \bar{Q}_m(q_m, \dot{q}_m) \\ M_s(q_s)\ddot{q}_s + C_s(q_s, \dot{q}_s)\dot{q}_s + G_s(q_s) &= \tau_s - F_e + \bar{Q}_s(q_s, \dot{q}_s) \end{aligned} \tag{11.1}$$

where $\bar{Q}_m(q_m, \dot{q}_m)$ and $\bar{Q}_s(q_s, \dot{q}_s)$ represent the unknown lumped system uncertainties such as system parameters uncertainties, friction forces and external disturbances.

In this chapter, the functions $M_i(q_i)$, $C_i(q_i, \dot{q}_i)$ and $G_i(q_i)$ are assumed to be known functions. In addition, following [3] we also assume that the nonlinear function $C_i(q_i, \dot{q}_i)\dot{q}_i$ is Lipschitz ($i = m, s$).

For finite time control, the following definition and lemmas are needed.

Definition 11.1 ([14])

$$sig(\xi)^{\alpha} = \left[|\xi_1|^{\alpha} sign(\xi_1), \ldots, |\xi_n|^{\alpha} sign(\xi_n)\right]^T, \tag{11.2}$$

where $\xi = [\xi_1, \xi_2, \ldots, \xi_n]^T \in R^n, \alpha > 0$ and $sign(\cdot)$ being the standard signum function.

Lemma 11.1 *Choose a vector system as follows*

$$\dot{e} = -\alpha sig(e)^{\gamma_1} - \beta sig(e)^{\gamma_2} \tag{11.3}$$

where $\alpha > 0$, $\beta > 0$, $\gamma_1 > 1$ *and* $0 < \gamma_2 < 1$. *The convergence time* T_1 *of* e *is about*

$$T_1 < \frac{1}{\alpha}\frac{1}{\gamma_1 - 1} + \frac{1}{\beta}\frac{1}{1 - \gamma_2} \tag{11.4}$$

Lemma 11.2 ([15]) *Let* $a_1, a_2, \ldots, a_n \geq 0$, *then if* $c > 0$

$$(a_1 + \cdots + a_n)^c \leq \max(n^{c-1}, 1)(a_1^c + \cdots + a_n^c) \tag{11.5}$$

In this chapter, the following RBF NNs are used to approximate a continuous function $f(X) : R^q \to R^p$ *[16–19]:*

$$f(X) = W^T\varphi(X) + \epsilon \tag{11.6}$$

where $X \in \Omega_X \in R^q$ *is the input vector,* Ω_X *is a compact set,* $W \in R^{n\times p}$ *denotes the weight matrix,* $n > 1$ *is the number of the neurons,* $\epsilon \in R^p$ *represents the estimation error and* $\varphi(X) = [\varphi_1(X), \varphi_2(X), \ldots, \varphi_n(X)]^T$, *with* $\varphi_i(X)$ *being the RBF functions, where the Gaussian RBF function is*

$$\varphi_i(X) = \exp(-\frac{\|X - C_i\|^2}{2b_i^2}), i = 1, 2, \ldots, n \tag{11.7}$$

where C_i, b_i *are the center and the width of the ith neuron.*

With the universal approximation property of NNs, for any continuous function $f(X)$, *there exists a NN such that*

$$f(X) = W^{*T}\varphi(X) + \epsilon^*, \left\|\epsilon^*\right\| \leq \epsilon_N \tag{11.8}$$

where W^* *denotes the optimal weight matrix of the NN,* ϵ^* *represents the bounded NN approximation error and* ϵ_N *is an upper bound of the approximation error* ϵ^*.

Remark 11.1 With the definition of Gaussian RBF function, it is easy to obtain that $|\varphi_i(X_1) - \varphi_i(X_2)| \leq \|X_1 - X_2\|$, which will be utilized to verify the finite-time convergence of the designed velocity observer presented in the following section.

11.3 Main Results

11.3.1 FTSM Based Finite-Time Velocity Observer Design

In this subsection, a new NNs-based FTSM finite-time velocity observer will be presented. Let us define $x_{i1} = q_i, \bar{x}_{i2} = \dot{q}_i, i = m, s$, then the dynamics model of the teleoperation system (11.1) can be transformed into

$$\begin{cases} \dot{x}_{i1} = \bar{x}_{i2} \\ \dot{\bar{x}}_{i2} = -M_i^{-1}(x_{i1})(C_i(x_{i1}, \bar{x}_{i2})\bar{x}_{i2} + G_i(x_{i1}) \\ \qquad - \tau_i - F_{ih,e} - \bar{Q}_i(x_{i1}, \bar{x}_{i2})) \end{cases} \tag{11.9}$$

where $F_{mh,e} = F_h$ and $F_{sh,e} = -F_e$.

Then we consider the following coordination transformation matrices

$$T_i = diag(a_{i1}, a_{i2}, \ldots, a_{in}), a_{ij} > 0, j = 1, 2, \ldots, n$$

With definition $x_{i2} = \bar{x}_{i2} - T_i x_{i1}$, then the system (11.9) can be rewritten as

$$\begin{cases} \dot{x}_{i1} = T_i x_{i1} + x_{i2} \\ \dot{x}_{i2} = -M_i^{-1}(x_{i1})(C_i(x_{i1}, \bar{x}_{i2})\bar{x}_{i2} + G_i(x_{i1})) \\ \qquad + M_i^{-1}(x_{i1})(\tau_i + F_{ih,e}) + Q_i(x_{i1}, x_{i2}) \ - T_i^2 x_{i1} - T_i x_{i2} \end{cases} \tag{11.10}$$

where $Q_i(x_{i1}, x_{i2}) = M_i^{-1}(x_{i1})\bar{Q}_i(x_{i1}, \bar{x}_{i2})$.

In this chapter, we assume that for master and the slave only the position information is available, i.e., $y_i = x_{i1}$. The FTSM based finite-time velocity observers proposed for the master and the slave are presented as follows

$$\begin{cases} \dot{\hat{x}}_{i1} = T_i\hat{x}_{i1} + \hat{x}_{i2} - L_{i1}\tilde{x}_{i1} + \nu_{i1} \\ \dot{\hat{x}}_{i2} = -M_i^{-1}(y_i)(C_i(y_i, \hat{\bar{x}}_{i2})\hat{\bar{x}}_{m2} + G_i(y_i)) + M_i^{-1}(y_i)(\tau_i + F_{ih,e}) \\ \qquad + \hat{Q}_i(y_i, \hat{x}_{i2}) - T_i^2\hat{x}_{i1} - T_i\hat{x}_{i2} - L_{i2}\tilde{x}_{i1} + \nu_{i2} + \chi_i sign(\nu_{i1}) \end{cases} \tag{11.11}$$

where $\hat{x}_{i1}$, $\hat{x}_{i2}$, $\hat{\bar{x}}_{i2}$ are the estimations for x_{i1}, x_{i2}, $\bar{x}_{i2}$, respectively. $\tilde{x}_{i1} = \hat{x}_{i1} - x_{i1}$, $L_{i1} = diag(l_{i11}, l_{i12}, \ldots, l_{i1n})$ and $L_{i2} = diag(l_{i21}, l_{i22}, \ldots, l_{i2n})$, each term of L_{i1} and L_{i2} is a positive constant. $\hat{Q}_i(y_i, \hat{x}_{i2})$ is the estimation of $Q_i(x_{i1}, x_{i2})$, $\nu_{i1} = -k_{i1} sign(\tilde{x}_{i1})$, $\nu_{i2} = k_{i2} |\nu_{i1}|^{r_{i1}} sign(\nu_{i1}) + k_{i3} |\nu_{i1}|^{r_{i2}} sign(\nu_{i1})$. For simplification, r_{i1} and r_{i2} are chosen as constants satisfying $r_{i1} > 1$, $0 < r_{i2} < 1$. In addition, k_{i1}, k_{i2} and k_{i3} being positive constants, $\chi_i = diag(\chi_{i1}, \chi_{i2}, \ldots, \chi_{in}) > 0$, $(i = m, s)$.

Let us define $\tilde{x}_{i2} = \hat{x}_{i2} - x_{i2}$ and $\tilde{\bar{x}}_{i2} = \hat{\bar{x}}_{i2} - \bar{x}_{i2}$, then we have

$$\begin{cases} \dot{\tilde{x}}_{i1} = (T_i - L_{i1})\tilde{x}_{i1} + \tilde{x}_{i2} + \nu_{i1} \\ \dot{\tilde{x}}_{i2} = -(T_i^2 + L_{i2})\tilde{x}_{i1} - T_i\tilde{x}_{i2} + \nu_{i2} + \Phi_i \\ \qquad + \hat{Q}_i(y_i, \hat{x}_{i2}) - Q_i(y_i, x_{i2}) + \chi_i sign(\nu_{i1}) \end{cases} \tag{11.12}$$

where $\Phi_i = -M_i^{-1}(y_i)C_i(y_i, \hat{\bar{x}}_{i2})\hat{\bar{x}}_{i2} + M_i^{-1}(y_i)C_i(y_i, \bar{x}_{i2})\bar{x}_{i2}$.

With the NN approximation ability, we have $\hat{Q}_i(y_i, \hat{x}_{i2}) = \hat{W}_i^T\varphi_i(y_i, \hat{x}_{i2})$, further we can obtain that

$$\hat{Q}_i(y_i, \hat{x}_{i2}) - Q_i(y_i, x_{i2}) = -\tilde{W}_i^T\varphi_i(y_i, \hat{x}_{i2}) + N_i - \epsilon_i^* \tag{11.13}$$

where $\tilde{W}_i = W_i^* - \hat{W}_i$, $N_i = W_i^{*T}\varphi_i(y_i, \hat{x}_{i2}) - W_i^{*T}\varphi_i(y_i, x_{i2})$.

11.3.2 Finite-Time Convergence of the Velocity Observer

In this subsection, the conditions for finite-time convergence of the velocity observer will be given. Let us define $\tilde{x}_i = [\tilde{x}_{i1}^T, \tilde{x}_{i2}^T]^T$, with Eq. (11.12), we have $\dot{\tilde{x}}_i = A_i\tilde{x}_i + H_i + \Delta Q_i + \nu_i$, where

$$A_i = \begin{bmatrix} T_i - L_{i1} & I_n \\ -T_i^2 - L_{i2} & -T_i \end{bmatrix}, H_i = \begin{bmatrix} 0_{n\times 1} \\ \Phi_i \end{bmatrix}$$
$$\Delta Q_i = \begin{bmatrix} 0_{n\times 1} \\ \hat{Q}_i(y_i, \hat{x}_{i2}) - Q_i(y_i, x_{i2}) \end{bmatrix}, \nu_i = \begin{bmatrix} \nu_{i1} \\ \nu_{i2} + \chi_i sign(\nu_{i1}) \end{bmatrix}$$

Proposition 11.1 *Consider the velocity observation error system (11.12), assume the human operator/environment insert forces are zero, i.e., $F_h = 0$, $F_e = 0$ and take the NNs parameter tuning law as*

$$\begin{aligned} \hat{W}_i &= \Omega_i\varphi_i(y_i, \hat{x}_{i2})y_i^T - \Omega_i\xi_i \\ \xi_i &= \frac{d\varphi_i(y_i, \hat{x}_{i2})}{dt}y_i^T + \varphi_i(y_i, \hat{x}_{i2})(\hat{x}_{i2} + T_iy_i)^T \end{aligned} \tag{11.14}$$

where Ω_i is a diagonal positive definite constant matrix. If the following conditions hold

$$
\begin{aligned}
&P_iA_i + A_i^T P_i + (\frac{1}{\varepsilon_i} + \frac{1}{z_i})P_iP_i + \Psi_i < 0 \\
&l_{i1j} - a_{ij} = \omega_i > 0 \\
&k_{i1} - \bar{R}_i = \varrho_i > 0 \\
&-2P_{i2}T_i + (\frac{1}{\varepsilon_i} + \frac{1}{z_i})P_iP_i + \varepsilon_i C_i^2 I_n + z_i b_i^2 I_n < 0 \\
&\chi_{ij} \geq F_i
\end{aligned} \tag{11.15}
$$

where $P_i = diag(P_{i1}, P_{i2}) > 0$, $P_{i1} = diag(p_{i11}, p_{i12}, \ldots, p_{i1n})$, $P_{i2} = diag(p_{i21}, p_{i22}, \ldots, p_{i2n})$; ε_i, z_i, b_i, ω_i, ϱ_i, $\bar{C}_i$ *are positive constants;* $\Psi_i = diag(2\varepsilon_i \bar{C}_i^2 \|T_i\|^2 I_n, 0_{n\times n})+ diag(0_{n\times n}, 2\varepsilon_i \bar{C}_i^2 I_n)+ diag(0_{n\times n}, z_i b_i^2 I_n)$; $\bar{R}_i$ *is the derived convergence bound of* $\|\tilde{x}_i\|$ *which will be given later;* $F_i = \left\| \hat{W}_i^T \varphi_i(y_i, \hat{x}_{i2}) - W_i^{*T} \varphi_i(y_i, \hat{x}_{i2}) - \epsilon_i^* \right\|$, *then the velocity estimation errors will converge to zero in finite time.*

The detailed proof for the Proposition 11.1 is given as Appendix.

Remark 11.2 Different from the TSM velocity observers proposed in [6, 20], the coordination transformation matrix T_i is utilized to design the new finite-time velocity observer in this chapter. With the coordinate transformation, one has $A_i = \begin{bmatrix} T_i - L_{i1} & I_n \\ -T_i^2 - L_{i2} & -T_i \end{bmatrix}$. The matrix A_i plays an important role in determining the stability of the velocity observer. With the added term T_i, more degrees of design freedom can be obtained.

Moreover, in order to provide higher convergence speed, the ν_{i2} term is designed with two parts: $k_{i2} |\nu_{i1}|^{r_{i1}} sign(\nu_{i1})$ and $k_{i3} |\nu_{i1}|^{r_{i2}} sign(\nu_{i1})$. When the velocity observer states stay at a long distance from the actual system states, $k_{i2} |\nu_{i1}|^{r_{i1}} sign(\nu_{i1})$ dominates over $k_{i3} |\nu_{i1}|^{r_{i2}} sign(\nu_{i1})$, thus a fast convergence rate can be guaranteed. Moreover, when the velocity observation errors are close to the region, the dominant term $k_{i3} |\nu_{i1}|^{r_{i2}} sign(\nu_{i1})$ determines finite-time convergence.

11.3.3 Finite-Time Controller Design

In this section, with the estimated velocity signals, the finite-time control method will be proposed by designing a novel NFITSM.

Define position and velocity synchronization errors between the master and the slave as $\hat{e}_{m1} = \hat{x}_{s1}(t - d_s) - \hat{x}_{m1}$, $\hat{e}_{m2} = \hat{\bar{x}}_{s2}(t - d_s) - \hat{\bar{x}}_{m2}$, $\hat{e}_{s1} = \hat{x}_{m1}(t - d_m) - \hat{x}_{s1}$, $\hat{e}_{s2} = \hat{\bar{x}}_{m2}(t - d_m) - \hat{\bar{x}}_{s2}$, we have

$$\begin{aligned}
\dot{\hat{e}}_{m1} &= \dot{\hat{x}}_{s1}(t-d_s) - \dot{\hat{x}}_{m1} \\
&= \dot{\hat{x}}_{s1}(t-d_s) - T_m\hat{x}_{m1} - \hat{x}_{m2} + L_{m1}\tilde{x}_{m1} - \nu_{m1} \\
\dot{\hat{e}}_{m2} &= \dot{\hat{\bar{x}}}_{s2}(t-d_s) - \dot{\hat{\bar{x}}}_{m2} \\
&= \dot{\hat{\bar{x}}}_{s2}(t-d_s) + M_m^{-1}(y_m)(C_m(y_m,\hat{\bar{x}}_{m2})\hat{\bar{x}}_{m2} + G_m(y_m)) \\
&\quad - M_m^{-1}(y_m)(\tau_m + F_h) - \hat{Q}_m(y_m,\hat{x}_{m2}) + L_{m2}\tilde{x}_{m1} - \nu_{m2} \\
&\quad - \chi_m sign(\nu_{m1}) + T_m L_{m1}\tilde{x}_{m1} - T_m\nu_{m1}
\end{aligned} \tag{11.16}$$

Similarly, we can also obtain that

$$\begin{aligned}
\dot{\hat{e}}_{s1} &= \dot{\hat{x}}_{m1}(t-d_m) - \dot{\hat{x}}_{s1} \\
&= \dot{\hat{x}}_{m1}(t-d_m) - T_s\hat{x}_{s1} - \hat{x}_{s2} + L_{s1}\tilde{x}_{s1} - \nu_{s1} \\
\dot{\hat{e}}_{s2} &= \dot{\hat{\bar{x}}}_{m2}(t-d_m) - \dot{\hat{\bar{x}}}_{s2} \\
&= \dot{\hat{\bar{x}}}_{m2}(t-d_m) + M_s^{-1}(y_s)(C_s(y_s,\hat{\bar{x}}_{s2})\hat{\bar{x}}_{s2} + G_s(y_s) \\
&\quad - M_s^{-1}(y_s)(\tau_s - F_e) - \hat{Q}_s(y_s,\hat{x}_{s2}) + L_{s2}\tilde{x}_{s1} - \nu_{s2} \\
&\quad - \chi_s sign(\nu_{s1}) + T_s L_{s1}\tilde{x}_{s1} - T_s\nu_{s1}
\end{aligned} \tag{11.17}$$

where d_m represents the unknown constant signal transmission time delay from the master side to the slave side and d_s stands for the constant transmission time delay in inverse direction. Here the buffer strategy is used to allow for the delivery of data at a constant rate even with an irregular transmission schedule in the network, which is a common approach [21].

With the defined synchronization errors between the master and the slave, new NFITSM surfaces for the master and the slave are designed as

$$\begin{aligned}
\hat{s}_i = \hat{e}_{i2} + \int_0^t &(\lambda_{i21} sig(\hat{e}_{i2})^{\alpha_{i21}} + \lambda_{i22} sig(\hat{e}_{i2})^{\alpha_{i22}} \\
&+ \lambda_{i11} sig(\hat{e}_{i1})^{\alpha_{i11}} + \lambda_{i12} sig(\hat{e}_{i1})^{\alpha_{i12}})d\tau
\end{aligned} \tag{11.18}$$

where $\lambda_{i21}, \lambda_{i22}, \lambda_{i11}, \lambda_{i12}$ are positive constants; $\alpha_{i21}, \alpha_{i22}, \alpha_{i11}$ and α_{i12} are positive constants with the bi-limit homogeneity degrees setting as $\alpha_{i11} = \dfrac{\alpha_{i21}}{2-\alpha_{i21}}$, $\alpha_{i12} = \dfrac{\alpha_{i22}}{2-\alpha_{i22}}$ and $1 < \alpha_{i21} < 2, 0 < \alpha_{i22} < 1$.

It is clear that by choosing proper initial value for the integral function, it has that $\hat{s}_i(0) = 0$. Moreover, the variable $\hat{s}_i$ will be kept at zero by designing effective control scheme, thus the system convergence time will be greatly reduced with the negligible reaching phase.

The derivative of the NFITSM is

$$\begin{aligned}\dot{\hat{s}}_i =&\dot{\hat{e}}_{i2} + \lambda_{i21} sig(\hat{e}_{i2})^{\alpha_{i21}} + \lambda_{i22} sig(\hat{e}_{i2})^{\alpha_{i22}} \\ &+ \lambda_{i11} sig(\hat{e}_{i1})^{\alpha_{i11}} + \lambda_{i12} sig(\hat{e}_{i1})^{\alpha_{i12}}\end{aligned} \tag{11.19}$$

Then the controllers with the estimated velocity signals proposed for the master and the slave are given by

$$\begin{aligned}\tau_m =& C_m(y_m, \hat{\bar{x}}_{m2})\hat{\bar{x}}_{m2} + G_m(y_m) + M_m(y_m)(\dot{\hat{\bar{x}}}_{s2}(t-d_s) + L_{m2}\tilde{x}_{m1} \\ &- \nu_{m2} - \chi_m sign(\nu_{m1}) - \hat{Q}_m(y_m, \hat{x}_{m2}) + \lambda_{m21} sig(\hat{e}_{m2})^{\alpha_{m21}} \\ &+ \lambda_{m22} sig(\hat{e}_{m2})^{\alpha_{m22}} + \lambda_{m11} sig(\hat{e}_{m1})^{\alpha_{m11}} + \lambda_{m12} sig(\hat{e}_{m1})^{\alpha_{m12}} \\ &+ T_m L_{m1}\tilde{x}_{m1} - T_m \nu_{m1} + K_{m1} sig(\hat{s}_m)^{\rho_1} + K_{m2} sig(\hat{s}_m)^{\rho_2}) \\ \tau_s =& C_s(y_s, \hat{\bar{x}}_{s2})\hat{\bar{x}}_{s2} + G_s(y_s) + M_s(y_s)(\dot{\hat{\bar{x}}}_{m2}(t-d_m) + L_{s2}\tilde{x}_{s1} - \nu_{s2} \\ &- \chi_s sign(\nu_{s1}) - \hat{Q}_s(y_s, \hat{x}_{s2}) + \lambda_{s21} sig(\hat{e}_{s2})^{\alpha_{s21}} + \lambda_{s22} sig(\hat{e}_{s2})^{\alpha_{s22}} \\ &+ \lambda_{s11} sig(\hat{e}_{s1})^{\alpha_{s11}} + \lambda_{s12} sig(\hat{e}_{s1})^{\alpha_{s12}} + T_s L_{s1}\tilde{x}_{s1} \\ &- T_s \nu_{s1} + K_{s1} sig(\hat{s}_s)^{\rho_1} + K_{s2} sig(\hat{s}_s)^{\rho_2})\end{aligned} \tag{11.20}$$

where K_{m1}, K_{m2}, K_{s1} and K_{s2} are diagonal positive-definite constant matrices, ρ_1 and ρ_2 are positive constants satisfying $\rho_1 > 1, 0 < \rho_2 < 1$.

The teleoperation system with the finite-time controller and FTSM finite-time velocity observer is shown in Fig. 11.1.

Remark 11.3 We define that $\Xi_m = C_m(y_m, \hat{\bar{x}}_{m2})\hat{\bar{x}}_{m2} + G_m(y_m) + M_m(y_m)(\dot{\hat{\bar{x}}}_{s2}(t - d_s) + L_{m2}\tilde{x}_{m1} - \nu_{m2} - \chi_m sign(\nu_{m1}) - \hat{Q}_m(y_m, \hat{x}_{m2}) + T_m L_{m1}\tilde{x}_{m1} - T_m \nu_{m1})$ and $\Xi_s = C_s(y_s, \hat{\bar{x}}_{s2})\hat{\bar{x}}_{s2} + G_s(y_s) + M_s(y_s)\,(\dot{\hat{\bar{x}}}_{m2}(t - d_m) + L_{s2}\tilde{x}_{s1} - \nu_{s2} - \chi_s\, sign(\nu_{s1}) - \hat{Q}_s (y_s, \hat{x}_{s2}) + T_s L_{s1}\tilde{x}_{s1} - T_s \nu_{s1})$. In this chapter, the RBF NNs are applied to design the velocity observer. Therefore, the terms of Ξ_m, Ξ_s used in controller design can be obtained from the velocity observer directly.

Remark 11.4 In this chapter, a novel NFITSM surface is designed. Compared with the TSM surface proposed in [14, 22, 23], the singularity problem is solved without any switching operation. Moreover, the high-degree terms $\lambda_{i21} sig(\hat{e}_{i2})^{\alpha_{i21}}$ and $\lambda_{i11} sig(\hat{e}_{i1})^{\alpha_{i11}}$ providing accelerated finite-time uniform convergence of the system errors are added. The convergence time will be bounded by a constant, which is independent of the initial conditions of the sliding mode.With the designed NFITSM surface, the corresponding finite-time control scheme (11.20) is proposed with high-degree term $K_{i1} sig(\hat{s}_i)^{\rho_1}$ and low-degree term $K_{i2} sig(\hat{s}_i)^{\rho_2}$. The accelerated convergence when the system initial states has long distance from the NFITSM surface can be guaranteed with the high-degree term. Furthermore, when the states in the neighborhood of the NFITSM, the low-degree term ensures the time to reach the NFITSM is finite.

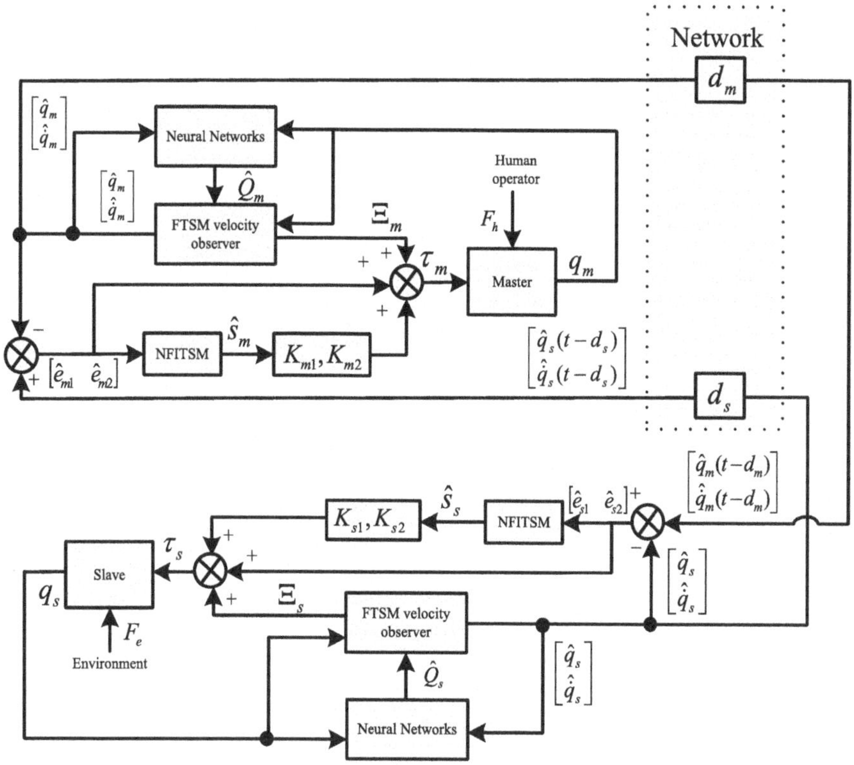

Fig. 11.1 Control block diagram of teleoperation system

With the controller (11.20) for teleoperation system (11.1), we have the following theorem.

Theorem 11.1 *Consider the teleoperator system (11.1) with the new control algorithm (11.20) in free motion, i.e.,* $F_h = 0$, $F_e = 0$, *the system trajectories will always stay on* $\hat{s}_i = 0$, *then the synchronization errors* e_{i1}, $\hat{e}_{i1}$, e_{i2} *and* $\hat{e}_{i2}$ *will converge to zero in finite time.*

Proof Consider Lyapunov function

$$V = \frac{1}{2}\hat{s}_m^T\hat{s}_m + \frac{1}{2}\hat{s}_s^T\hat{s}_s \tag{11.21}$$

Differentiating V along the system trajectories yields

$$\dot{V} = \hat{s}_m^T\dot{\hat{s}}_m + \hat{s}_s^T\dot{\hat{s}}_s \tag{11.22}$$

with the definitions of $\hat{s}_m$ and $\hat{s}_s$, we have

$$\begin{aligned}\dot{V} =& \hat{s}_m^T(\dot{\widehat{\bar{x}}}_{s2}(t-d_s) + M_m^{-1}(y_m)(C_m(y_m,\widehat{\bar{x}}_{m2})\widehat{\bar{x}}_{m2} + G_m(y_m)) \\ & - M_m^{-1}(y_m)\tau_m - \hat{Q}_m(y_m,\hat{x}_{m2}) + T_mL_{m1}\tilde{x}_{m1} - T_m\nu_{m1} + L_{m2}\tilde{x}_{m1} \\ & - \nu_{m2} - \chi_m sign(\nu_{m1}) + \lambda_{m21}sig(\hat{e}_{m2})^{\alpha_{m21}} + \lambda_{m22}sig(\hat{e}_{m2})^{\alpha_{m22}} \\ & + \lambda_{m11}sig(\hat{e}_{m1})^{\alpha_{m11}} + \lambda_{m12}sig(\hat{e}_{m1})^{\alpha_{m12}}) + \hat{s}_s^T(\dot{\widehat{\bar{x}}}_{m2}(t-d_m) \\ & + M_s^{-1}(y_s)(C_s(y_s,\widehat{\bar{x}}_{s2})\widehat{\bar{x}}_{s2} + G_s(y_s)) - M_s^{-1}(y_s)\tau_s - \hat{Q}_s(y_s,\hat{x}_{s2}) \\ & + T_sL_{s1}\tilde{x}_{s1} - T_s\nu_{s1} + L_{s2}\tilde{x}_{s1} - \nu_{s2} - \chi_s sign(\nu_{s1}) + \lambda_{s21}sig(\hat{e}_{s2})^{\alpha_{s21}} \\ & + \lambda_{s22}sig(\hat{e}_{s2})^{\alpha_{s22}} + \lambda_{s11}sig(\hat{e}_{s1})^{\alpha_{s11}} + \lambda_{s12}sig(\hat{e}_{s1})^{\alpha_{s12}})\end{aligned} \tag{11.23}$$

Substituting controller (11.20) into (11.23) gives

$$\begin{aligned}\dot{V} =& -\hat{s}_m^TK_{m1}sig(\hat{s}_m)^{\rho_1} - \hat{s}_m^TK_{m2}sig(\hat{s}_m)^{\rho_2} \\ & - \hat{s}_s^TK_{s1}sig(\hat{s}_s)^{\rho_1} - \hat{s}_s^TK_{s2}sig(\hat{s}_s)^{\rho_2} \\ \leq & -\lambda_{\min}(K_{m1})n^{1-\rho_1/2}(\hat{s}_m^T\hat{s}_m)^{1+\rho_1/2} \\ & - \lambda_{\min}(K_{s1})n^{1-\rho_1/2}(\hat{s}_s^T\hat{s}_s)^{1+\rho_1/2} \\ & - \lambda_{\min}(K_{m2})(\hat{s}_m^T\hat{s}_m)^{1+\rho_2/2} \\ & - \lambda_{\min}(K_{s2})(\hat{s}_s^T\hat{s}_s)^{1+\rho_2/2} \\ \leq & -\alpha(2V)^{1+\rho_1/2} - \beta(2V)^{1+\rho_2/2}\end{aligned} \tag{11.24}$$

where

$$\begin{aligned}\alpha =& 2^{1-\rho_1/2}\min\{\lambda_{\min}(K_{m1})n^{1-\rho_1/2}, \lambda_{\min}(K_{s1})n^{1-\rho_1/2}\}, \\ \beta =& \min\{\lambda_{\min}(K_{m2}), \lambda_{\min}(K_{s2})\},\end{aligned}$$

$\lambda_{\min}(K_{ij})$ denotes the minimum eigenvalue of matrix K_{ij}, $i = m, s, j = 1, 2$. Choose $\vartheta = \sqrt{2V}$, then the inequality (11.24) can be rewritten as $\dot{\vartheta} = -\alpha\vartheta^{\rho_1} - \beta\vartheta^{\rho_2}$. Based on Lemma 11.2, the system states will always stay on the pre-described NFITSM $\hat{s}_i = 0$. With the definition of $\hat{s}_i$, it is obvious that the synchronization errors between the master and the slave, i.e., $\hat{e}_{m1}$, $\hat{e}_{m2}$, $\hat{e}_{s1}$ and $\hat{e}_{s2}$ will also converge to zero in finite time. When $t \geq T_2$, we have that

$$\begin{aligned}s_m =& e_{m2} + \int_0^t(\lambda_{m21}sig(e_{m2})^{\alpha_{m21}} + \lambda_{m22}sig(e_{m2})^{\alpha_{m22}} \\ & + \lambda_{m11}sig(e_{m1})^{\alpha_{m11}} + \lambda_{m12}sig(e_{m1})^{\alpha_{m12}})d\tau \\ s_s =& e_{s2} + \int_0^t(\lambda_{s21}sig(e_{s2})^{\alpha_{s21}} + \lambda_{s22}sig(e_{s2})^{\alpha_{s22}} \\ & + \lambda_{s11}sig(e_{s1})^{\alpha_{s11}} + \lambda_{s12}sig(e_{s1})^{\alpha_{s12}})d\tau\end{aligned} \tag{11.25}$$

where $e_{m1} = x_{s1}(t - d_s) - x_{m1}$, $e_{m2} = \bar{x}_{s2}(t - d_s) - \bar{x}_{m2}$, $e_{s1} = x_{m1}(t - d_m) - x_{s1}$, $e_{m2} = \bar{x}_{m2}(t - d_m) - \bar{x}_{s2}$. Hence, we can also obtain that the variables e_{m1}, e_{m2}, e_{s1}, e_{m2} converge to zero in finite time. This completes the proof.

Remark 11.5 The analysis on how the parameters of the FTSM velocity observer, the NFITSM and the finite-time controller influence the system performance is also given. Firstly, consider the parameters of the finite-time velocity observer, with the Proposition 11.1, the parameters of the velocity observer should be chosen to make the conditions shown in Eq. (11.15) hold. Moreover, parameters k_{i2}, k_{i3}, r_{i1} should be chosen appropriately large and r_{i2} should be chosen appropriately small to make the velocity estimation time T to satisfy the practical velocity estimation demand. Secondly, consider the parameters of the NFITSM surface, λ_{i21}, λ_{i22}, λ_{i11} and λ_{i12} are chosen as positive constants. Based on Lemma 2, appropriately large values of $\lambda_{i21}, \lambda_{i11}, \lambda_{i22}, \lambda_{i12}$ should be set to reduce the convergence time. Moreover, based on bi-limit homogeneity theory, parameters α_{i21}, α_{i22}, α_{i11}, α_{i12} should satisfy $\alpha_{i11} = \dfrac{\alpha_{i21}}{2 - \alpha_{i21}}, \alpha_{i12} = \dfrac{\alpha_{i22}}{2 - \alpha_{i22}}, 1 < \alpha_{i21} < 2, 0 < \alpha_{i22} < 1$. Additionally, appropriately large values of α_{i11}, α_{i21} and appropriately small values of α_{i12}, α_{i22} should be set to reduce the convergence time. Finally, consider the parameters of the finite-time controller. According to the inequality (11.24) and Lemma 11.2, appropriately large values of K_{i1} and K_{i2} should to be set. In addition, appropriately large values of ρ_1 and appropriately small values of ρ_2 are required to guarantee the NFITSM converges to zero in finite time with high speed.

Remark 11.6 For practical teleoperation design, the synchronization accuracy and synchronization time are very important. However, the existing teleoperation control design methods can only guarantee that the synchronization error converges to zero asymptotically/exponetically. It is well known that the state variables reach zero when $t \to \infty$ for the asymptotical/exponetical stability. Therefore, the synchronization performance can not suit the requirements from some demanding applications. For example, for the underwater exploration, the finite-time control can raise the working efficiency and reduce the energy consumption. Moreover, the strong robustness with respect to uncertain dynamics and disturbances of TSM control will provide some satisfactory performances for the teleoperation system under complex ocean environment.

Remark 11.7 In this chapter, a finite-time velocity observer and finite-time control approach are proposed for the teleoperation system, simultaneously. With the proposed control schemes, the synchronization performance can be achieved in finite time by choosing proper parameters. Then high accuracy, fast convergence rate and strong robustness performances can be realized. Moreover, the control approaches given in (11.20) are realized only with the position variables q_m and q_s. Therefore, the cost related to the hardware on the teleoperation system can be effectively reduced with the proposed control scheme.

11.4 Simulation and Experimental Results

In this section, simulation results are first presented to verify the effectiveness of the main result. Then, the experiment on a teleoperation system consisting of two Phantom Premium 1.5A robot manipulators is also performed.

11.4.1 Simulation Results

The simulations are performed on 2-degree-of-freedom manipulators. For the sake of simplicity, two identical manipulators are chosen as the master and the slave. We choose the system parameters as $m_1 = 0.5\,\text{kg}$, $m_2 = 1\,\text{kg}$, $L_1 = 0.7\,\text{m}$, $L_2 = 0.5\,\text{m}$ and $g = 9.81\,\text{m/s}^2$. The specific definitions for $M(q)$, $C(q,\dot{q})\dot{q}$ and $G(q)$ are same as above chapters. The time delays are set as $d_m = 600\,\text{ms}$ and $d_s = 600\,\text{ms}$. The controller parameters are chosen as $K_{i1} = diag(10, 10)$, $K_{i2} = diag(8, 8)$, $\rho_1 = 3$, $\rho_2 = 9/11$. The sliding mode surfaces parameters are set as $\lambda_{i11} = 5, \lambda_{i22} = 5, \lambda_{i12} = 5$, $\lambda_{i21} = 5$, $\alpha_{i11} = 3$, $\alpha_{i21} = \frac{3}{2}$, $\alpha_{i12} = \frac{3}{7}$, $\alpha_{i22} = \frac{3}{5}$. The velocity observer parameters are chosen as $T_i = diag(2, 2)$, $k_{i1} = diag(5, 5)$, $k_{i2} = diag(0.5, 0.5)$, $k_{i3} = diag(0.8670, 0.8670)$, $L_{i1} = diag(446.4864, 446.4864)$, $L_{i2} = diag(-14.3050, -14.3050)$, $r_{i1} = 3$, $r_{i2} = 9/11$, $\Omega_i = 0.5$. The unknown lumped system uncertainties are set as $\bar{Q}_m(q_m, \dot{q}_m) = 0.2\dot{q}_m^2 + 0.3q_m + 0.5$ and $\bar{Q}_s(q_s, \dot{q}_s) = 0.3\dot{q}_s^2 + 0.2q_s + 0.6$. We apply a human force F to the master site in Y-direction, which is shown in Fig. 11.2. The human-input force is 0 at 0 s and then increases to 20 N and it decreases to zero from 1 s to 2 s. The simulation results are used to verify the following: (1) when we move the master robot, does the velocity observer estimate the actual velocity in high speed? (2) When we move the master robot, does the slave follow the master? (3) When the human-input force decreases to zero, does the synchronization errors between the master and slave converge to zero in finite time?

Firstly, the position and velocity estimation results are presented to verify the effectiveness of the designed finite-time velocity observer. The actual joint positions and the estimated joint positions are shown in Fig. 11.3 for the master and the slave. The actual joint velocities and the estimated joint velocities are shown in Fig. 11.4. In Figs. 11.3 and 11.4, $j-$ denotes the estimated position/velocity. The joint velocity estimation errors for the master and the slave are shown in Fig. 11.5. The velocity estimation results with the linear sliding velocity observer are shown in Fig. 11.6. Compared with Figs. 11.4 and 11.6, it is clear that the finite-time velocity observer designed in this chapter provides higher estimation speed and precision. Then the position synchronization results with the proposed controllers are presented. Non-zero initial values of the NFITSM are chosen to show the higher convergence speed with the proposed control scheme clearly. The values of the sliding mode surfaces are shown in Fig. 11.7. It is obvious that when there are human insert forces, the values of the sliding mode are bounded. Then when the human insert forces disappear, the values of the sliding mode converge to zero in high speed. In Fig. 11.8, the position

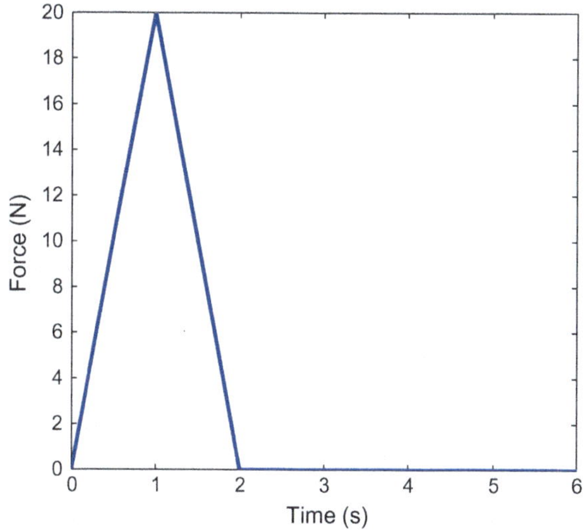

Fig. 11.2 Human input force

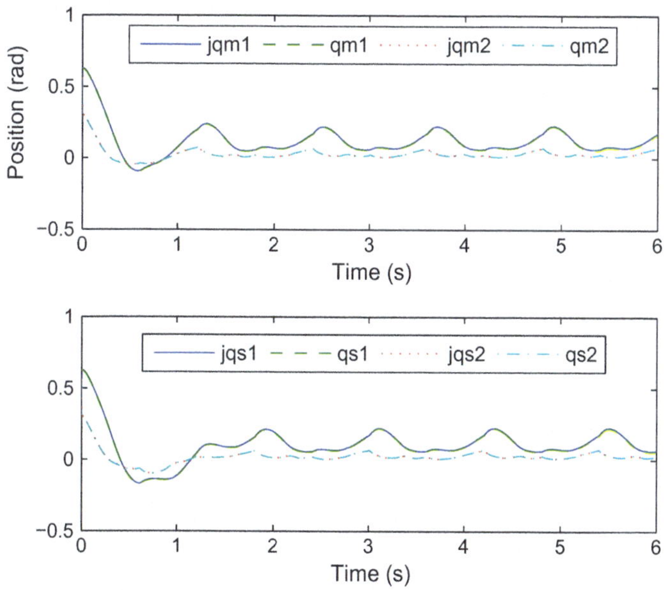

Fig. 11.3 The actual positions and the estimated positions for the master and the slave

synchronization errors are also given. From Figs. 11.7 and 11.8, we can see that the slave joint follows the movement of the master joint even though with complex system uncertainties. Moreover, when the human insert forces decrease to zero, the synchronization errors converge to zero in finite time.

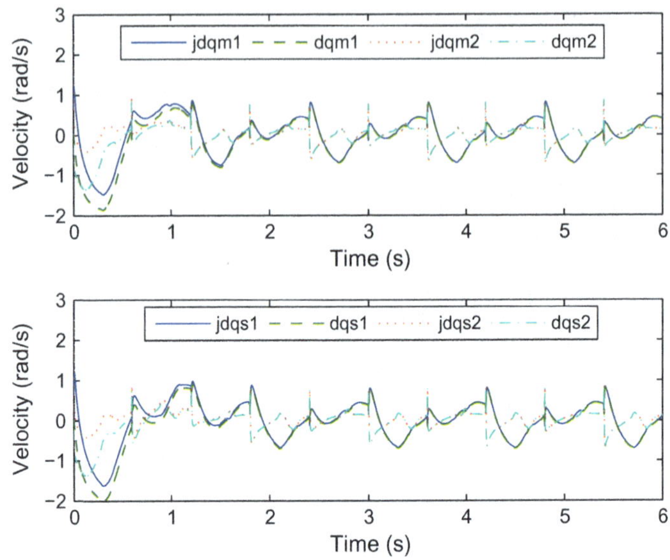

Fig. 11.4 The actual velocities and the estimated velocities for the master and the slave

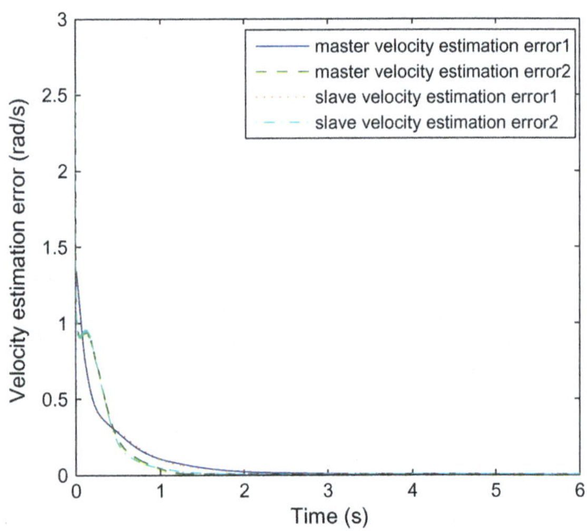

Fig. 11.5 The velocity estimation errors at the master side and the slave side

To illustrate the superior control performance of NFITSM controller, the comparisons with the previous LSM controller have been made. The LSM is designed as follows

$$\hat{s}_m = \hat{e}_{m2} + b_m \hat{e}_{m1};\ \hat{s}_s = \hat{e}_{s2} + b_s \hat{e}_{s1}$$

where b_m and b_s are positive constants, which are set as $b_m = 10$ and $b_s = 10$.

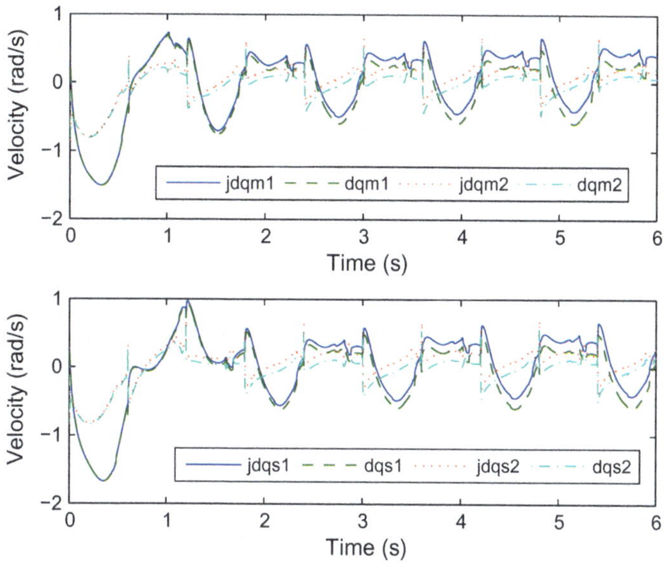

Fig. 11.6 The actual velocities and the estimated velocities for the master and the slave with LSM velocity observer

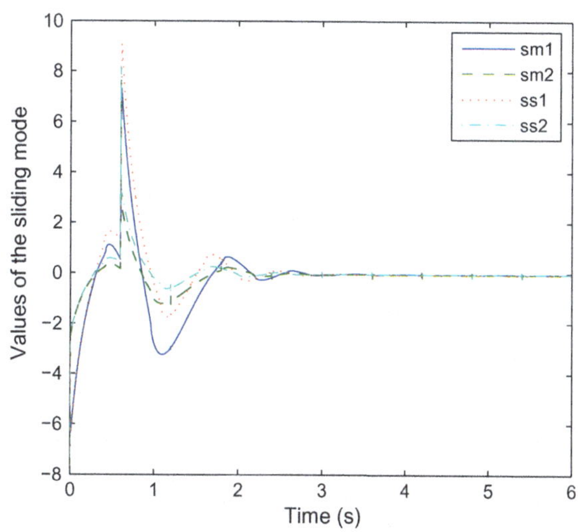

Fig. 11.7 The values of the sliding mode surface for the master and the slave

The position synchronization errors are shown in Figs. 11.9 and 11.10. In Figs. 11.9, 11.10, $l-$ is LSM and $n-$ denotes NFITSM. As we can see from these two figures, the position error e_m converges to zero at about $t = [3.7s; 1.8s]$ with the NFITSM controller and at about $t = [4s; 4.2s]$ with the LSM controller. In Fig. 11.10, the e_s converges to zero at about $t = [2.7s; 2.9s]$ with the NFITSM controller and

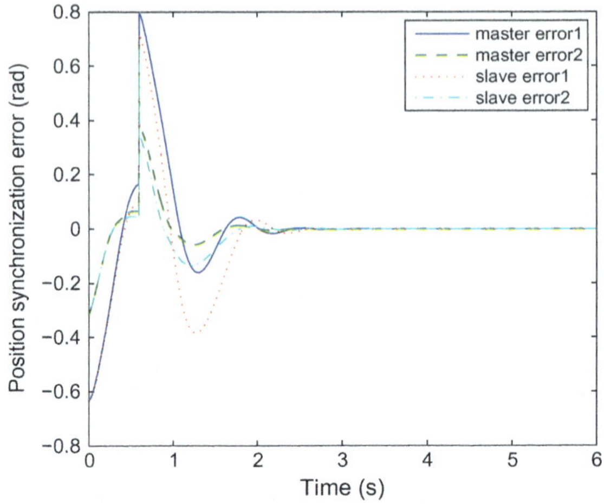

Fig. 11.8 The position synchronization errors at the master side and the slave side

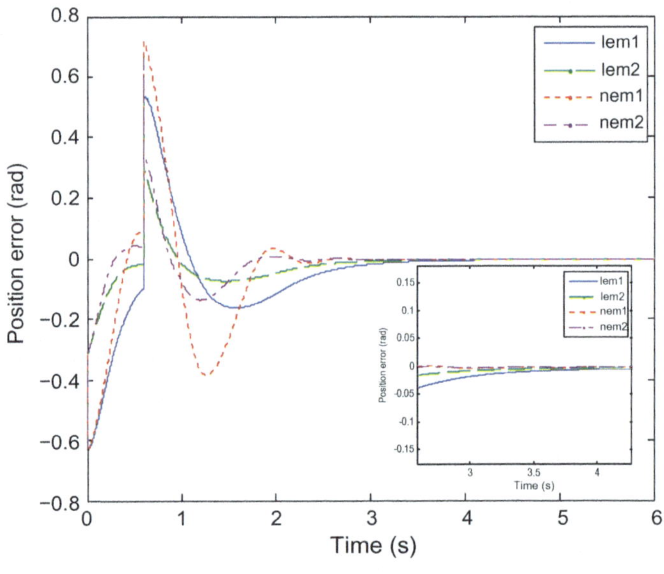

Fig. 11.9 The position synchronization errors at the master side with LSM and NFITSM

at about $t = [3.1s; 3.3s]$ with the LSM controller. It is obvious that faster synchronization rate and higher synchronization precision are obtained with the proposed control algorithm.

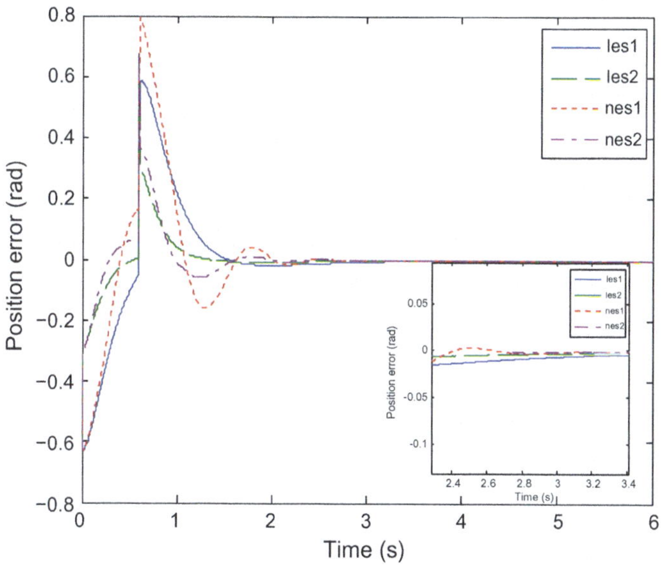

Fig. 11.10 The position synchronization errors at the slave side with LSM and NFITSM

11.4.2 Experiment Results

To demonstrate the effectiveness of the proposed finite-time velocity observer and the controller, the experiment on the teleoperation platform will be conducted. The controller parameters are chosen as $K_{i1} = diag(0.3, 0.3, 0.3)$, $K_{i2} = diag(0.4, 0.4, 0.4)$, $\rho_1 = 3$, $\rho_2 = 9/11$. The sliding mode surfaces parameters are set as $\lambda_{i11} = 5$, $\lambda_{i22} = 5$, $\lambda_{i12} = 5$, $\lambda_{i21} = 5$, $\alpha_{i11} = 3$, $\alpha_{i21} = \frac{3}{2}$, $\alpha_{i12} = \frac{3}{7}$, $\alpha_{i22} = \frac{3}{5}$. The velocity observer parameters are chosen as $T_i = diag(0.1, 0.1, 0.1)$, $k_{i1} = diag(2.3084, 2.3084, 2.3084)$, $k_{i2} = diag(0.5, 0.5, 0.5)$, $k_{i3} = diag(0.8670, 0.8670, 0.867)$, $L_{i1} = diag(2, 2, 2)$, $L_{i2} = diag(-0.14, -0.14, -0.14)$, $r_{i1} = 3$, $r_{i2} = 9/11$, $\Omega_i = 0.5$.

To prove the effectiveness of the proposed velocity observer and the controller, first the time delays are set as $d_m = 100\,\text{ms}$ and $d_s = 100\,\text{ms}$. The experimental results of the estimated velocity and the actual velocity are shown in Fig. 11.11. The joint position synchronization trajectories between master and slave robots are presented in Fig. 11.12. Then the time delays are increased to $d_m = 500\,\text{ms}$ and $d_s = 500\,\text{ms}$. The experimental results of the velocity estimation and the joint position synchronization trajectories are shown in Fig. 11.13 and Fig. 11.14, respectively. Finally, time delays are set as $d_m = 1000\,\text{ms}$ and $d_s = 1000\,\text{ms}$. The experimental results of the velocity estimation and the joint position trajectories are given in Fig. 11.15 and Fig. 11.16, respectively. In Figs 11.11, 11.12, 11.13, 11.14, 11.15 and 11.16, the experimental results for joint 1, joint 2 and joint 3 are presented in top,

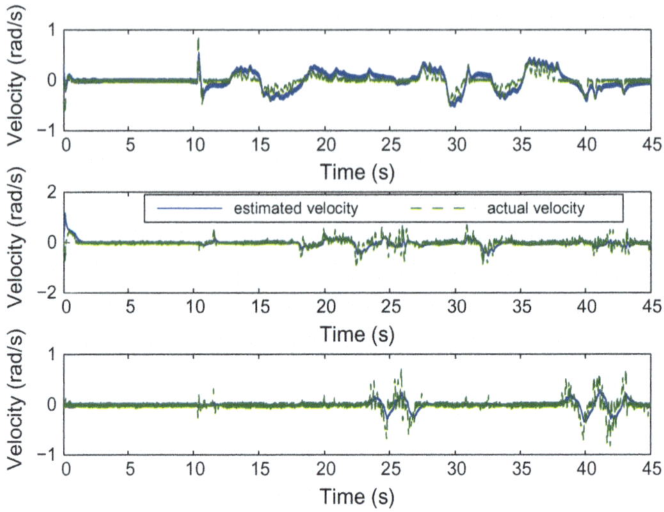

Fig. 11.11 The estimated velocities and the actual velocities with time delays $d_m = 100\,\text{ms}$ and $d_s = 100\,\text{ms}$

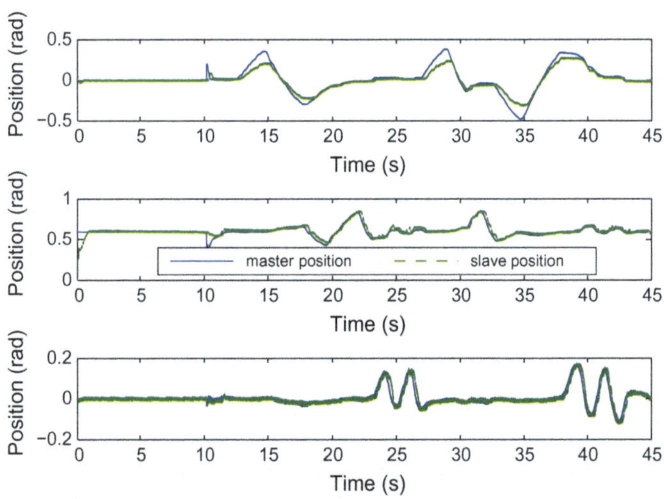

Fig. 11.12 The position synchronization trajectories between the master and the slave with time delays $d_m = 100\,\text{ms}$ and $d_s = 100\,\text{ms}$

middle and bottom, respectively. As we can see from Figs. 11.11, 11.13 and 11.15, the FTSM based finite-time velocity observer can provide good velocity estimation performance. Moreover, from Figs. 11.12, 11.14 and 11.16, when we move the master robot, because of the time delay, the slave tracks the delayed position signals of the master, even though there are synchronization errors, the trajectories of the master and

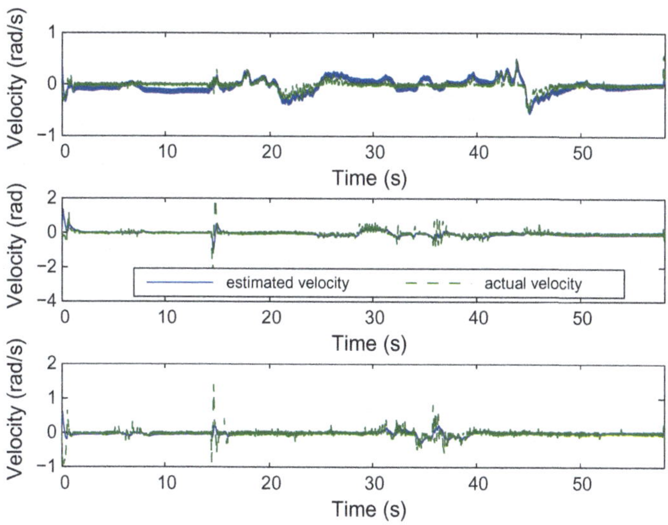

Fig. 11.13 The estimated velocities and the actual velocities with time delays $d_m = 500\,\text{ms}$ and $d_s = 500\,\text{ms}$

the slave are same. Moreover, when the master stops, the position synchronization errors between the master and the slave converge to zero in high speed. It is obvious that with the estimated velocity signals, the NFITSM based control scheme proposed in this chapter can guarantee good synchronization performance.

Next the contact motion between slave robot and the environment will be considered. The teleoperation system with contact force is shown in Fig. 11.17. We put a rigid box at 0.2 m in Y-direction of the slave manipulator which represents a stiff wall. We apply the human-input force to the master manipulator and push the last joint to move in Y-direction. The slave joint follows the movement of the master joint, and then, the slave joint touches the stiff box and could move no more. We move the master joint further and can feel the feedback force and then move the master joint in inverse direction. With the proposed controller and same parameters above, the movements of the master and slave joints in Y-direction are shown in Fig. 11.18. In Fig. 11.19, the trajectories in task-space i.e., x_m and x_s are presented. We can see that when the slave joint reaches 0.2 m, it could no longer move. With the master joint back to −0.2 m, the slave manipulator follows the movement of the master manipulator and also reaches −0.2 m. In spite of there is no force signal transformation, with the bilateral communication channel, when the slave contacts with the rigid box, a big feedback force will be felt by the human operator. The experimental results further show the effectiveness and correctness of the proposed main results.

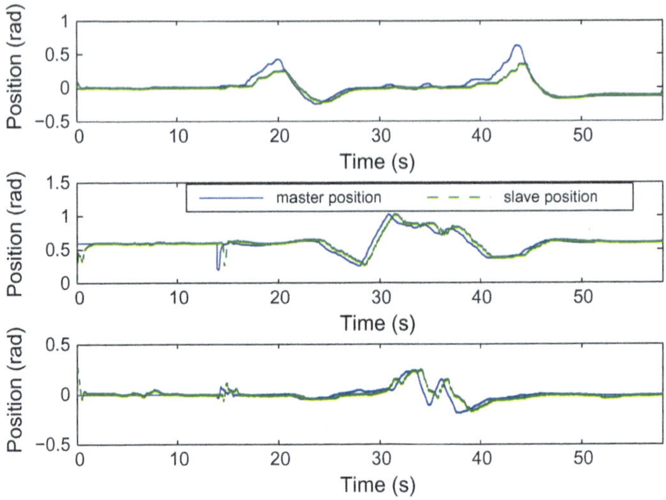

Fig. 11.14 The position synchronization trajectories between the master and the slave with time delays $d_m = 500\,\text{ms}$ and $d_s = 500\,\text{ms}$

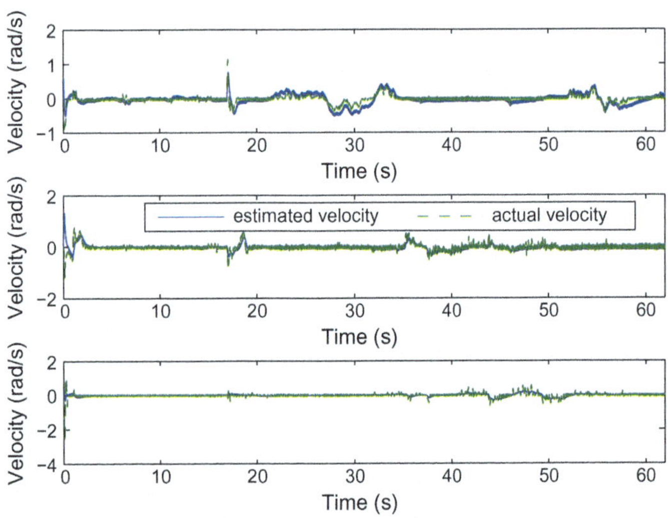

Fig. 11.15 The estimated velocities and the actual velocities with time delays $d_m = 1000\,\text{ms}$ and $d_s = 1000\,\text{ms}$

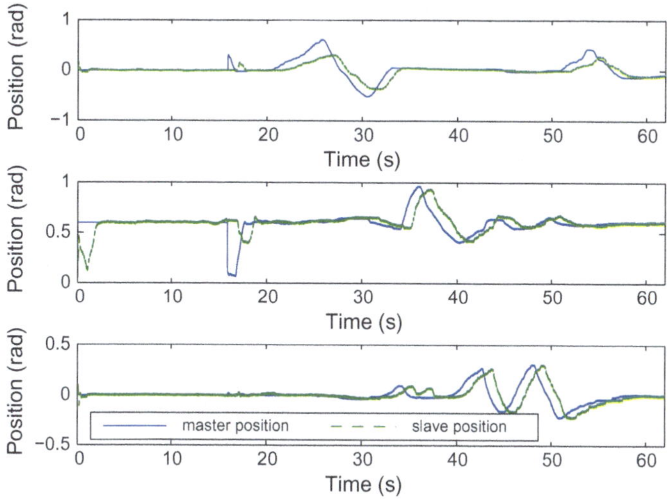

Fig. 11.16 The position synchronization trajectories between the master and the slave with time delays $d_m = 1000\,\text{ms}$ and $d_s = 1000\,\text{ms}$

Fig. 11.17 Teleoperation system with contact force

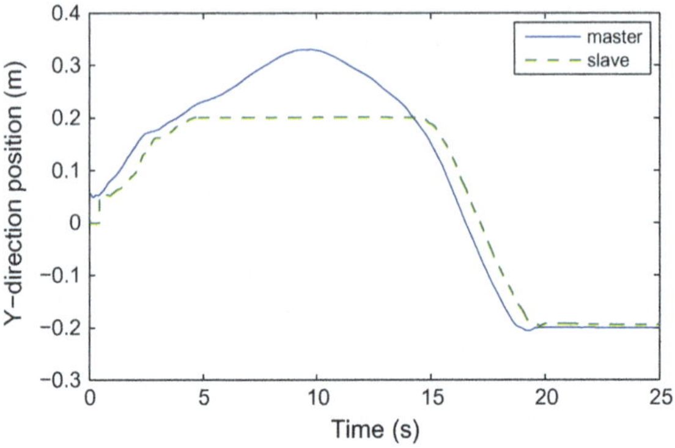

Fig. 11.18 The Y-direction positions of the master and the slave

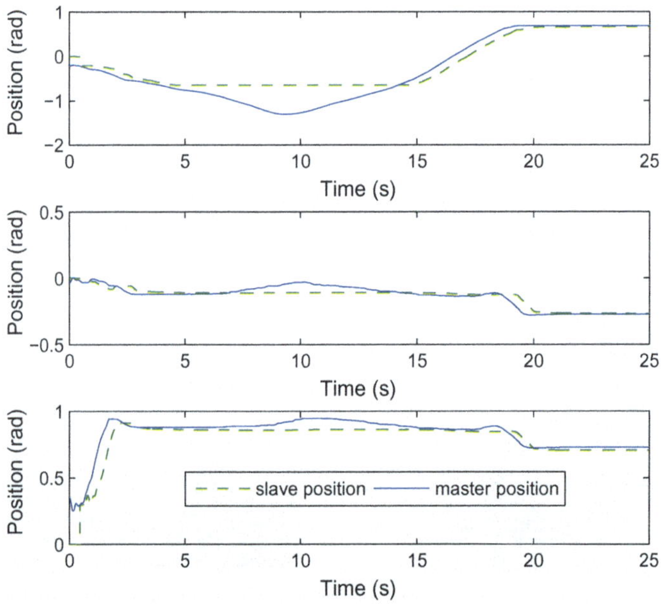

Fig. 11.19 The joint positions of the master and the slave

11.5 Summary

This chapter addresses the finite-time output feedback synchronization control problem for a bilateral teleoperation system in the presence of system uncertainties. To estimate the velocity, the new NNs based FTSM finite-time velocity observers are proposed. Furthermore, the corresponding NFITSM based finite-time controller is designed with the estimated velocity signals. Based on Lyapunov stability theory, it is strictly proved that the finite-time convergence is achieved for the closed-loop teleoperation system. The simulation and experimental results are both presented to demonstrate the effectiveness of the proposed control technology.

Appendix

Proof: The specific prove procedure for the FTSM finite-time velocity observer will be divided into three steps.

Step1: Let us consider the following Lyapunov candidate

$$U_1 = \tilde{x}_i^T P_i \tilde{x}_i + \tilde{W}_i^T P_{i2} \Omega_i^{-1} \tilde{W}_i \tag{A.1}$$

Differentiating U_1 with respect to time yields

$$\begin{aligned} \dot{U}_1 \le & \tilde{x}_i^T (P_i A_i + A_i^T P_i)\tilde{x}_i + 2\tilde{x}_i^T P_i H_i \\ & + 2\tilde{x}_i^T P_i \Delta Q_i + 2\tilde{x}_i^T P_i \nu_i - 2\tilde{W}_i P_{i2} \Omega_i^{-1} \dot{\hat{W}}_i \end{aligned} \tag{A.2}$$

With the Young inequality, we have $2\tilde{x}_i^T P_i H_i \le \frac{1}{\varepsilon_i}\tilde{x}_i^T P_i P_i \tilde{x}_i + \varepsilon_i H_i^T H_i$, ε_i is a positive constant. Since $C_i(y_i, \bar{x}_{i2})\bar{x}_{i2}$ is Lipschitz, thus $\varepsilon_i H_i^T H_i \le \varepsilon_i \bar{C}_i^2 \left\|\tilde{\bar{x}}_{i2}\right\|^2$, where $\bar{C}_i$ is a positive constant. Furthermore, with the transformation $\tilde{x}_{i2} = \tilde{\bar{x}}_{i2} - T_i \tilde{x}_{i1}$, we have $\bar{C}_i^2 \left\|\tilde{\bar{x}}_{i2}\right\|^2 \le 2\bar{C}_i^2 \|\tilde{x}_{i2}\|^2 + 2\bar{C}_i^2 \|T_i\|^2 \|\tilde{x}_{i1}\|^2$.

According to Remark 1 yields $\left\|W_i^{*T}\varphi_i(y_i, \hat{x}_{i2}) - W_i^{*T}\varphi_i(y_i, x_{i2})\right\| \le b_i \|\tilde{x}_{i2}\|$, b_i is a positive constant. Additionally, one has that $2\tilde{x}_i^T P_i N_i \le \dfrac{1}{z_i}\tilde{x}_i^T P_i P_i \tilde{x}_i + z_i N_i^T N_i \le \dfrac{1}{z_i}\tilde{x}_i^T P_i P_i \tilde{x}_i + z_i b_i^2 \|\tilde{x}_{i2}\|^2$, z_i is a positive constant.

Afterwards, substituting the NNs adaptive tuning law $\dot{\hat{W}}_i = -\Omega_i \varphi_i(y_i, \hat{x}_{i2})$ $\tilde{x}_{i2}^T$ into (A.2) yields

$$\begin{aligned} \dot{U}_1 \le & \tilde{x}_i^T (P_i A_i + A_i^T P_i + \frac{1}{\varepsilon_i} P_i P_i + \frac{1}{z_i} P_i P_i \\ & + \Psi_i)\tilde{x}_i + 2\tilde{x}_{i2}^T P_{i2} \epsilon_i^* + 2\tilde{x}_i^T P_i \nu_i \end{aligned} \tag{A.3}$$

With $O_i = P_i A_i + A_i^T P_i + \dfrac{1}{\varepsilon_i} P_i P_i + \dfrac{1}{z_i} P_i P_i + \Psi_i < 0$, $\Lambda_i = -\lambda_{\max}(O_i)$, $\Psi_i = diag(2\varepsilon_i \bar{C}_i^2 \|T_i\|^2 I_n, 0_{n\times n}) + diag(0_{n\times n}, 2\varepsilon_i \bar{C}_i^2 I_n) + diag(0_{n\times n}, z_i b_i^2 I_n)$, it has that

$$\begin{aligned}\dot{U}_1 <& -\Lambda_i\|\tilde{x}_i\|^2 - 2k_{i1}\sum_{j=1}^{n}p_{i1j}\left|\tilde{x}_{i1j}\right| - 2P_{ik1}\tilde{x}_i \\ &- 2k_{i1}^{r_{i1}}P_{ik2}\tilde{x}_i - 2k_{i1}^{r_{i2}}P_{ik3}\tilde{x}_i + 2\tilde{x}_{i2}^T P_{i2}\epsilon_i^* \\ &\leq -\|\tilde{x}_i\|\,(\Lambda_i\|\tilde{x}_i\| - 2k_{i1}^{r_{i1}}\sqrt{\sum_{j=1}^{n}k_{i2}^2p_{i2j}^2} - 2k_{i1}^{r_{i2}}\sqrt{\sum_{j=1}^{n}k_{i3}^2p_{i2j}^2} \\ &- 2\sqrt{\sum_{j=1}^{n}p_{i2j}^2\chi_{ij}^2} - 2\sqrt{\sum_{j=1}^{n}p_{i2j}^2\epsilon_{ij}^{*2}})\end{aligned} \tag{A.4}$$

where

$$\begin{aligned}P_{ik1} &= [0_{1\times n}, p_{i21}\chi_{i1}sign(\tilde{x}_{i11})\cdots p_{i2n}\chi_{in}sign(\tilde{x}_{i1n})],\\ P_{ik2} &= [0_{1\times n}, p_{i21}k_{i2}sign(\tilde{x}_{i11})\cdots p_{i2n}k_{i2}sign(\tilde{x}_{i1n})],\\ P_{ik3} &= [0_{1\times n}, p_{i21}k_{i3}sign(\tilde{x}_{i11})\cdots p_{i2n}k_{i3}sign(\tilde{x}_{i1n})].\end{aligned}$$

Furthermore, we can obtain that $U_1 < 0$ with $\|\tilde{x}_i\| > R_i$, where

$$R_i = \frac{2k_{i1}^{r_{i1}}\sqrt{\sum_{j=1}^{n}k_{i2}^2p_{i2j}^2} + 2k_{i1}^{r_{i2}}\sqrt{\sum_{j=1}^{n}k_{i3}^2p_{i2j}^2} + 2\sqrt{\sum_{j=1}^{n}p_{i2j}^2\chi_{ij}^2} + 2\sqrt{\sum_{j=1}^{n}p_{i2j}^2\epsilon_{ij}^{*2}}}{\Lambda_i}$$

This implies that if $\|\tilde{x}_i\|$ is outside the ball with radius R_i, then $U_1 < 0$, the $\|\tilde{x}_i\|$ will reduce, resulting in $\|\tilde{x}_i\|$ being bounded by $\|\tilde{x}_i\| \leq R_i$. Therefore, there exists a moment t_0 such that $\|\tilde{x}_i\| = \bar{R}_i$, where $\bar{R}_i > R_i$.

Step2: Consider Lyapunov function $U_2 = \frac{1}{2}p_{i1j}\tilde{x}_{i1j}^2$, differentiating U_2, we have

$$\dot{U}_2 = p_{i1j}\dot{\tilde{x}}_{i1j} \tag{A.5}$$

Applying Eq. (11.14), we can obtain that

$$\dot{U}_2 \leq -p_{i1j}(l_{i1j} - a_{i1j})\tilde{x}_{i1j}^2 - p_{i1j}\left|\tilde{x}_{i1j}\right|(k_{i1} - \left|\tilde{x}_{i2j}\right|) \tag{A.6}$$

Set $l_{i1j} - a_{i1j} = \omega_i > 0$ and $k_{i1} - \bar{R}_i = \varrho_i > 0$, we have $\dot{U}_2 \leq -p_{i1j}\omega_i\tilde{x}_{i1j}^2 - p_{i1j}\varrho_i\left|\tilde{x}_{i1j}\right| = -2\omega_i U_2 - \sqrt{2p_{i1j}}\varrho_i U_2^{1/2}$. Hence the $\tilde{x}_{i1}$ and $\dot{\tilde{x}}_{i1}$ will converge to zero in finite time with the convergence time $t_j \leq \frac{1}{\omega_i}\ln(1 + \dfrac{2\omega_i U_2^{1/2}(\bar{R}_i)}{\sqrt{2p_{i1j}}\varrho_i}) + t_0$ $(j = 1, 2, \ldots, n)$.

Step3: When the sliding mode $\tilde{x}_{i1} = \tilde{x}_{i1} = 0$, which causes $\nu_{i1} = -\tilde{x}_{i2}$, then $\nu_{i2} = -k_{i2} |\tilde{x}_{i2}|^{r_{i1}} sign(\tilde{x}_{i2}) - k_{i3} |\tilde{x}_{i2}|^{r_{i2}} sign(\tilde{x}_{i2})$. Consider the Lyapunov function $U_3 = \tilde{x}_{i2}^T P_{i2}\tilde{x}_{i2}$, with the results obtained in Step1, the boundedness of $\hat{W}_i^T \varphi_i(y_i, \hat{x}_{i2}) - W_i^{*T} \varphi_i(y_i, \hat{x}_{i2})$ can be achieved. Differentiating U_3 with Lemma 2 yields

$$
\begin{aligned}
\dot{U}_3 <& \tilde{x}_{i2}^T(-2P_{i2}T_i + \frac{1}{\varepsilon_i}P_{i2}P_{i2} + \frac{1}{z_i}P_{i2}P_{i2} + \varepsilon_i\bar{C}_i^2 I_n + z_i b_i^2 I_n)\tilde{x}_{i2} \\
&+ 2\sum_{j=1}^{n} p_{i2j}\tilde{x}_{i2j}F_i - 2\sum_{j=1}^{n} \chi_{ij}p_{i2j}\left|\tilde{x}_{i2j}\right| \\
&- 2k_{i2}\lambda_{\min}(P_{i2})n^{1-r_{i1}/2}(\tilde{x}_{i2}^T\tilde{x}_{i2})^{r_{i1}+1/2} \\
&- 2k_{i3}\lambda_{\min}(P_{i2})(\tilde{x}_{i2}^T\tilde{x}_{i2})^{r_{i2}+1/2} \\
\le& \tilde{x}_{i2}^T(-2P_{i2}T_i + \frac{1}{\varepsilon_i}P_{i2}P_{i2} + \frac{1}{z_i}P_{i2}P_{i2} + \varepsilon_i C_i^2 I_n + z_i b_i^2 I_n)\tilde{x}_{i2} \\
&+ 2\sum_{j=1}^{n} p_{i2j}\tilde{x}_{i2j}F_{ij} - 2\sum_{j=1}^{n} \chi_{ij}p_{i2j}\left|\tilde{x}_{i2j}\right| \\
&- 2k_{i2}n^{1-r_{i1}/2}\frac{\lambda_{\min}(P_{i2})}{\lambda_{\max}(P_{i2})^{r_{i1}+1/2}}U_3^{r_{i1}+1/2} \\
&- 2k_{i3}\frac{\lambda_{\min}(P_{i2})}{\lambda_{\max}(P_{i2})^{r_{i2}+1/2}}U_3^{r_{i2}+1/2}
\end{aligned}
\tag{A.7}
$$

When $-2P_{i2}T_i + \frac{1}{\varepsilon_i}P_iP_i + \frac{1}{z_i}P_iP_i + \varepsilon_i\bar{C}_i^2 I_n + z_i b_i^2 I_n < 0$ and $\chi_{ij} \ge F_i$, then we have $\dot{U}_3 < -2k_{i2}n^{1-r_{i1}/2}\frac{\lambda_{\min}(P_{i2})}{\lambda_{\max}(P_{i2})^{r_{i1}+1/2}}U_3^{r_{i1}+1/2} - 2k_{i3}\frac{\lambda_{\min}(P_{i2})}{\lambda_{\max}(P_{i2})^{r_{i2}+1/2}}U_3^{r_{i2}+1/2}$ $= -\psi_{i1}U_3^{r_{i1}+1/2} - \psi_{i2}U_3^{r_{i2}+1/2}$, where $\psi_{i1} = 2k_{i2}n^{1-r_{i1}/2}\frac{\lambda_{\min}(P_{i2})}{\lambda_{\max}(P_{i2})^{r_{i1}+1/2}}$ and $\psi_{i2} = 2k_{i3}\frac{\lambda_{\min}(P_{i2})}{\lambda_{\max}(P_{i2})^{r_{i2}+1/2}}$. Set $\psi = \sqrt{U_3}$, with Eq. (A.7) we have $\dot{\psi} = \frac{1}{2\sqrt{U_3}}\dot{U}_3 = -\psi_{i1}\psi^{r_{i1}} - \psi_{i2}\psi^{r_{i2}}$. According to Lemma 1, it has that the ψ will converge to zero within the settling time $T = \frac{1}{\psi_{i1}}\frac{1}{r_{i1}-1} + \frac{1}{\psi_{i2}}\frac{1}{1-r_{i2}}$. Based on the definitions of ψ and U_3, we also can obtain that $\tilde{x}_{i2}$ will converge to zero in finite time within the settling time T. Therefore the settling time of the velocity observer is $T_2 = T + \max(t_1, \ldots, t_n)$. This completes the proof.

References

1. J. Artigas, C. Preusche, G. Hirzinger et al., Bilateral energy transfer in delayed teleoperation on the time domain, in *IEEE International Conference on Robotics and Automation*, pp. 19–23 (2008)
2. I. Sarras, E. Nuño, M. Kinnaert, L. Basañez, Output-feedback control of nonlinear bilateral teleoperators, in *American Control Conference (ACC)*, pp. 27–29 (2012)

3. C.P. Tan, X.H. Yu, Z.H. Man, Terminal sliding mode observers for a class of nonlinear systems. Automatica **46**(8), 1401–1404 (2010)
4. B. Zhang, Y.M. Jia, F. Matsuno, Finite-time observers for multi-agent systems without velocity measurements and with input saturation. Syst. Control Lett. **68**(1), 86–94 (2014)
5. J. Davila, L. Fridman, A. Levant, Second-order sliding-mode observer for mechanical systems. IEEE Trans. Autom. Control **50**(11), 1785–1789 (2005)
6. J.M. Daly, D.W.L. Wang, Time-delayed output feedback bilateral teleoperation with force estimation for n-dof nonlinear manipulators. IEEE Trans. Control Syst. Technol. **22**(1), 299–306 (2014)
7. D.Y. Zhao, S.Y. Li, Q.M. Zhu, Output feedback terminal sliding mode control for a class of second order nonlinear systems. Asian J. Control **15**(1), 237–247 (2013)
8. Q.L. Hu, B. Xiao, D.W. Wang, Spacecraft attitude fault tolerant control with terminal sliding mode observer. J. Aerosp. Eng. **15**(1), 1–15 (2013)
9. J. Wang, C. Zhang, S. Li et al., Finite-time output feedback control for pwm-based dc-dc buck power converters of current sensorless mode. IEEE Trans. Control Syst. Technol. **25**(4), 1359–1371 (2017)
10. H.B. Sun, S.H. Li, C.Y. Sun, Finite time integral sliding mode control of hypersonic vehicles. IEEE Trans. Fuzzy Syst. **73**(1–2), 229–244 (2013)
11. S. Laghrouche, F. Plestan, A. Glumineau, Higher order sliding mode control based on integral sliding mode. Automatica **43**(3), 531–537 (2007)
12. S.H. Yu, X.J. Long, Finite-time consensus for second-order multi-agent systems with disturbances by integral sliding mode. Automatica **54**, 158–165 (2015)
13. S. Kamal, A. Raman, B. Bandyopadyay, Finite-time stabilization of fractional order uncertain chain of integrator: an integral sliding mode approach. IEEE Trans. Autom. Control **58**(6), 1597–1602 (2013)
14. S.H. Yu, X.H. Yu, B. Shirinzadeh, Z.H. Man, Continuous finite-time control for robotic manipulators with terminal sliding mode. Automatica **41**(11), 1957–1964 (2005)
15. Y.N. Yang, C.C. Hua, J.P. Li, X.P. Guan, Fixed-time coordination control for bilateral telerobotics system with asymmetric time-varying delays. J. Intell. Robot. Syst. **86**(3–4), 447–466 (2017)
16. F.L. Lewis, A. Yesildirak, S. Jagannathan, *Neural Network Control of Robot Manipulators and Nonlinear Systems* (Taylor & Francis, 1998)
17. A. Forouzantabar, H.A. Talebi, A.K. Sedigh, Adaptive neural network control of bilateral teleoperation with constant time delay. Nonlinear Dyn. **67**(2), 1123–1134 (2012)
18. Y. J. Liu, J. Li, S. C. Tong, C.L. Philip Chen, Neural network control-based adaptive learning design for nonlinear systems with full-state constraints. IEEE Trans. Neural Netw. Learn. Syst. **27**(7), 1562–1571 (2016)
19. C.C. Hua, Y.N. Yang, X.P. Guan, Neural network-based adaptive position tracking control for bilateral teleoperation under constant time delay. Neurocomputing **113**(3), 204–212 (2013)
20. M. Farza, M.M. Saad, M. Triki et al., High gain observer for a class of non-triangular systems. Syst. Control Lett. **60**(1), 27–35 (2011)
21. N. Chopra, P. Berestesky, M.W. Spong, Bilateral teleoperation over unreliable communication networks. IEEE Trans. Control Syst. Technol. **16**(2), 304–313 (2008)
22. Y. Feng, X.H. Yu, Z.H. Man, Non-singular terminal sliding mode control of rigid manipualtors. Automatica **38**(12), 2159–2167 (2002)
23. L. Yang, J.Y. Yang, Nonsingular fast terminal sliding-mode control for nonlinear dynamical systems. Int. J. Robust Nonlinear Control **21**(16), 1865–1879 (2011)

Part IV
Constrained Input for Teleoperation System

Chapter 12
On Exploring the Domain of Attraction for Bilateral Teleoperator

Abstract The domain of attraction problem is investigated for networked teleoperation system subject to actuator saturation in this chapter. The forward and backward time-varying communication delays are assumed to be interval and asymmetric, which is the case for network-based teleoperation system. We propose a novel Lyapunov-Krasovskii functional for the closed-loop teleoperation system with the consideration of the interval values of the time delays. The delay-dependent estimation of the domain of attraction is presented using linear matrix inequality (LMI) technique. The problem of designing P+d control law such that the domain of attraction is enlarged is formulated and solved as an optimization problem with LMI constraints. Experiments are performed to verify the effectiveness of the proposed approach.

12.1 Introduction

As we cited in above chapters with the introduction of communication networks into the teleoperation system, time delays are unavoidable while data are exchanged between the master and slave sites. Moreover, it is well known that the existence of even a small delay in the communication channel may cause instability of a teleoperation system. Many methods have been proposed to deal with the problem [1–5]. Recently, the stability conditions in the form of linear matrix inequalities (LMIs) have been presented in [6] and [7]. The above cited references consider that the delays vary from zero. However, it is well known that the communication delays often vary in an interval in practice, i.e., the lower bound determined by the physical distance and the upper bound determined by the congestion level. Therefore, it is of great significance to consider the interval time-varying delay in teleoperation system.

On the other hand, actuator saturation is inevitable in practical feedback control systems. The ignorance of actuator saturation in control design possibly results in degraded performance or even instability, see [8–10] for linear systems and [11, 12] for nonlinear systems. When the system state is far from the equilibrium point, the actuator with bounded gain cannot render the system state to the equilibrium point. How to estimate the domain of attraction (DOA) is still an open problem. The

C. Hua et al., *Analysis and Design for Networked Teleoperation System*,
https://doi.org/10.1007/978-981-13-7936-9_12

physical restrictions exist in robot motors, thus the saturation problem should be considered in the torque design of master-slave robots in teleoperation. Reference [13] was a first work on actuator saturation of teleoperation system and proposed a control design method. However, the DOA problem was not discussed in [13], which was challenging in networked nonlinear teleoperation system.

In this chapter, we focus on exploring the DOA for teleoperation system with constrained input and interval time delays. The main works of the chapter lies: (i) A new delay-dependent stability condition is presented for networked teleoperation system subject to actuator saturation, which involves both the lower and upper bounds of the delays. The Lyapunov-Krasovskii functional constructed in this chapter makes full use of delay information and the resulted stability criterion is less conservative. (ii) The estimation of the DOA is explored for bilateral teleoperator for the first time. The relationship is established among the control parameters, the delay bounds, the saturation levels and the initial values of system state. The optimization of the allowable bounds for the time delays and the estimation of the DOA can be performed easily.

Notations: $diag\{\ldots\}$ denotes a block-diagonal matrix. $He(A)$ is defined as $He(A) = A + A^T$. The subscript of a matrix $n \times m$ means the dimension of the matrix is $n \times m$. $\lambda_{\max}(A)$ represents the maximum eigenvalue of matrix A.

12.2 Problem Statement

Consider a bilateral teleoperation system described by

$$\begin{cases} M_m(q_m)\ddot{q}_m + C_m(q_m, \dot{q}_m)\dot{q}_m + G_m(q_m) = \tau_m + F_h \\ M_s(q_s)\ddot{q}_s + C_s(q_s, \dot{q}_s)\dot{q}_s + G_s(q_s) = \tau_s - F_e \end{cases} \tag{12.1}$$

where $\tau_i \in \mathbb{R}^n$ is the applied torque to be designed. We know that there exist upper bound for τ_i. For simplicity, it is assumed that the upper bounds of τ_m and τ_s are the same. We use $T = [T_1, T_2, \ldots, T_n]^T$ as the bound, that is each element τ_{ij} of τ_i satisfies $|\tau_{ij}| \leq T_j$, where T_j is the given actuator saturation parameter.

The property of gravity torque are revisited below with $i = m, s$.

Property 12.1 For a manipulator with revolute joints, each element $G_{ij}(q_i)$ of the gravity vector $G_i(q_i)$ is bounded.

For simplicity, we assume that the bounds of the gravity functions are the same for the master and slave. The parameter γ_j is used to bound $G_{ij}(q_i)$, i.e. $|G_{ij}(q_i)| \leq \gamma_j$. The bound vector is $\gamma = [\gamma_1, \gamma_2, \ldots, \gamma_n]^T$.

The proportional position error plus damping injection (P+d) scheme with gravity compensator is designed as

$$\begin{cases} \tau_m = sat(-K_m(q_m - q_s(t - d_s(t))) - \alpha_m\dot{q}_m) + G_m(q_m) \\ \tau_s = sat(K_s(q_m(t - d_m(t)) - q_s) - \alpha_s\dot{q}_s) + G_s(q_s) \end{cases} \tag{12.2}$$

where K_m, α_m, K_s and α_s are gain matrices to be designed, $d_m(t)$ is denoted as forward delay and $d_s(t)$ is denoted as backward delay, $sat(\cdot)$ is the saturation function. With the bound vector T of τ_i and the bound vector γ of gravity g_i, we have the bound vector M of $sat(\cdot)$ based on (12.2), and each element of M satisfies $M_j \leq T_j - \gamma_j$.

For the master-slave system (12.1) with controller (12.2), the following assumptions are imposed.

Assumption 12.1 For time-varying delays $d_m(t)$ and $d_s(t)$, there exist positive scalars d_{i1}, d_{i2}, μ_{i1} and μ_{i2} such that $d_{i1} \leq d_i(t) \leq d_{i2} < \infty$ and $\mu_{i1} \leq \dot{d}_i(t) \leq \mu_{i2}$ for $i = m, s$. The bounds μ_{m2} and μ_{s2} satisfy $\mu_{m2} + \mu_{s2} < 1$.

Assumption 12.2 The master and the slave are both non-redundant, and the human operator and environment satisfy the passive condition, i.e., $-\int_0^t \dot{q}_m^T(\sigma)F_h(\sigma)d\sigma \geq 0$ and $\int_0^t \dot{q}_s^T(\sigma)F_e(\sigma)d\sigma \geq 0$.

Let D be the set of $n \times n$ diagonal matrices D_i with 0 or 1 as its diagonal entries, i.e., $D = \{D_i : i = 1, \ldots, 2^n\}.D_i^- = I - D_i$. The following lemma will be used to derive our main results.

Lemma 12.1 ([9]) *Let matrices F and $H \in R^{n\times n}$ be given, h_j denotes the jth row of H. For $x \in R^n$, if $|h_j x| \leq M_j$, $j = 1, 2, \ldots, n$ then*

$$sat(Fx) \in co\{D_i Fx + D_i^- Hx, i = 1, \ldots, 2^n\}$$

where $co\{\cdot\}$ denotes the convex hull of a set.

Remark 12.1 This chapter considers the interval time delay case. The delays are assumed to be lower and upper bounded. Assumption 12.2 is a passive condition in the joint space. In fact, it is more realistic to assume that in the task space. When the master and slave are both non-redundant, the two passive conditions are equivalent. Assumption 12.2 is commonly used in analyzing stability of teleoperation system, see [4, 6, 13].

Lemma 12.1 places $sat(Fx)$ into a convex combination of an actual feedback and an auxiliary counterpart. Define the set $L(H) = \{x \in R^n : |h_i x| \leq M_i, i = 1, 2, \ldots, n\}$. A subset of the set $L(H)$ is chosen to be an ellipsoid of the form $\varepsilon(Q, 1) = \{x \in R^n : x^T Qx \leq 1\}$ with $Q > 0$ to be determined. Denote the state trajectory of system as $\psi(t, x_0)$, then the domain of attraction of the origin is $\Gamma = \{x_0 \in C^1[-d_2, 0] : \lim_{t\to\infty} \psi(t, x_0) = 0\}$.

The control objective of this chapter is as follows: develop new stability criterion for teleoperation system (12.1) in the presence of time-varying delay and saturation, and propose the estimation of the domain of attraction.

12.3 Main Results

In this section, a Lyapunov-Krasovskii functional is constructed and the delay-dependent stability condition is proposed for teleoperation system (12.1) with controller (12.2). Furthermore, the estimation of the DOA is presented and the optimization of the DOA is discussed.

To demonstrate the domain of attraction, we define the bounds of initial values as follows: $\max\limits_{\theta\in[-d_2,0]} \|\dot{q}_{m0}(\theta)\| \le \delta_{m1}$, $\max\limits_{\theta\in[-d_2,0]} \|\dot{q}_{s0}(\theta)\| \le \delta_{s1}$, $\max\limits_{\theta\in[-d_2,0]} \|q_{m0}(\theta)-q_{s0}(\theta)\| \le \delta_{e1}$ and $\max\limits_{\theta\in[-d_2,0]} \|\dot{q}_{m0}(\theta)-\dot{q}_{s0}(\theta)\| \le \delta_{e2}$, where δ_{m1}, δ_{s1}, δ_{e1} and δ_{e2} are positive parameters.

Theorem 12.1 *Consider teleoperation system (12.1) with controller (12.2), define* $d_2 = d_{m2} + d_{s2}$, *if there exist positive definite matrices* W, R_m, R_s, R, S, U, Z *and matrices* X, N, M, $H_{\alpha m}$, H_{km}, $H_{\alpha s}$, H_{ks} *with proportional dimensions such that the following matrix inequalities hold*

$$\Phi_\kappa = \Phi_{0,\kappa} - \frac{1}{d_2}\Upsilon^T \Theta \Upsilon < 0, \kappa = 1, \ldots, 2^n \tag{12.3}$$

$$\Theta = \begin{bmatrix} \tilde{R} & X \\ * & \tilde{R} \end{bmatrix} > 0 \tag{12.4}$$

and $\varepsilon(Q, 1)$ *belongs to the intersection of the sets* $L_m(H_{\alpha m}, H_{km})$ *and* $L_s(H_{\alpha s}, H_{ks})$, *i.e.,* $\varepsilon(Q, 1) \subset (L_m(H_{\alpha m}, H_{km}) \cap L_s(H_{\alpha s}, H_{ks}))$, *then the asymptotic stability of teleoperation system is guaranteed and the estimation of the DOA for teleoperation system (12.1) is given by* $\Gamma = \{\dot{q}_{m0}, \dot{q}_{s0}, q_{m0} - q_{s0} : \Gamma_\delta \le 1\}$. *The matrices and sets used in the theorem are defined at the bottom of next page.*

Proof Choose the following Lyapunov-Krasovskii functional $V = \sum\limits_{i=1}^{5} V_i$ with

$$V_1 = \sum_{r=m,s} \dot{q}_r^T M_r(q_r)\dot{q}_r - 2\int_0^t (\dot{q}_m^T(\sigma) F_h(\sigma) + \dot{q}_s^T(\sigma) F_e(\sigma))d\sigma$$

$$V_2 = e^T(t)We(t),\ V_3 = \int_{t-d_2}^{t-d(t)} e^T(\sigma)Se(\sigma)d\sigma$$

$$V_4 = \sum_{r=m,s} \int_{-d_{r2}}^{0}\int_{t+\theta}^{t} \dot{q}_r^T(\sigma) R_r \dot{q}_r(\sigma)\, d\sigma d\theta + \int_{-d_2}^{0}\int_{t+\theta}^{t} \dot{e}^T(\sigma)R\dot{e}(\sigma)d\sigma d\theta$$

$$V_5 = \int_{t-d(t)}^{t} e^T(\sigma)d\sigma U \int_{t-d(t)}^{t} e(\sigma)d\sigma + \int_{t-d_2}^{t-d(t)} e^T(\sigma)d\sigma Z \int_{t-d_2}^{t-d(t)} e(\sigma)d\sigma \tag{12.5}$$

$$Q = diag\{\varrho_{m1}I, \varrho_{s1}I, W\}, \tilde{R} = diag\{R, 3R\}, \mu_i = \mu_{mi} + \mu_{si}, i = 1, 2,$$

$$\Upsilon^T = \left[\Upsilon_1^T \ \Upsilon_2^T \ \Upsilon_3^T \ \Upsilon_4^T\right]^T \ \Phi_{0,\kappa} = \hat{S} + \hat{W} + \hat{N} + \hat{M} + \hat{N}^T + \hat{M}^T + \hat{R}_m + \hat{R}_s + \hat{R} + \Phi_{3,\kappa}$$

in which $\Upsilon_1 = \left[0_{n\times 10n} \ I\right]$, $\Upsilon_2 = \left[0_{n\times 5n} \ I \ I \ 0 \ -2I \ 0_{n\times 2n}\right]$, $\Upsilon_3 = \left[0_{n\times 9n} \ I \ 0\right]$,

$$\Upsilon_4 = \left[0_{n\times 4n} \ I \ I \ 0 \ -2I \ 0_{n\times 3n}\right].\hat{W} = [\hat{W}_{i,j}]_{11n\times 11n} \text{ with } \hat{W}_{1,5} = W, \hat{W}_{2,5} = -W, \text{ others are 0.}$$

$$\hat{R} = d_2[\hat{R}_{i,j}]_{11n\times 11n} \text{ with } \hat{R}_{1,1} = \hat{R}_{2,2} = R, \hat{R}_{1,2} = \hat{R}_{2,1} = -R, \text{ others are 0.}$$

$$\hat{R}_m = diag\{d_{m2}R_m, 0, -\frac{1}{d_{m2}}R_m, 0_{8n}\},$$

$$\hat{R}_s = diag\{0_n, d_{s2}R_s, 0_n, -\frac{1}{d_{s2}}R_s, 0_{7n}\}, \hat{S} = diag\{0_{5n}, (1-\dot{d}(t))S, -S, 0_{4n}\},$$

$$\tilde{R} = diag\{R, 3R\}.$$

$$\hat{N} = NJ_2, \hat{M} = MJ_3, \text{ with } J_2 = \left[0_{n\times 4n} \ I \ -I \ 0_{n\times 3n} \ -I \ 0\right],$$

$$J_3 = \left[0_{n\times 5n} \ I \ -I \ 0_{n\times 3n} \ -I\right].$$

$$\Phi_{3,\kappa} = [\Phi_{i,j}^{3\kappa}]_{11n\times 11n} \text{ with } \Phi_{1,1}^{3\kappa} = He(-D_\kappa\alpha_m + D_\kappa^- H_{\alpha m}),$$

$$\Phi_{2,2}^{3\kappa} = He(-D_\kappa\alpha_s + D_\kappa^- H_{\alpha s}),$$

$$\Phi_{2,3}^{3\kappa} = -(D_\kappa K_s + D_\kappa^- H_{ks}), \Phi_{1,4}^{3\kappa} = \Phi_{1,5}^{3\kappa} = -D_\kappa K_m + D_\kappa^- H_{km},$$

$$\Phi_{2,5}^{3\kappa} = D_\kappa K_s + D_\kappa^- H_{ks}, \Phi_{5,8}^{3\kappa} = d(t)U,$$

$$\Phi_{6,8}^{3\kappa} = -(1-\dot{d}(t))d(t)U, \Phi_{6,9}^{3\kappa} = (1-\dot{d}(t))(d_2 - d(t))Z,$$

$$\Phi_{7,9}^{3\kappa} = -(d_2 - d(t))Z, \text{ others are 0.}$$

$$L_m(H_{\alpha m}, H_{km}) = \{\dot{q}_m, \dot{q}_s, q_m, q_s \in R^n : |h_{kmi}(q_m - q_s) + h_{\alpha mi}\dot{q}_m| \le M_i\},$$

$$L_s(H_{\alpha s}, H_{ks}) = \{\dot{q}_m, \dot{q}_s, q_m, q_s \in R^n : |h_{ksi}(q_m - q_s) + h_{\alpha si}\dot{q}_s| \le M_i\},$$

where $i = 1, \ldots, n$, h_{kmi} is the ith row of H_{km}, the same with $h_{ksi}, h_{\alpha mi}, h_{\alpha si}$.

$$\Gamma_\delta = (\varrho_{m2} + \frac{1}{2}d_{m2}^2\lambda_{\max}(R_m))\delta_{m1}^2 + (\varrho_{s2} + \frac{1}{2}d_{s2}^2\lambda_{\max}(R_s))\delta_{s1}^2 + (\lambda_{\max}(W)$$

$$+(d_2 - d_1)\lambda_{\max}(S) + d_2^2\lambda_{\max}(U) + (d_2 - d_1)^2\lambda_{\max}(Z))\delta_{e1}^2 + \frac{1}{2}d_2^2\lambda_{\max}(R)\delta_{e2}^2.$$

where $e(t) = q_m(t) - q_s(t)$, $d(t) = d_m(t) + d_s(t)$. Obviously, $V > 0$.

By Property 1.2, we have the time derivative of V_1

$$\begin{aligned}\dot{V}_1 =& 2\dot{q}_m^T[sat(-K_m(q_m - q_s(t - d_s(t))) - \alpha_m\dot{q}_m)] \\ &+ 2\dot{q}_s^T[sat(K_s(q_m(t - d_m(t)) - q_s) - \alpha_s\dot{q}_s)]\end{aligned} \tag{12.6}$$

Using Lemma 12.1 gives

$$\begin{aligned}\dot{V}_1 =& 2\sum_{\kappa=1}^{2^n}\eta_\kappa\{\dot{q}_m^T[(-D_\kappa K_m + D_\kappa^- H_{km}) \times (q_m - q_s + \int_{t-d_s(t)}^t \dot{q}_s(\sigma)d\sigma) \\ &+ (-D_\kappa\alpha_m + D_\kappa^- H_{\alpha m})\dot{q}_m] + \dot{q}_s^T[(D_\kappa K_s + D_\kappa^- H_{ks}) \\ &\times (q_m - q_s - \int_{t-d_m(t)}^t \dot{q}_m(\sigma)d\sigma) + (-D_\kappa\alpha_s + D_\kappa^- H_{\alpha s})\dot{q}_s]\}\end{aligned} \tag{12.7}$$

The time derivative of V_4 is

$$\begin{aligned}\dot{V}_4 =& \sum_{r=m,s} d_{r2}\dot{q}_r^T R_r\dot{q}_r - \int_{t-d_{r2}}^t \dot{q}_r(\sigma)R_r\dot{q}_r(\sigma)d\sigma \\ &+ d_2\dot{e}^T R\dot{e} - \int_{t-d_2}^t \dot{e}^T(\sigma)R\dot{e}(\sigma)d\sigma\end{aligned} \tag{12.8}$$

in which $-\int_{t-d_2}^t \dot{e}^T(\sigma)R\dot{e}(\sigma)d\sigma$ can be written as $-\int_{t-d_2}^{t-d(t)} \dot{e}^T(\sigma)R\dot{e}(\sigma)d\sigma - \int_{t-d(t)}^t \dot{e}^T(\sigma)R\dot{e}(\sigma)d\sigma$. By employing Wirtinger inequality [14], one obtains $-\int_{t-d(t)}^t \dot{e}^T(\sigma)R\dot{e}(\sigma)d\sigma \le -\frac{1}{d(t)}(\int_{t-d(t)}^t \dot{e}(\sigma)d\sigma)^T R(\int_{t-d(t)}^t \dot{e}(\sigma)d\sigma) - \frac{3}{d(t)}(e(t) + e(t-d(t)) - \frac{2}{d(t)}\int_{t-d(t)}^t e(\sigma)d\sigma)^T R(e(t) + e(t-d(t)) - \frac{2}{d(t)}\int_{t-d(t)}^t e(\sigma)d\sigma)$. We deal with $-\int_{t-d_2}^{t-d(t)} \dot{e}^T(\sigma)R\dot{e}(\sigma)d\sigma$ in the same way. Then, if there exists a matrix X such that $\Theta > 0$, we finally get $-\int_{t-d_2}^t \dot{e}^T(\sigma)R\dot{e}(\sigma)d\sigma \le -\frac{1}{d_2}\xi^T\Upsilon^T\Theta\Upsilon\xi$. Based on Jensen inequality, we obtain $-\int_{t-d_{r2}}^t \dot{q}_r^T(\sigma)R_r\dot{q}_r(\sigma)d\sigma \le -\frac{1}{d_{r2}}(\int_{t-d_r(t)}^t \dot{q}_r(\sigma)d\sigma)^T R_r(\int_{t-d_r(t)}^t \dot{q}_i(\sigma)d\sigma)$, $r = m, s$. Finally,

$$\begin{aligned}\dot{V}_4 \le& \sum_{r=m,s} d_{r2}\dot{q}_r^T R_r\dot{q}_r - \frac{1}{d_{r2}}(\int_{t-d_r(t)}^t \dot{q}_r(\sigma)\,d\sigma)^T R_r(\int_{t-d_r(t)}^t \dot{q}_i(\sigma)\,d\sigma) \\ &+ d_2\dot{e}^T R\dot{e} - \frac{1}{d_2}\xi^T\Upsilon^T\Theta\Upsilon\xi\end{aligned} \tag{12.9}$$

where $\xi^T = \Big[\dot{q}_m^T(t)\ \dot{q}_s^T(t)\ \int_{t-d_m(t)}^{t} \dot{q}_m^T(\sigma)\,d\sigma\ \int_{t-d_s(t)}^{t} \dot{q}_s^T(\sigma)\,d\sigma\ e^T(t)$
$e^T(t-d(t))\ e^T(t-d_2)\ \frac{1}{d(t)}\int_{t-d(t)}^{t} e^T(\sigma)d\sigma$
$\frac{1}{d_2-d(t)}\int_{t-d_2}^{t-d(t)} e^T(\sigma)d\sigma\ \int_{t-d(t)}^{t} \dot{e}^T(\sigma)d\sigma\ \int_{t-d_2}^{t-d(t)} \dot{e}^T(\sigma)d\sigma\Big]$. The following free-weighting matrices are considered.

$f_1 = 2\xi^T N[e(t) - e(t-d(t)) - \int_{t-d(t)}^{t} \dot{e}(\sigma)d\sigma] = 0$

$f_2 = 2\xi^T M[e(t-d(t)) - e(t-d_2) - \int_{t-d_2}^{t-d(t)} \dot{e}(\sigma)d\sigma] = 0$

and therefore, one finally obtains $\dot{V} = \sum_{i=1}^{5} \dot{V}_i + \sum_{j=1}^{2} f_j \leq \sum_{\kappa=1}^{2^n} \eta_\kappa \xi^T \Phi_\kappa \xi$. In view of (12.3) and (12.4), we have $\dot{V} < 0$ for all $x \in \varepsilon(Q, 1)\backslash\{0\}$ with $x^T = [\dot{q}_m^T\ \dot{q}_s^T\ q_m^T - q_s^T]$ and accordingly, $x^T Q x \leq V(x_t) < V(x_0) \leq (\varrho_{m2} + \frac{1}{2}d_{m2}^2\lambda_{\max}(R_m))\delta_{m1}^2 + (\varrho_{s2} + \frac{1}{2}d_{s2}^2\lambda_{\max}(R_s))\delta_{s1}^2 + (\lambda_{\max}(W) + (d_2 - d_1)\lambda_{\max}(S) + d_2^2\lambda_{\max}(U) + (d_2 - d_1)^2 \lambda_{\max}(Z))\delta_{e1}^2 + \frac{1}{2}d_2^2\lambda_{\max}(R)\delta_{e2}^2 = \Gamma_\delta$.

Then, the estimation of the DOA can be obtained from $\Gamma_\delta \leq 1$. This completes the proof. □

Remark 12.2 $\dot{V}$ is negative when $\Theta > 0$ and $\Phi_\kappa < 0$ with $\kappa = 1, \ldots, 2^n$ for all $(d(t), \dot{d}(t)) \in [d_1, d_2] \times [\mu_1, \mu_2]$. Since the matrix Φ_κ is convex, with respect to $d(t)$ and $\dot{d}(t)$, it is necessary and sufficient to ensure that $\Phi_\kappa < 0$ at the vertices of the interval $[d_1, d_2] \times [\mu_1, \mu_2]$. Then, condition (12.3) is in the form of LMI with the delay bounds. By solving the inequalities, the matrices W, R_m, R, S, U, Z, X, N, M, R_s and the upper bounds of the delays can all be obtained. The matrices $H_{\alpha m}$, $H_{\alpha s}$, H_{km} and H_{ks} are also calculated.

Remark 12.3 In this chapter, the interval time delay is considered and the new stability conditions are presented in the LMI form. When the lower bound is chosen as zero, the corresponding stability result can be obtained directly from Theorem 12.1. In the proof, we can find that the velocities and synchronization error appear in ξ. It is different from the classic proof procedure [6, 7] that only the velocities appear in ξ. Thus the asymptotical synchronization of the position can be directly obtained with the new proposed method. When the teleoperation system is free of time delays, one can also obtain the corresponding control design condition and the estimated DOA.

Remark 12.4 In the proof, Wirtinger inequality [14] is used, but we find that the LMI (Φ_κ in (12.3) without matrices N and M) is not feasible for the reason that some terms in inequality are restricted strongly. Then, we introduce some free weighting matrices in f_1 and f_2 to relax the restriction. An example is given below to show the less conservativeness of the proposed stability condition. Choose $K_m = K_s = I$, $\alpha_m = \alpha_s = 3I$ in (12.2). For the symmetric time delay case, we solve LMIs (12.3) and (12.4) and obtain the allowable maximum time delays $d_{m2} = d_{s2} = 2.1990$ when $d_{m1} = d_{s1} = 0$. Using the same parameters, we have $d_{m2} = d_{s2} = 0.75$ in [13]. Thus the proposed result is less conservative.

Remark 12.5 Based on Lemma 12.1, if $|h_{kmi}(q_m - q_s) + h_{\alpha mi}\dot{q}_m| \leq M_i$ and $|h_{ksi}(q_m - q_s) + h_{\alpha si}\dot{q}_s| \leq M_i$, $i = 1, \ldots, n$ for all $\{\dot{q}_m, \dot{q}_s, q_m - q_s\} \in \varepsilon(Q, 1)$, i.e.,

$\varepsilon(Q,1) \subset (L_m(H_{\alpha m}, H_{km}) \cap L_s(H_{\alpha s}, H_{ks}))$, the saturation function can be presented as a convex combination of the actual feedback and the auxiliary counterpart. With these expressions, if the conditions (12.3) and (12.4) are satisfied, the asymptotical stability of teleoperation system (12.1) can be guaranteed and the estimation of the DOA is obtained.

It is interesting to come up with a solution such that the estimation of the DOA is maximized. The following optimization problem is obtained

min ϖ

subject to

(a) $\omega_i > 0, i = 1, 2, ..., 7, W > 0, R_m > 0, R_s > 0, R > 0, S > 0, U > 0, Z > 0$

(b) $\omega_1 I - W \geq 0, \omega_2 I - S \geq 0, \omega_3 I - R_m \geq 0, \omega_4 I - R_s \geq 0, \omega_5 I - R \geq 0, \omega_6 I - U \geq 0, \omega_7 I - Z \geq 0$

(c) $\varepsilon(Q,1) \subset (L_m(H_{\alpha m}, H_{km}) \cap L_s(H_{\alpha s}, H_{ks}))$

(d) LMIs (12.3) and (12.4)

where $\varpi = \omega_1 + (d_2 - d_1)\omega_2 + \frac{1}{2}d_{m2}^2\omega_3 + \frac{1}{2}d_{s2}^2\omega_4 + 2d_2^2\omega_5 + d_2^2\omega_6 + (d_2 - d_1)^2 \omega_7$, $d_1 = d_{m1} + d_{s1}$. The constraint b) is equivalent to $\lambda_{\max}(W) \leq \omega_1$. The same holds for $\omega_i, i = 2, 3, ..., 7$. Hence, we get $\lambda_{\max}(W) + (d_2 - d_1)\lambda_{\max}(S) + \frac{1}{2}d_{m2}^2\lambda_{\max}(R_m) + \frac{1}{2}d_{s2}^2\lambda_{\max}(R_s) + \frac{1}{2}d_2^2\lambda_{\max}(R) + d_2^2\lambda_{\max}(U) + (d_2 - d_1)^2\lambda_{\max}(Z) \leq \varpi$. Then a maximized estimation of the DOA can be designed as $\delta_{\max} = \frac{1}{\sqrt{\varpi}}$. Furthermore, if the lower bound of delay is fixed, the estimation of the DOA varies inversely with the size of time delay, i.e., the larger the range of delay is, the smaller the estimation of the DOA is. For simplicity, the optimization is considered for the case $\delta_{m1} = \delta_{s1} = \delta_{e1} = \delta_{e2}$. In fact, we may fix some parameters and optimize others by the same way to achieve the optimal values.

In this chapter, the commonly assumed condition is used that the human input force F_h and the environment force F_e are passive. With this condition, the asymptotical stability criterion is achieved. For the non-passive case (including bounded input case), the input to state stability (ISS) and the small gain idea could be used, see [15–18] and the references therein. We can use the proposed method to derive the ISS condition for the non-passive case.

12.4 Experiment

In this section, experiments are done to verify the effectiveness of the proposed main results. The teleoperation system for the experiments includes two Phantom Premium 1.5A robotic arms with three degrees of freedom positional sensing, see Fig. 12.1. The master robot and slave robot are controlled by two different PCs, which are connected by communication network. Our goal is to evaluate the effectiveness of the estimation of the DOA. The tracking performance will be shown for the two cases that the initial conditions are inside the estimation of the DOA and out of the DOA.

Fig. 12.1 Experimental setup

In [13], the new stability conditions are presented without the discussion of DOA. We also show that the DOA is very important for the saturation control problem.

The master and slave devices are controlled by (12.2). The controller parameters are $\alpha_m = \alpha_s = diag\{0.3, 0.3, 0.3\}$, $K_m = K_s = diag\{1, 1, 1\}$. The saturation level is ± 2. The maximum of estimate of the DOA can be obtained as $\delta_{\max} = 1.1405$.

Case I: inside the estimation of the DOA. The initial positions of master and slave are $q_m(0) = \left[-0.1710\ 0.3214\ 0.3805\right]^T$, $q_s(0) = \left[0\ 0.314\ 0.314\right]^T$, and the initial velocities are $\dot{q}_m(0) = \dot{q}_s(0) = \left[0\ 0\ 0\right]^T$. A human operator moves the master manipulator, and the slave follows the motion of the master. In Fig. 12.2, the positions of master and slave in experiment are presented. It can be seen that the teleoperation system demonstrates stable behavior no matter how the master manipulator moves. The solid line represents the master and the dashed line represents the slave. Results shown in Figs. 12.3 and 12.4 correspond to the velocities of master and slave, respectively, and Figs. 12.5 and 12.6 correspond to the input torques of master and slave, respectively. Tracking performance can be achieved in this case both for our chapter and [13]. Figures for [13] are omitted due to the space limitation.

Case II: outside the estimation of the DOA. The initial positions of master and slave are $q_m(0) = \left[-0.5319\ 0.8531\ -0.7588\right]^T$, $q_s(0) = \left[0\ 0.314\ 0.314\right]^T$, and the initial velocities are $\dot{q}_m(0) = \dot{q}_s(0) = \left[0\ 0\ 0\right]^T$. As shown in Fig. 12.7, the teleoperation system clearly demonstrates unstable dynamics. The positions of three joints are shown in Fig. 12.7. The solid line represents the master and the dashed line represents the slave. For the first joint, the slave only tracks the motion of master in a certain range. It remains motionless when the position of the master is less than $-1.513rad$. For the second and third joints, the slave hardly tracks the motion of the master manipulator.

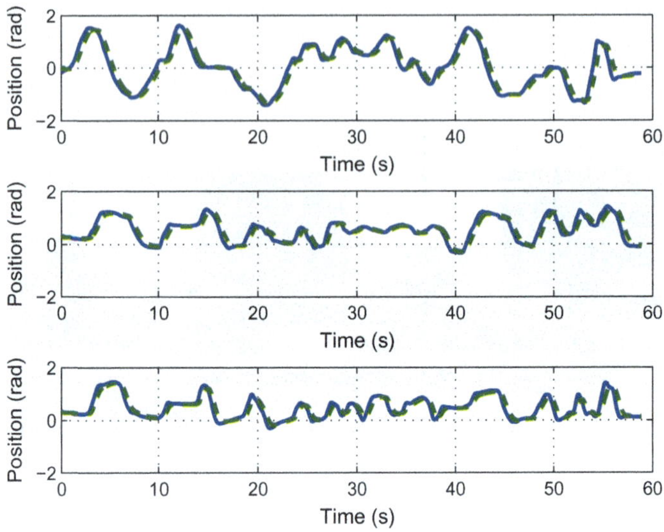

Fig. 12.2 Positions (Case I)

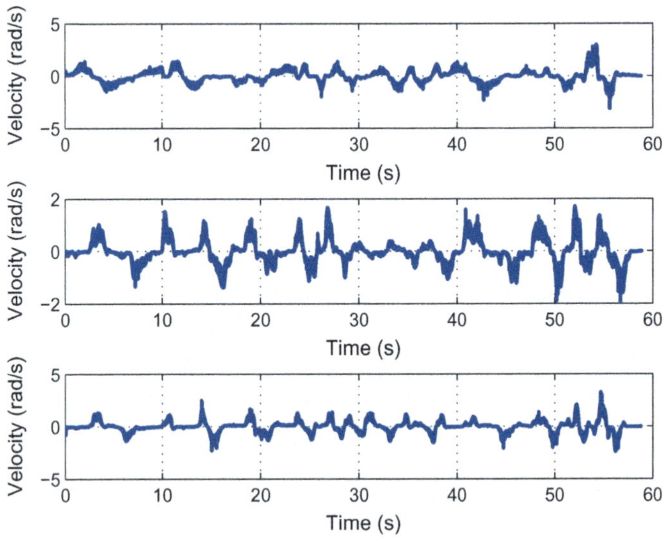

Fig. 12.3 Master velocity (Case I)

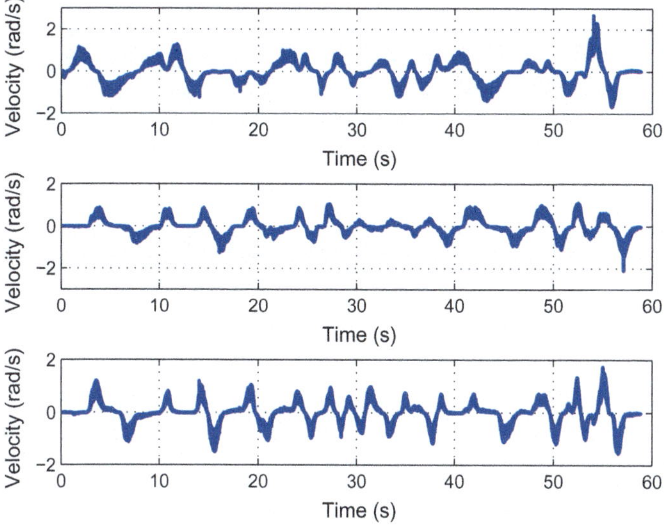

Fig. 12.4 Slave velocity (Case I)

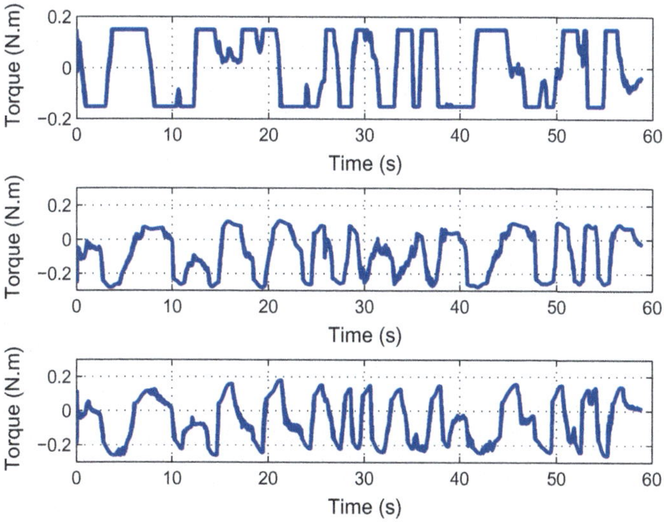

Fig. 12.5 Master torque (Case I)

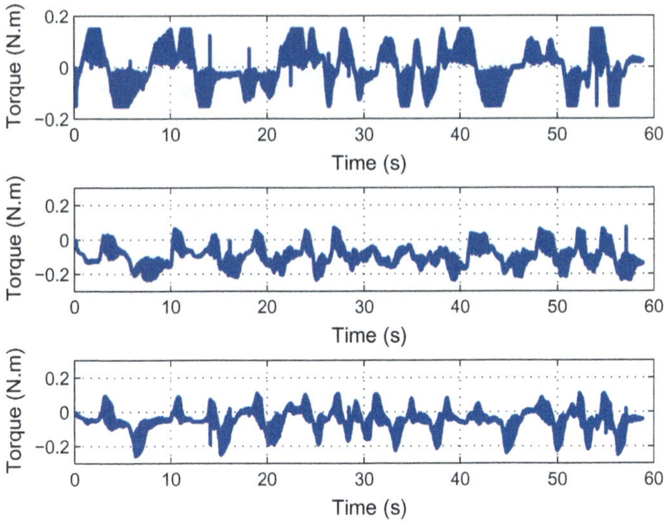

Fig. 12.6 Slave torque (Case I)

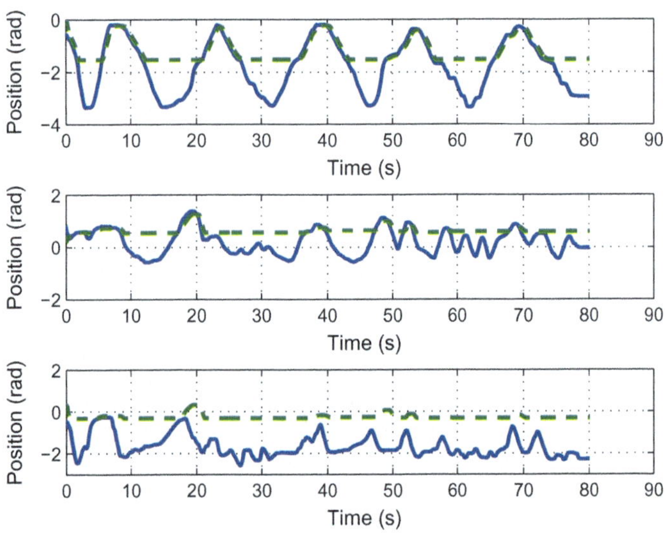

Fig. 12.7 Positions (Case II)

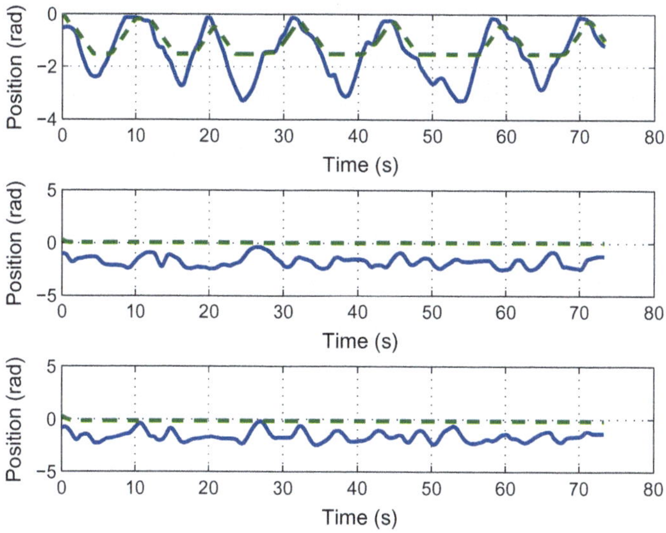

Fig. 12.8 Positions [13]

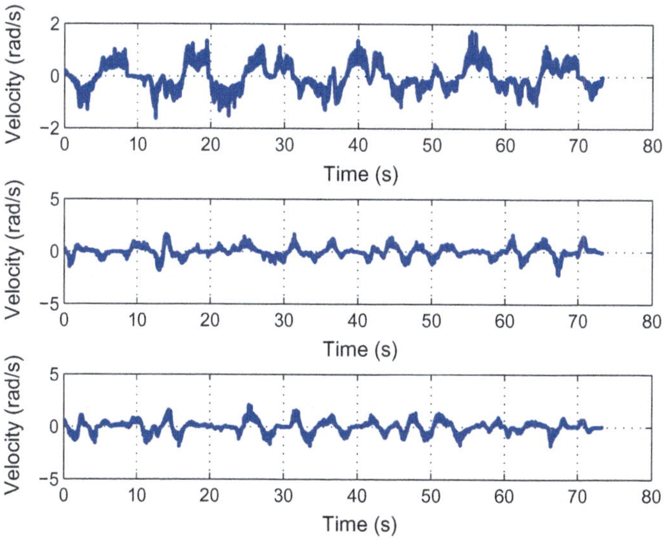

Fig. 12.9 Master velocity [13]

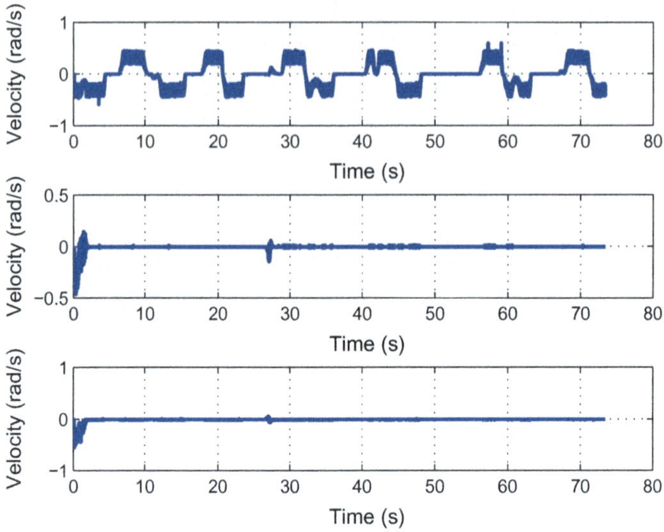

Fig. 12.10 Slave velocity [13]

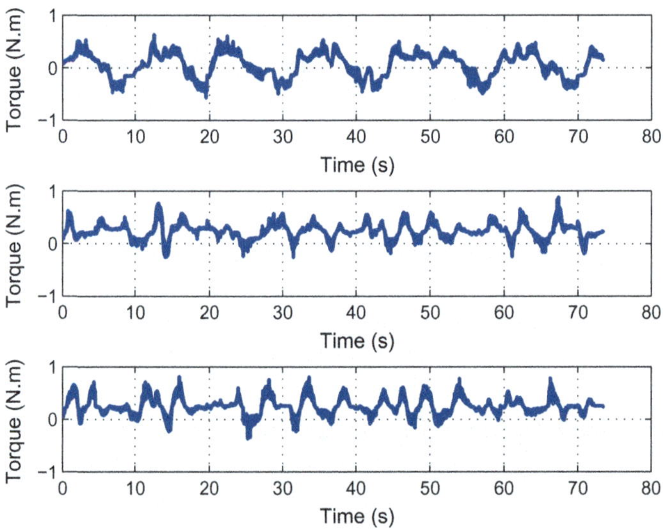

Fig. 12.11 Master torque [13]

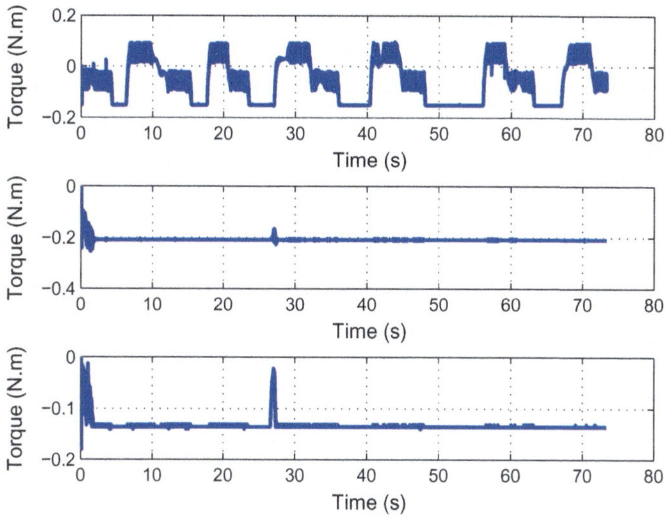

Fig. 12.12 Slave torque [13]

In the next set of experiments, we address the case in [13] where the initial conditions are chosen outside the estimation of the DOA. Our experiments indicate that, although the control algorithm in [13] is effective, the tracking performance is poor when the initial conditions are chosen outside the estimation of the DOA. The domain of attraction was not discussed in [13]. In this case, the teleoperation system demonstrates unstable behavior. The positions of master and slave are shown in Fig. 12.8. The solid line represents the master and the dashed line represents the slave. The qualitative figure is similar with the above case. The first joint of slave tracks the movement of the master in a certain range, while the second and third joints of slave do not track the master manipulator at all. Results shown in Figs. 12.9 and 12.10 correspond to the velocities of master and slave, respectively, and Figs. 12.11 and 12.12 correspond to the input torques of master and slave. Overall, the experimental investigations indicate that it is vitally important to discuss DOA for saturated networked teleoperation system and the method proposed in this chapter to estimate the DOA of teleoperation system is effective.

12.5 Summary

In this chapter, we have discussed the problem on exploring the domain of attraction for bilateral teleoperator with interval delays and actuator saturation. The novel Lyapunov-Krasovskii functional is proposed and the domain of attraction is analyzed. The delay-dependent estimation of the domain of attraction is presented using linear matrix inequality technique. The derived condition involves the control gains, the allowable bounds of the delays, the initial values of system state and the parameters of actuator saturation. Experiments are performed and the results show the effectiveness of the proposed method. In the future, it is possible to propose the output feedback controller and search for a less conservative method to estimate the domain of attraction for teleoperation system.

References

1. H.J. Yang, Y.Q. Xia, P. Shi, M.Y. Fu, Stability analysis for high frequency networked control system. IEEE Trans. Autom. Control **57**(10), 2694–2700 (2012)
2. A. Suzuki, K. Ohnishi, Frequency-domain damping design for timedelayed bilateral teleoperation system based on modal space analysis. IEEE Trans. Ind. Electron. **60**(1), 177–190 (2013)
3. H.J. Gao, T.W. Chen, New results on stability of discrete-time systems with time-varying state delay. IEEE Trans. Autom. Control **52**(2), 328–334 (2007)
4. S. Islam, P.X. Liu, A.E. Saddik, Y.B. Yang, Bilateral control of teleoperation systems with time delay. IEEE/ASME Trans. Mechatron. **20**(1), 1–12 (2015)
5. I.G. Polushin, S.N. Dashkovskiy, A. Takhmar, R.V. Patel, A small gain framework for networked cooperative force-reflecting teleoperation. Automatica **49**(2), 338–348 (2013)
6. C.C. Hua, X.P. Liu, Delay-dependent stability criteria of teleoperation systems with asymmetric time-varying delays. IEEE Trans. Robot. **26**(5), 925–932 (2010)
7. C.C. Hua, X.P. Liu, Teleoperation over the internet with/without velocity signal. IEEE Trans. Instrum. Measurement **60**, 4–13 (2011)
8. Y.L. Li, Z.L. Lin, Design of saturation-based switching antiwindup gains for the enlargement of the domain of attraction. IEEE Trans. Autom. Control **28**(7), 1810–1816 (2013)
9. T.S. Hu, Z.L. Lin, B.M. Chen, An analysis and design method for linear systems subject to actuator saturation and disturbance. Automatica **38**(2), 351–359 (2002)
10. Z.Y. Meng, Z.Y. Zhao, Z.L. Lin, On global leader-following consensus of identical linear dynamic systems subject to actuator saturation. Syst. Control Lett. **62**(2), 132–142 (2013)
11. Z. Zhu, Y.Q. Xia, M.Y. Fu, Adaptive sliding mode control for attitude stabilization with actuator saturation. IEEE Trans. Ind. Electron. **58**(10), 4898–4907 (2011)
12. A.M. Zou, K.D. Kumar, Neural network-based distributed attitude coordination control for spacecraft formation flying with input saturation. IEEE Trans. Neural Netw. Learn. Syst. **23**(7), 1155–1162 (2012)
13. F. Hashemzadeh, I. Hassanzadeh, M. Tavakoli, Teleoperation in the presence of varying time delays and sandwich linearity in actuators. Automatica **49**(9), 2813–2821 (2013)
14. A. Seuret, F. Gouaisbaut, Wirtinger-based integral inequality: application to time-delay systems. Automatica **49**(9), 2860–2866 (2013)
15. S. Dashkovskiy, D. Efimov, E.D. Sontag, Input to state stability and allied system properties. Autom. Remote Control **72**(8), 1579–1614 (2011)
16. D. Efimov, A. Loria, E. Panteley, Robust output stabilization: Improving performance via supervisory control. Int. J. Robust Nonlinear Control **21**(10), 1219–1236 (2011)
17. I.G. Polushin, H.J. Marquez et al., A multichannel ios small gain theorem for systems with multiple time-varying communication delays. IEEE Trans. Autom. Control **54**(2), 404–409 (2009)
18. S. Dashkovskiy, F.R. Wirth, M. Kosmykov, A small-gain condition for interconnections of iss systems with mixed iss characterizations. IEEE Trans. Autom. Control **56**(6), 1247–1258 (2011)

Chapter 13
Adaptive Controller Design for Teleoperation System with Actuator Saturation

Abstract The output-feedback based controller design problem is investigated for the networked teleoperation system in this chapter. A new control scheme is proposed to guarantee the global asymptotic stability of the bilateral teleoperation system with time-varying delays and bounded inputs. Firstly, a new fast terminal sliding mode velocity observer is proposed to estimate the unknown velocity signals for the teleoperation system. Then, by considering the unknown gravity term, an adaptive SP+Sd-type (saturated proportion plus saturated damping) controller is designed based on the estimated velocity. In the new controllers, the specific sigmoidal function is not used, and any one on a set of saturation functions can be applied. Furthermore, by choosing Lyapunov-Krasovskii functional, we show that the master-slave teleoperation system is stable under specific linear matrix inequality (LMI) conditions. With the given controller design parameters and the upper bound of the input, the allowable maximal transmission delay can be computed by using the proposed stability criteria. Finally, both simulation and experiment are performed to show the effectiveness of the proposed methods.

13.1 Introduction

As we cited in above chapters in recent years various teleoperation strategies have been reported in the literature to deal with the problem associated with time delays [1–6]. Like many applications of control systems, the actuator output has a limited amplitude, i.e., is subject to saturation. Controllers that ignore actuator saturation may cause undesirable responses and even closed-loop system instability. Therefore, it is highly desirable to develop control methods that take any actuator saturation into account at the design outset, therefore, allow for efficient and stable control with small-size actuators that inevitably possess a limited output capacity.

In order to deal with the actuator saturation problem for the control of manipulator, many effective approaches have been presented. In [7], a bounded dynamical extension of the PD+ (proportional and derivation) algorithm was proposed. Two globally

C. Hua et al., *Analysis and Design for Networked Teleoperation System*,
https://doi.org/10.1007/978-981-13-7936-9_13

stabilizing bounded control schemes for the tracking control of robot manipulators with saturating inputs were designed in [8]. An output-feedback adaptive SP+SD-type control scheme for the global position stabilization of robot manipulator with bounded inputs was proposed in [9]. To deal with the actuator saturation problem in teleoperation system with the time-varying delay, a novel control scheme was designed in [10].

Although there are many achievements on the time-delayed teleoperation system control design, to the best of the authors' knowledge, there have been no works on output-feedback adaptive saturation control for the teleoperation system to date. Considering the time-varying delay and actuator saturation, we propose a new finite-time velocity observer based SP+Sd-type controller. The main contents of this chapter can be summarized as follows: (i) A new fast terminal sliding mode based finite-time velocity observer is proposed to estimate the unknown velocity. (ii) The gravity term is not required to be known and the adaptive method is used to estimate it online. (iii) The SP+Sd-type control approach is proposed for the teleoperation system with time-varying delay and bounded inputs, and the delay-dependent stability criteria are derived. The simulations and the experiments are performed and the results show the effectiveness of our developed methods.

13.2 Preliminaries

Let us consider a master-slave bilateral teleoperation system given by the following model:

$$\begin{aligned} M_m(q_m)\ddot{q}_m + C_m(q_m, \dot{q}_m)\dot{q}_m + G_m(q_m) &= \tau_m + F_h \\ M_s(q_s)\ddot{q}_s + C_s(q_s, \dot{q}_s)\dot{q}_s + G_s(q_s) &= \tau_s - F_e \end{aligned} \tag{13.1}$$

where $\tau_m, \tau_s \in R^{n\times 1}$ are the applied torques restricted by the input saturation bound.

The well known properties for robot system are recalled here:

Property 13.1 *The gravity vector $G(q) = [g_1(q), g_2(q), ..., g_n(q)]^T$ is bounded as it consists of sinusoidal function q, i.e., every element $g_i(q)$ of the gravity vector $G(q)$, satisfies $|g_i(q)| \le \bar{g}_i$, $\forall q \in R^n$ for some positive constants $\bar{g}_i$, $i = 1, 2, ..., n$.*

Property 13.2 *The gravity vector can be written as $G(q) = Y(q)\theta$, where $\theta \in R^p$ is a constant vector whose elements depend on the system parameters and $Y(q) \in R^{n\times p}$ is the regression matrix whose elements depend on the configuration variables and do not involve any of the system parameters. Equivalently, the potential energy function of the robot can be written as $E(q, \theta) = \Upsilon(q)\theta$, where $\Upsilon(q) \in R^{1\times p}$ (the regression vector) is a continuous row vector function whose elements depend exclusively on the configuration variables and do not involve any the system parameters. Actually, $Y(q) = \frac{\partial}{\partial q}\Upsilon(q)$.*

Property 13.3 *Consider the gravity vector $G(q)$. Let θ_{Mi} represents any upper bound of $|\theta_i|$, such that $|\theta_i| \leq \theta_{Mi}, \forall i = (1, 2, ..., p)$. Let $\theta_M = [\theta_{M1}, ..., \theta_{Mp}]^T$. By property 5 and property 6, there exists positive constant $\bar{g}_i^{\theta_M} \geq \bar{g}_i$.*

Remark 13.1 In this chapter, we consider the teleoperation design problem with the following challenging issues: (i) The joint velocities are not available; (ii) The gravity term are unknown; (iii) There exists energy constraints for the input torques; (iv) The communication delay is time-varying. To deal with these problems, firstly a fast terminal sliding mode based finite time velocity observer is proposed to estimate the velocity signal. Secondly, the adaptive method is applied to approximate the gravity term online. Then by considering the bounded input and the time-varying delay, we propose a SP+Sd-type controller in which the upper bounds of such saturation functions are explicitly used. Finally, the LMI based stability conditions are presented for the closed-loop system.

13.3 Velocity Observer

13.3.1 Velocity Observer Design

In this section, a new fast terminal sliding mode velocity observer will be designed. We define $x_{m1} = q_m$, $\bar{x}_{m2} = \dot{q}_m$, $x_{s1} = q_s$, $\bar{x}_{s2} = \dot{q}_s$. Then the dynamics model of teleoperation system can be transformed into

$$\begin{aligned}
\dot{x}_{m1} &= \bar{x}_{m2} \\
\dot{\bar{x}}_{m2} &= -M_m^{-1}(x_{m1})(C_m(x_{m1}, \bar{x}_{m2})\bar{x}_{m2} \\
&\quad + G_m(x_{m1}) - \tau_m - F_h) \\
\dot{x}_{s1} &= \bar{x}_{s2} \\
\dot{\bar{x}}_{s2} &= -M_s^{-1}(x_{s1})(C_s(x_{s1}, \bar{x}_{s2})\bar{x}_{s2} \\
&\quad + G_s(x_{s1}) - \tau_s + F_e)
\end{aligned} \tag{13.2}$$

Consider the following coordination transformation

$$\begin{aligned}
T_m &= diag(a_{m1}, a_{m2}, ..., a_{mn}), a_{mi} > 0 \\
T_s &= diag(a_{s1}, a_{s2}, ..., a_{sn}), a_{si} > 0
\end{aligned}$$

Define $x_{m2} = \bar{x}_{m2} - T_m x_{m1}$ and $x_{s2} = \bar{x}_{s2} - T_s x_{s1}$, then we have

$$\begin{aligned}
\dot{x}_{m1} &= T_m x_{m1} + x_{m2} \\
\dot{x}_{m2} &= F_m(y_m, x_{m2}) + h_m(y_m)(-G_m(q_m) + \tau_m \\
&\quad + F_h) - T_m^2 x_{m1} - T_m x_{m2} \\
\dot{x}_{s1} &= T_s x_{s1} + x_{s2} \\
\dot{x}_{s2} &= F_s(y_s, x_{s2}) + h_s(y_s)(-G_s(q_s) + \tau_s \\
&\quad - F_e) - T_s^2 x_{s1} - T_s x_{s2}
\end{aligned} \tag{13.3}$$

where $F_i(y_i, x_{i2}) = -M_i^{-1}(y_i)(C_i(y_i, \bar{x}_{i2})\bar{x}_{i2})$, $h_i(y_i) = M_i^{-1}(y_i)$, $(i = m, s)$.

We assume that only the position information is available, i.e., $y_m = x_{m1}$ and $y_s = x_{s1}$. y_m, y_s represent the system outputs of the master and the slave, respectively. The fast terminal sliding mode velocity observers designed in this chapter are as follows

$$\begin{aligned}
\dot{\hat{x}}_{m1} &= T_m \hat{x}_{m1} + \hat{x}_{m2} - L_{m1} e_{m1} + \nu_{m1} \\
\dot{\hat{x}}_{m2} &= F_m(y_m, \hat{x}_{m2}) + h_m(y_m)(-Y_m(y_m)\hat{\theta}_m + \tau_m + F_h) \\
&\quad - T_m^2 \hat{x}_{m1} - T_m \hat{x}_{m2} - L_{m2} e_{m1} + \nu_{m2} + \chi_m sign(\nu_{m1}) \\
\dot{\hat{x}}_{s1} &= T_s \hat{x}_{s1} + \hat{x}_{s2} - L_{s1} e_{s1} + \nu_{s1} \\
\dot{\hat{x}}_{s2} &= F_s(y_s, \hat{x}_{s2}) + h_s(y_s)(-Y_s(y_s)\hat{\theta}_s + \tau_s - F_e) \\
&\quad - T_s^2 \hat{x}_{s1} - T_s \hat{x}_{s2} - L_{s2} e_{s1} + \nu_{s2} + \chi_s sign(\nu_{s1})
\end{aligned} \tag{13.4}$$

where $\hat{x}_{i1}$, $\hat{x}_{i2}$ are the estimations of x_{i1}, x_{i2}. $e_{i1} = \hat{x}_{i1} - x_{i1}$; $\hat{\theta}_i$ is the estimation of the θ_i, $L_{i1} = diag\{l_{i11}, l_{i12}, ..., l_{i1n}\}$ and $L_{i2} = diag\{l_{i21}, l_{i22}, ..., l_{i2n}\}$, each term of L_{i1} and L_{i2} is a constant. χ_i is a positive definite diagonal matrix i.e., $\chi_i = diag\{\chi_{i1}, \chi_{i2}, ..., \chi_{in}, \} > 0$. $\nu_{i1} = -k_{i1} sign(e_{i1})$, $\nu_{i2} = k_{i2}\nu_{i1} + k_{i3}|\nu_{i1}|^{\gamma_i} sign$ (v_{i1}) and $0 < \gamma_i < 1$. k_{i1}, k_{i2} and k_{i3} are positive constants $(i = m, s)$. Therefore, the velocity estimation error functions are

$$\begin{aligned}
\dot{e}_{m1} &= (T_m - L_{m1})e_{m1} + e_{m2} + \nu_{m1} \\
\dot{e}_{m2} &= -(T_m^2 + L_{m2})e_{m1} + \varphi_m(y_m)\tilde{\theta}_m - T_m e_{m2} + \nu_{m2} \\
&\quad + F_m(y_m, \hat{x}_{m2}) - F_m(y_m, x_{m2}) + \chi_m sign(\nu_{m1}) \\
\dot{e}_{s1} &= (T_s - L_{s1})e_{s1} + e_{s2} + \nu_{s1} \\
\dot{e}_{s2} &= -(T_s^2 + L_{s2})e_{s1} + \varphi_s(y_s)\tilde{\theta}_s - T_s e_{s2} + \nu_{s2} \\
&\quad + F_s(y_s, \hat{x}_{s2}) - F_s(y_s, x_{s2}) + \chi_s sign(\nu_{s1})
\end{aligned} \tag{13.5}$$

where $e_{i2} = \hat{x}_{i2} - x_{i2}$, $\varphi_i(y_i)\tilde{\theta}_i = \varphi_i(y_i)\theta_i - \varphi_i(y_i)\hat{\theta}_i$, $h_i(y_i)Y_i(y_i) = \varphi_i(y_i)$ $(i = m, s)$.

Remark 13.2 A new fast terminal sliding mode based finite time velocity observer is designed in this section. Compared with the high-gain observer and the velocity observer used in [11, 12], the finite time velocity observer can provide faster estimation speed and higher estimation precision. In comparison with the terminal sliding mode velocity observers proposed in [13–15], the application of T_m and T_s will provide more degrees of design freedom. Moreover, with the consideration of the convergence speed problem, the ν_{i2} term are designed to be composed of two parts: $k_{i2}\nu_{i1}$ and $k_{i3}\,|\nu_{i1}|^{\gamma_i}\, sign(\nu_{i1})$. When the velocity observer states stay at a distance from the actual system states, $k_{i2}\nu_{i1}$ dominates over $k_{i3}\,|\nu_{i1}|^{\gamma_i}\, sign(\nu_{i1})$, thus a fast convergence rate can be guaranteed; when the system states are close to the region, the dominant term $k_{i3}\,|\nu_{i1}|^{\gamma_i}\, sign(\nu_{i1})$ determines finite time convergence.

13.3.2 Stability Analysis

In this section, the stability analysis for the observer will be given. The following lemma is needed for the stability analysis.

Lemma 13.1 [16] *Consider the dynamics model $\dot{x} = f(x)$, $f(0) = 0$ and $x \in R^n$. If there exists a positive definite scalar function $V(x)$ such that*

$$\dot{V}(x) \le -\alpha V(x) - \beta V(x)^{\delta} \tag{13.6}$$

where α, $\beta > 0$, $0 < \delta < 1$, then the system is finite-time stable. Furthermore, the settling time is given by

$$T \le \frac{1}{\alpha(1-\delta)} \ln \frac{\alpha V^{1-\delta}(x_0) + \beta}{\beta} \tag{13.7}$$

Similar with [16], we assume that the nonlinear functions $F_m(y_m, x_{m2})$ and $F_m(y_m, \hat{x}_{m2})$ are Lipschitz, i.e., $\left\| F_m(y_m, \hat{x}_{m2}) - F_m(y_m, x_{m2}) \right\| \le \bar{C}_m \left\| e_{m2} \right\|$, where $\bar{C}_m$ is a positive constant. Let us define $e_m = [e_{m1}, e_{m2}]^T$, with equation (13.5) we have $\dot{e}_m = A_m e_m + \Delta F_m + H_m + \bar{\nu}_m$, where

$$A_m = \begin{bmatrix} T_m - L_{m1} & I_n \\ -T_m^2 - L_{m2} & -T_m \end{bmatrix}$$

$$\Delta F_m = \begin{bmatrix} 0_{n\times 1} \\ F_m(y_m, \hat{x}_{m2}) - F_m(y_m, x_{m2}) \end{bmatrix}$$

$$H_m = \begin{bmatrix} 0_{n\times 1} \\ \varphi_m(y_m)\tilde{\theta}_m \end{bmatrix}$$

$$\bar{\nu}_m = \begin{bmatrix} \nu_{m1} \\ \nu_{m2} + \chi_m sign(\nu_{m1}) \end{bmatrix}$$

With the above knowledge, we have the following Proposition for the master velocity observer (to save the space, only the proposition and proof for master are presented, the same conditions can be derived for the slave in similar way.).

Proposition 13.1 *Assuming the human operator insert forces are zero, i.e., ($F_h =$ 0), with the adaptive parameter tuning law*

$$\begin{aligned} \hat{\theta}_m &= \Omega_m \xi_m - \Omega_m \varphi_m(y_m) y_m \\ \dot{\xi}_m &= \frac{d\varphi_m(y_m)}{dt} y_m + \varphi_m(y_m)(\hat{x}_{m2} + T_m y_m) \end{aligned} \tag{13.8}$$

and with the following conditions hold, the velocity observer estimation errors will converge to zero in finite time.

$$\begin{aligned} P_m A_m + A_m^T P_m + \frac{P_m P_m}{\epsilon_m} + \Psi_m < 0 \\ l_{m1i} > a_{mi} \\ k_{m1} - R_m = \eta_m > 0 \\ -2P_{m2} T_m + \frac{P_{m2} P_{m2}}{\epsilon_m} + \epsilon_m \bar{C}_m^2 I_n < 0 \\ \chi_{mi} \geq F_m \end{aligned} \tag{13.9}$$

where $P_m = diag\{P_{m1}, P_{m2}\} > 0$, $P_{m1} = diag\{p_{m11}, p_{m12}, ..., p_{m1n}\}$, $P_{m2} = diag\{p_{m21}, p_{m22}, ..., p_{m2n}\}$. $\epsilon_m, \eta_m, \bar{C}_m$ *are positive constants,* R_m *is the deduced boundedness of the* $\|e_m\|$, $F_m = \left\| \varphi_m(y_m)\hat{\theta}_m - \varphi_m(y_m)\theta_m \right\|$, $\Psi_m = diag(0_{n\times n}, \epsilon_m \bar{C}_m^2 I_n)$.

Proof The proof procedure will be divided into three steps.

Step 1: Choose the following Lyapunov function

$$V_1' = e_m^T P_m e_m + \tilde{\theta}_m^T P_{m2} \Omega_m^{-1} \tilde{\theta}_m \tag{13.10}$$

where Ω_m is a positive definite constant matrix.

By differentiating V_1', we have

$$\begin{aligned} \dot{V}_1' \leq\ & e_m^T (P_m A_m + A_m^T P_m) e_m + 2e_m^T P_m \Delta F_m \\ & + 2e_m^T P_m H_m + 2e_m^T P_m \bar{\nu}_m - 2\tilde{\theta}_m P_{m2} \Omega_m^{-1} \dot{\hat{\theta}}_m \end{aligned} \tag{13.11}$$

Note that $2e_m^T P_m \Delta F_m \leq \dfrac{e_m^T P_m P_m e_m}{\epsilon_m} + \epsilon_m \Delta F_m^T \Delta F_m$. Since $F_m(y_m, x_{m2})$ and $F_m(y_m, \hat{x}_{m2})$ are Lipschitz, then we have $\Delta F_m^T \Delta F_m \leq \bar{C}_m^2 \|e_{m2}\|^2$. Therefore, the above inequality can be transformed into

$$\dot{V}_1' \leq e_m^T(P_m A_m + A_m^T P_m)e_m + \frac{e_m^T P_m P_m e_m}{\epsilon_m} + \epsilon_m \bar{C}_m^2 \|e_{m2}\|^2 + 2e_{m2}^T P_{m2}\varphi_m(y_m)\tilde{\theta}_m + 2e_m^T P_m \bar{\nu}_m - 2\tilde{\theta}_m P_{m2}\Omega_m^{-1}\dot{\hat{\theta}}_m \tag{13.12}$$

With the adaptive parameter tuning law (13.8), it is easy to achieve $\dot{\hat{\theta}}_m = \Omega_m \varphi_m(y_m)e_{m2}$. Then we have

$$\dot{V}_1' \leq e_m^T(P_m A_m + A_m^T P_m + \frac{P_m P_m}{\epsilon_m} + \Psi_m)e_m + 2e_m^T P_m \bar{\nu}_m \tag{13.13}$$

where $\Psi_m = diag(0_{n\times n}, \epsilon_m \bar{C}_m^2 I_n)$.

If $Q_m = P_m A_m + A_m^T P_m + \frac{P_m P_m}{\epsilon_m} + \Psi_m < 0$, $\Lambda_m = -\lambda_{\max}(Q_m)$, we have

$$\begin{aligned}\dot{V}_1' &< -\Lambda_m \|e_m\|^2 - 2k_{m1}\sum_{i=1}^{n} p_{m1i}|e_{m1i}| \\ &\quad - 2P_{mk1}e_m - 2k_{m1}P_{mk2}e_m - 2k_{m1}^{\gamma_m}P_{mk3}e_m \\ &\leq -\|e_m\| \left(\begin{array}{c} \Lambda_m \|e_m\| - 2k_{m1}\sqrt{\sum_{i=1}^{n} k_{m2}^2 p_{m2i}^2} \\ -2k_{m1}^{\gamma_m}\sqrt{\sum_{i=1}^{n} k_{m3}^2 p_{m2i}^2} - 2\sqrt{\sum_{i=1}^{n} p_{m2i}^2 \chi_{mi}^2} \end{array} \right)\end{aligned} \tag{13.14}$$

where

$$\begin{aligned} P_{mk1} &= [0_{1\times n}, p_{m21}\chi_{m1}sign(e_{m11}), \ldots, p_{m2n}\chi_{mn}sign(e_{m1n})] \\ P_{mk2} &= [0_{1\times n}, p_{m21}k_{m2}sign(e_{m11}), \ldots, p_{m2n}k_{m2}sign(e_{m1n})] \\ P_{mk3} &= [0_{1\times n}, p_{m21}k_{m3}sign(e_{m11}), \ldots, p_{m2n}k_{m3}sign(e_{m1n})] \end{aligned}$$

Furthermore, we can obtain that $\dot{V}_1' < 0$ if $\|e_m\| > R_m$, where $R_m = \frac{2k_{m1}\sqrt{\sum_{i=1}^{n} k_{m2}^2 p_{m2i}^2} + 2k_{m1}^{\gamma_m}\sqrt{\sum_{i=1}^{n} k_{m3}^2 p_{m2i}^2} + 2\sqrt{\sum_{i=1}^{n} p_{m2i}^2 \chi_{mi}^2}}{\Lambda_m}$. This implies that if $\|e_m\|$ is outside the ball of radius R_m then $\dot{V}_1' < 0$ and hence $\|e_m\|$ will reduce, resulting in $\|e_m\|$ being bounded by $\|e_m\| \leq R_m$ in finite time, and the proof is completed.

Step 2: Consider the Lyapunov function $V_2' = \frac{1}{2}e_{m1i}^T p_{m1i} e_{m1i}$. Differentiating V_2', yields

$$\dot{V}_2' \leq -p_{m1i}(l_{m1i} - a_{mi})e_{m1i}^2 - p_{m1i}|e_{m1i}|(k_{m1} - |e_{m2i}|) \tag{13.15}$$

Set $l_{m1i} > a_{mi}$ and $k_{m1} - R_m = \eta_m > 0$, where η_m is a positive constant, we have $\dot{V}_2' \leq -p_{m1i}\eta_m |e_{m1i}| = -\sqrt{2p_{m1i}}\eta_m V_2'^{1/2}$. Hence e_{m1} and $\dot{e}_{m1}$ will converge to zero in finite time.

Step 3: When the sliding mode $e_{m1} = \dot{e}_{m1} = 0$, which causes to $\nu_{m1} = -e_{m2}$, then $\nu_{m2} = -k_{m2}e_{m2} - k_{m3}|e_{m2}|^{\gamma_m} sign(e_{m2})$. Consider the Lyapunov function $V_3' = e_{m2}^T P_{m2} e_{m2}$. With the results obtained in Step 1, the boundedness of $\varphi_m(y_m)\theta_m - \varphi_m(y_m)\hat{\theta}_m$ can be achieved. Differentiating V_3', yields

$$\begin{aligned}\dot{V}_3' \leq\ & e_{m2}^T(-2P_{m2}T_m + \frac{P_{m2}P_{m2}}{\epsilon_m} + \epsilon_m \bar{C}_m^2 I_n)e_{m2} \\ & + 2\sum_{i=1}^{n} F_m p_{m2i}|e_{m2i}| - 2\sum_{i=1}^{n} \chi_{mi} p_{m2i}|e_{m2i}| \\ & - 2\sum_{i=1}^{n} k_{m2} p_{m2i} e_{m2i}^2 - 2\sum_{i=1}^{n} k_{m3} p_{m2i}|e_{m2i}|^{1+\gamma_m}\end{aligned} \tag{13.16}$$

When $-2P_{m2}T_m + \frac{P_{m2}P_{m2}}{\epsilon_m} + \epsilon_m \bar{C}_m^2 I_n < 0$ and $\chi_{mi} \geq F_m$, then we have $\dot{V}_3' \leq -2\sum_{i=1}^{n} k_{m2} p_{m2i} e_{m2i}^2 - 2\sum_{i=1}^{n} k_{m3} p_{m2i}|e_{m2i}|^{1+\gamma_m} = -\psi_{m1}V_3 - \psi_{m2}V_3^{(1+\gamma_m)/2}$, where $\psi_{m1} = 2k_{m2}$ and $\psi_{m2} = 2k_{m3}\sum_{i=1}^{n} p_{m2i}^{(1-\gamma_m)/2}$, hence yields the e_{m2} will converges to zero in finite time. This completes the proof.

Remark 13.3 In this section, a new fast terminal sliding mode velocity observer is designed for the teleoperation system. With the velocity observer, the position and velocity estimation errors will converge to zero in finite time. The exact convergence time can be computed as $t_i = \frac{2V_2'^{1/2}(e_{m1i}(0))}{\sqrt{2p_{m1i}}\eta_m}$ $(i = 1, 2, ..., n)$ and $T_2 = \frac{1}{\psi_{m1}(1-\gamma_m)} \ln \frac{\psi_{m1}U^{(1-\gamma_m)}(e_{m2}(T_1)) + \psi_{m2}}{\psi_{m2}}$, where $T_1 = \max\{t_1, t_2, ..., t_n\}$.

13.4 New Control Scheme

In this section, the SP+Sd-type controller will be designed with the estimated velocity signals (Fig. 13.1).

13.4.1 Controller Design

Let us suppose that the absolute value of each input τ_i (the ith element of the input vector τ) is constrained to be smaller than a given saturation bound $U_i > 0$, i.e.,

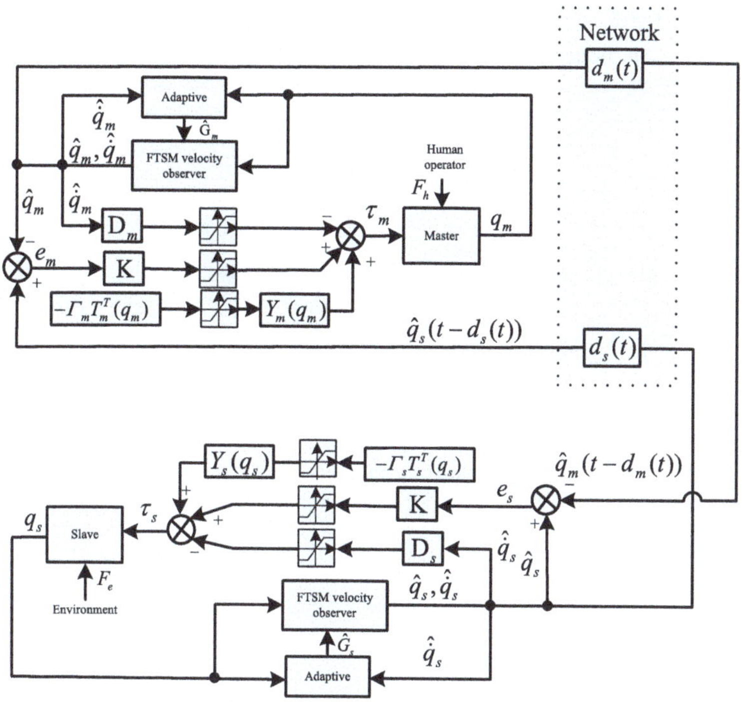

Fig. 13.1 Teleoperation system with new controller

$|\tau_i| \leq U_i, i = 1, ..., n$. So we have

$$\tau_i = U_i sat(\frac{u_i}{U_i})$$

where $sat(\cdot)$ is the standard saturation function, i.e., $sat(x) = sign(x) \min(|x|, 1)$.

Assumption 13.1 $U_i > \bar{g}_i, i = 1, ..., n$.

Assumption 13.2 The time delay $d_m(t)$ and $d_s(t)$ are bounded, i.e., there exist positive scalars $\bar{d}_m$ and $\bar{d}_s$ such that $d_m(t) \leq \bar{d}_m$ and $d_s(t) \leq \bar{d}_s$.

Definition 13.1 Given a positive constant $\bar{\sigma}$, $\sigma(x)$ is strictly increasing, bounded, continuous, passing through the origin, concave for positive x and convex for negative x, and have next properties.

P-I: $|\sigma(x)| \leq |x|$ and $\sigma(-x) = -\sigma(x)$.
P-II: For any $x, y \in R$, if $x < y$ then $\sigma(x) < \sigma(y)$.
P-III: For any $x, y \in R$, $|\sigma(x) - \sigma(y)| \leq 2\sigma(|x - y|)$.
P-IV: For any $x, y \in R_{\geq 0}$, $\sigma(x + y) \leq \sigma(x) + \sigma(y)$.
P-V: For any $x \in R$, $|\sigma(x)| \leq \min(|x|, \bar{\sigma})$.

Let $M_a = [M_{a1}, ..., M_{ap}]^T$ and $\Theta_a = [-M_{a1}, M_{a1}] \times ... \times [-M_{ap}, M_{ap}]$, with M_{aj}, $j = 1, ..., p$, being positive constant, such that

$$\left|\theta_j\right| < M_{aj} \tag{13.17}$$

and

$$\bar{g}_i^{M_a} < U_i \tag{13.18}$$

New controllers proposed in this chapter are as follows

$$\begin{aligned} u_m &= Y_m(q_m)\bar{\theta}_m + S_P(K(q_s(t - d_s(t)) - q_m)) - S_d(D_m\dot{\hat{q}}_m) \\ u_s &= Y_s(q_s)\bar{\theta}_s + S_P(K(q_m(t - d_m(t)) - q_s)) - S_d(D_s\dot{\hat{q}}_s) \end{aligned} \tag{13.19}$$

where K, D_m, $D_s \in R^{n\times n}$ are positive constants.

Remark 13.4

$$\begin{aligned} S_P(x) &: R^n \to R^n \\ & x \longmapsto [\sigma_{P1}(x_1), \sigma_{P2}(x_2), ..., \sigma_{Pn}(x_n)]^T \end{aligned}$$

and

$$\begin{aligned} S_d(x) &: R^n \to R^n \\ & x \longmapsto [\sigma_{d1}(x_1), \sigma_{d2}(x_2), ..., \sigma_{dn}(x_n)]^T \end{aligned}$$

with $\sigma_{Pi}(\cdot)$ and $\sigma_{di}(\cdot)$, $i = 1, ..., n$, being generalized saturation functions with bounds $\bar{S}_{Pi}$ and $\bar{S}_{di}$ such that

$$\bar{S}_{Pi} + \bar{S}_{di} < U_i - \bar{g}_i^{M_a} \tag{13.20}$$

Remark 13.5 The control scheme (13.19) consists of three parts. The first term $Y_i(q_i)\bar{\theta}_i$ is used to estimate the unknown gravity function. Compared with the classical controllers designed for teleoperation system assuming that the gravity term can be obtained directly, more realistic phenomenon is considered in this chapter. Because of the actuator saturation, the estimation parameter $\bar{\theta}_i$ should be bounded as (13.22). The second term is used to deal with the position synchronization errors between the master and the slave. The third part $-S_d(D_i\dot{\hat{q}}_i)$ applying the estimated velocity signal provides damping force. In the designed controller, the control gains K and D_i are not tied to satisfy any saturation-avoidance inequality and may consequently take any positive values, which may be considered beneficial for performance-adjustment/improvement purposes ($i = m, s$).

Remark 13.6 Assumption 13.1 is a conventional assumption in the literature about the input saturation. This general assumption is the precondition of the system stability. Assumption 13.2 is imposed on the communication time delays. The backward and forward time delays are time varying and bounded, which is very common for the practical communication channel and assumption 13.2 also can be found in [1, 10, 11, 17].

Remark 13.7 In designing the controller, the adaptive method is used to approximate the unknown gravity term. Compared with the literatures about teleoperation system [1, 5, 6, 17], the gravity functions were directly used in the torque design to compensate gravity term. Unknown gravity case is considered in this chapter. The adaptive method is used to estimate the gravity term online. Moreover, compared with our previous work [11], the constrained input torque is considered. As the result of the input saturation, it is very difficult to design the adaptive parameter tuning law. In this chapter, a new adaptive parameter tuning law is used by defining the upper bound of the adaptive parameter, this problem is solved successfully.

13.4.2 Stability Analysis

With the new controllers, the closed-loop dynamics model of teleoperation system is

$$
\begin{aligned}
&M_m(q_m)\ddot{q}_m + C_m(q_m, \dot{q}_m)\dot{q}_m = Y_m(q_m)\tilde{S}_{am}(\phi_m)\\
&+ S_P(K(q_s(t-d_s(t)) - q_m)) - S_d(D_m\dot{\hat{q}}_m) + F_h\\
&M_s(q_s)\ddot{q}_s + C_s(q_s, \dot{q}_s)\dot{q}_s = Y_s(q_s)\tilde{S}_{as}(\phi_s)\\
&+ S_P(K(q_m(t-d_m(t)) - q_s)) - S_d(D_s\dot{\hat{q}}_s) - F_e
\end{aligned} \tag{13.21}
$$

where $\tilde{S}_{ai}(\phi_i) = S_{ai}(\phi_i) - S_{ai}(\phi_i^*)$, $\phi_i = -\Gamma_i\Upsilon_i^T(q_i)$, Γ_i is a positive definite diagonal matrix. $S_{ai}(\phi_i^*) = \theta_i$, $S_{ai}(\cdot)$ stands for the saturation function like

$$
\begin{aligned}
S_{ai}(x) : R^n &\rightarrow R^n\\
x &\longmapsto [\sigma_{ai1}(x_1), \sigma_{ai2}(x_2), \ldots, \sigma_{ain}(x_n)]^T
\end{aligned}
$$

with $\sigma_{aij}(\cdot)$, $i = m/s$, $j = 1, \ldots, n$, being generalized saturation functions with bounds M_{ai}.

Theorem 13.1 *Assuming the human operator and the environment are passive, in the bilateral teleoperation (13.1) controlled by the new controller (13.19) proposed in this chapter, with the estimation adaptive parameters*

$$
\begin{aligned}
\bar{\theta}_m &= S_{am}(-\Gamma_m\Upsilon_m^T(q_m))\\
\bar{\theta}_s &= S_{as}(-\Gamma_s\Upsilon_s^T(q_s))
\end{aligned} \tag{13.22}
$$

and the fast terminal sliding mode velocity observer (13.4), if there exist positive-definite matrices Z and Q such that the following LMI holds, the position errors between the master manipulator and the slave manipulator are bounded.

$$\begin{bmatrix} \Psi_{11} & \Psi_{12} & 0 & 0 \\ * & \Psi_{22} & 0 & 0 \\ * & * & \Psi_{33} & \Psi_{34} \\ * & * & * & \Psi_{44} \end{bmatrix} < 0, \tag{13.23}$$

$$\begin{bmatrix} \bar{\Psi}_{11} & \bar{\Psi}_{12} & 0 & 0 \\ * & \bar{\Psi}_{22} & 0 & 0 \\ * & * & \bar{\Psi}_{33} & \bar{\Psi}_{34} \\ * & * & * & \bar{\Psi}_{44} \end{bmatrix} < 0, \tag{13.24}$$

$$\begin{bmatrix} \Psi_{11} & \Psi_{12} & 0 & 0 \\ * & \Psi_{22} & 0 & 0 \\ * & * & \bar{\Psi}_{33} & \bar{\Psi}_{34} \\ * & * & * & \bar{\Psi}_{44} \end{bmatrix} < 0, \tag{13.25}$$

$$\begin{bmatrix} \bar{\Psi}_{11} & \bar{\Psi}_{12} & 0 & 0 \\ * & \bar{\Psi}_{22} & 0 & 0 \\ * & * & \Psi_{33} & \Psi_{34} \\ * & * & * & \Psi_{44} \end{bmatrix} < 0, \tag{13.26}$$

where $\Psi_{11} = -D_m + 2\bar{d}_m ZK + \frac{\mu_m^2}{\upsilon_m}$, $\Psi_{12} = 2\bar{d}_s K$, $\Psi_{22} = -2\bar{d}_s QK$, $\Psi_{33} = -D_s + 2\bar{d}_s QK + \frac{\mu_s^2}{\upsilon_s}$, $\Psi_{34} = 2\bar{d}_m K$, $\Psi_{44} = -2\bar{d}_m ZK$. $\bar{\Psi}_{11} = -\bar{S}_{dm} + 2\bar{S}_{Pm}\bar{d}_m Z + \frac{\mu_m^2}{\upsilon_m}$, $\bar{\Psi}_{12} = 2\bar{d}_s \bar{S}_{Pm}$, $\bar{\Psi}_{22} = -2\bar{d}_s Q\bar{S}_{Pm}$, $\bar{\Psi}_{33} = -\bar{S}_{ds} + 2\bar{S}_{Ps}\bar{d}_s Q + \frac{\mu_s^2}{\upsilon_s}$, $\bar{\Psi}_{34} = 2\bar{d}_m \bar{S}_{Ps}$, $\bar{\Psi}_{44} = -2\bar{d}_m Z\bar{S}_{Ps}$. μ_m, μ_s, υ_m *and* υ_s *will be defined below.*

Proof Next, we choose the following Lyapunov functional $V = V_1 + V_2 + V_3 + V_4 + V_5$ with

$$\begin{aligned} V_1 &= \frac{1}{2}\dot{q}_m^T M_m \dot{q}_m + \frac{1}{2}\dot{q}_s^T M_s \dot{q}_s \\ &\quad + \int_0^t -\dot{q}_m^T(\delta) F_h(\delta) d\delta + \int_0^t \dot{q}_s^T(\delta) F_e(\delta) d\delta \\ V_2 &= \sum_{i=1}^{n} \int_0^{q_{mi}(t)-q_{si}(t)} S_{Pi}(K\zeta_i) d\zeta_i \\ V_3 &= \int_{0_n}^{\phi_m} \tilde{S}_{am}^T(r)\Gamma_m^{-1} dr + \int_{0_n}^{\phi_s} \tilde{S}_{as}^T(r)\Gamma_s^{-1} dr \end{aligned}$$

$$
\begin{aligned}
V_4 &= 2\int_{-\bar{d}_m}^{0}\int_{t+\eta}^{t}\dot{q}_m^T(\xi)ZS_P(K\dot{q}_m(\xi))d\xi d\eta \\
&\quad + 2\int_{-\bar{d}_s}^{0}\int_{t+\eta}^{t}\dot{q}_s^T(\xi)QS_P(K\dot{q}_s(\xi))d\xi d\eta \\
V_5 &= D_m e_{m2}^T P_{m2} e_{m2} + D_s e_{s2}^T P_{s2} e_{s2}
\end{aligned}
\tag{13.27}
$$

where P_{m2}, P_{s2} have been defined in Sect. 13.3.

The time derivative of V_1 is given by

$$
\begin{aligned}
\dot{V}_1 &= \dot{q}_m^T(\tau_m + F_h - G_m(q_m)) + \dot{q}_s^T(\tau_s - F_e \\
&\quad - G_s(q_s)) - \dot{q}_m^T F_h + \dot{q}_s^T F_e \\
&= \dot{q}_m^T(Y_m(q_m)\tilde{S}_{am}(\phi_m) + S_P(K(q_s(t-d_s(t)) - q_m)) \\
&\quad - S_d(D_m\dot{q}_m)) + \dot{q}_s^T(Y_s(q_s)\tilde{S}_{as}(\phi_s) + S_P(K(q_m(t \\
&\quad - d_m(t)) - q_s)) - S_d(D_s\dot{q}_s)) - \dot{q}_m^T(S_d(D_m\dot{\hat{q}}_m) \\
&\quad - S_d(D_m\dot{q}_m)) - \dot{q}_s^T(S_d(D_s\dot{\hat{q}}_s) - S_d(D_s\dot{q}_s))
\end{aligned}
\tag{13.28}
$$

The time derivatives of V_2, V_3, V_4 and V_5 are

$$
\dot{V}_2 = (\dot{q}_m - \dot{q}_s)^T S_P(K(q_m - q_s)) \tag{13.29}
$$

$$
\begin{aligned}
\dot{V}_3 &= \tilde{S}_{am}^T(\phi_m)\Gamma_m^{-1}\dot{\phi}_m + \tilde{S}_{as}^T(\phi_s)\Gamma_s^{-1}\dot{\phi}_s \\
&= -\tilde{S}_{am}^T(\phi_m)Y_m^T(q_m)\dot{q}_m - \tilde{S}_{as}^T(\phi_s)Y_s^T(q_s)\dot{q}_s
\end{aligned}
\tag{13.30}
$$

$$
\begin{aligned}
\dot{V}_4 &= 2\bar{d}_m\dot{q}_m^T ZS_P(K\dot{q}_m) - 2\int_{t-\bar{d}_m}^{t}\dot{q}_m^T(\xi)ZS_P(K\dot{q}_m(\xi))d\xi \\
&\quad + 2\bar{d}_s\dot{q}_s^T QS_P(K\dot{q}_s) - 2\int_{t-\bar{d}_s}^{t}\dot{q}_s^T(\xi)QS_P(K\dot{q}_s(\xi))d\xi \\
&\le 2\bar{d}_m\dot{q}_m^T ZS_P(K\dot{q}_m) - 2\int_{t-d_m(t)}^{t}\dot{q}_m^T(\xi)ZS_P(K\dot{q}_m(\xi))d\xi \\
&\quad + 2\bar{d}_s\dot{q}_s^T QS_P(K\dot{q}_s) - 2\int_{t-d_s(t)}^{t}\dot{q}_s^T(\xi)QS_P(K\dot{q}_s(\xi))d\xi
\end{aligned}
\tag{13.31}
$$

$$
\begin{aligned}
\dot{V}_5 &< D_m e_{m2}^T\left(\frac{P_{m2}P_{m2}}{\epsilon_m} + \epsilon_m\bar{C}_m^2 I_n - 2P_{m2}T_m\right)e_{m2} \\
&\quad + D_s e_{s2}^T\left(\frac{P_{s2}P_{s2}}{\epsilon_s} + \epsilon_s\bar{C}_s^2 I_n - 2P_{s2}T_s\right)e_{s2} \\
&\le -\lambda_m D_m \|e_{m2}\|^2 - \lambda_s D_s \|e_{s2}\|^2
\end{aligned}
\tag{13.32}
$$

where $-\lambda_m = \lambda_{\max}(\frac{P_{m2}P_{m2}}{\epsilon_m} + \epsilon_m \bar{C}_m^2 I_n - 2P_{m2}T_m)$ and $-\lambda_s = \lambda_{\max}(\frac{P_{s2}P_{s2}}{\epsilon_s} + \epsilon_s \bar{C}_s^2 I_n - 2P_{s2}T_s)$, the specific prove process for V_5 has been shown in the stability analysis of velocity observer.

With P-III, we know that $\dot{q}_m^T S_P(K(q_s(t-d_s(t)) - q_m)) + \dot{q}_m^T S_P(K(q_m - q_s)) \le 2\left|\dot{q}_m^T\right| S_P(K\left|q_s(t-d_s(t)) - q_s\right|)$ and $\dot{q}_s^T S_P(K(q_m(t-d_m(t)) - q_s)) + \dot{q}_s{}^T S_P(K(q_s - q_m)) \le 2\left|\dot{q}_s^T\right| S_P(K\left|q_m(t-d_m(t)) - q_m\right|)$. Moreover, $-\dot{q}_m^T(S_d(D_m\hat{q}_m) - S_d(D_m\dot{q}_m)) \le 2\left|\dot{q}_m\right|^T S_d(D_m\left|e_{m2}\right|)$ and $-\dot{q}_s^T(S_d(D_s\hat{q}_s) - S_d(D_s\dot{q}_s)) \le 2\left|\dot{q}_s\right|^T S_d(D_s\left|e_{s2}\right|)$.

Then with the P-IV, we can obtain that

$$2\dot{q}_m^T S_P(K\left|q_s(t-d_s(t)) - q_s\right|) \le 2\left|\dot{q}_m\right|^T \int_{t-d_s(t)}^{t} S_P(K\left|\dot{q}_s(\xi)\right|)d\xi$$

and

$$2\dot{q}_s^T S_P(K\left|q_m(t-d_m(t)) - q_m\right|) \le 2\left|\dot{q}_s\right|^T \int_{t-d_m(t)}^{t} S_P(K\left|\dot{q}_m(\xi)\right|)d\xi$$

furthermore, we have

$$\begin{aligned}
&\dot{V}_1 + \dot{V}_2 + \dot{V}_3 + \dot{V}_4 \\
\le & -\dot{q}_m^T S_d(D_m\dot{q}_m) - \dot{q}_s^T S_d(D_s\dot{q}_s) \\
&+ 2\left|\dot{q}_m\right|^T S_d(D_m\left|e_{m2}\right|) + 2\left|\dot{q}_s\right|^T S_d(D_s\left|e_{s2}\right|) \\
&+ 2\left|\dot{q}_m\right|^T \int_{t-d_s(t)}^{t} S_P(K\left|\dot{q}_s(\xi)\right|)d\xi \\
&+ 2\left|\dot{q}_s\right|^T \int_{t-d_m(t)}^{t} S_P(K\left|\dot{q}_m(\xi)\right|)d\xi \\
&+ 2\bar{d}_m\dot{q}_m^T Z S_P(K\dot{q}_m) - 2\int_{t-d_m(t)}^{t} \dot{q}_m^T(\xi) Z S_P(K\dot{q}_m(\xi))d\xi \\
&+ 2\bar{d}_s\dot{q}_s^T Q S_P(K\dot{q}_s) - 2\int_{t-d_s(t)}^{t} \dot{q}_s^T(\xi) Q S_P(K\dot{q}_s(\xi))d\xi \\
\le & -\dot{q}_m^T S_d(D_m\dot{q}_m) - \dot{q}_s^T S_d(D_s\dot{q}_s) \\
&+ 2\bar{d}_m\dot{q}_m^T Z S_P(K\dot{q}_m) + 2\bar{d}_s\dot{q}_s^T Q S_P(K\dot{q}_s)
\end{aligned} \tag{13.33}$$

$$\begin{aligned}
&+ 2\bar{d}_s\dot{q}_m^T Q^{-1} S_P(K\dot{q}_m) + 2\bar{d}_m\dot{q}_s^T Z^{-1} S_P(K\dot{q}_s) \\
&+ 2\left|\dot{q}_m\right|^T S_d(D_m\left|e_{m2}\right|) + 2\left|\dot{q}_s\right|^T S_d(D_s\left|e_{s2}\right|)
\end{aligned}$$

Define $\bar{\varphi}_m = \frac{\dot{q}_m^T S_d(D_m\dot{q}_m)}{D_m\|\dot{q}_m\|^2}$, $\bar{\varphi}_s = \frac{\dot{q}_s^T S_d(D_s\dot{q}_s)}{D_s\|\dot{q}_s\|^2}$, $\bar{\eta}_m = \frac{\dot{q}_m^T S_P(K\dot{q}_m)}{K\|\dot{q}_m\|^2}$, $\bar{\eta}_s = \frac{\dot{q}_s^T S_P(K\dot{q}_s)}{K\|\dot{q}_s\|^2}$, $\mu_m = \frac{\left|\dot{q}_m\right|^T S_d(D_m\left|e_{m2}\right|)}{D_m\|\dot{q}_m\|\|e_{m2}\|}$ and $\mu_s = \frac{\left|\dot{q}_s\right|^T S_d(D_s\left|e_{s2}\right|)}{D_s\|\dot{q}_s\|\|e_{s2}\|}$. Then we can have

$$\begin{aligned}
\dot{V} \leq & -\bar{\varphi}_m D_m \left\|\dot{q}_m\right\|^2 + 2\mu_m D_m \left\|\dot{q}_m\right\| \left\|e_{m2}\right\| \\
& - \bar{\varphi}_s D_s \left\|\dot{q}_s\right\|^2 + 2\mu_s D_s \left\|\dot{q}_s\right\| \left\|e_{s2}\right\| \\
& + \bar{\eta}_m (2\bar{d}_m Z K + 2\bar{d}_s K Q^{-1}) \left\|\dot{q}_m\right\|^2 - \lambda_m D_m \left\|e_{m2}\right\|^2 \\
& + \bar{\eta}_s (2\bar{d}_s Q K + 2\bar{d}_m K Z^{-1}) \left\|\dot{q}_s\right\|^2 - \lambda_s D_s \left\|e_{s2}\right\|^2 \\
\leq & -\bar{\varphi}_m D_m \left\|\dot{q}_m\right\|^2 + \frac{\mu_m^2}{\upsilon_m} \left\|\dot{q}_m\right\|^2 - \lambda_m D_m \left\|e_{m2}\right\|^2 \\
& + \upsilon_m D_m^2 \left\|e_{m2}\right\|^2 + \bar{\eta}_m (2\bar{d}_m Z K + 2\bar{d}_s K Q^{-1}) \left\|\dot{q}_m\right\|^2 \\
& - \bar{\varphi}_s D_s \left\|\dot{q}_s\right\|^2 + \frac{\mu_s^2}{\upsilon_s} \left\|\dot{q}_s\right\|^2 - \lambda_s D_s \left\|e_{s2}\right\|^2 \\
& + \upsilon_s D_s^2 \left\|e_{s2}\right\|^2 + \bar{\eta}_s (2\bar{d}_s Q K + 2\bar{d}_m K Z^{-1}) \left\|\dot{q}_s\right\|^2 \\
= & -(\varphi_m D_m - \bar{\eta}_m (2\bar{d}_m Z K + 2\bar{d}_s K Q^{-1}) - \frac{\mu_m^2}{\upsilon_m}) \left\|\dot{q}_m\right\|^2 \\
& - (\varphi_s D_s - \bar{\eta}_s (2\bar{d}_s Q K + 2\bar{d}_m K Z^{-1}) - \frac{\mu_s^2}{\upsilon_s}) \left\|\dot{q}_s\right\|^2 \\
& - (\lambda_m D_m - \upsilon_m D_m^2) \left\|e_{m2}\right\|^2 - (\lambda_s D_s - \upsilon_s D_s^2) \left\|e_{s2}\right\|^2
\end{aligned} \tag{13.34}$$

where υ_m is a positive constant. If $\lambda_m D_m \geq \upsilon_m D_m^2$ and $\lambda_s D_s \geq \upsilon_s D_s^2$, then we have the following LMI

$$\begin{bmatrix} \acute{\Psi}_{11} & \acute{\Psi}_{12} & 0 & 0 \\ * & \acute{\Psi}_{22} & 0 & 0 \\ * & * & \acute{\Psi}_{33} & \acute{\Psi}_{34} \\ * & * & * & \acute{\Psi}_{44} \end{bmatrix} < 0 \tag{13.35}$$

where $\acute{\Psi}_{11} = -\bar{\varphi}_m D_m + \frac{\mu_m^2}{\upsilon_m} + 2\bar{\eta}_m \bar{d}_m Z K$, $\acute{\Psi}_{12} = 2\bar{\eta}_m \bar{d}_s K$, $\acute{\Psi}_{22} = -2\bar{\eta}_m \bar{d}_s Q K$, $\acute{\Psi}_{33} = -\bar{\varphi}_s D_s + \frac{\mu_s^2}{\upsilon_s} + 2\bar{\eta}_s \bar{d}_s Q K$, $\acute{\Psi}_{34} = 2\bar{\eta}_s \bar{d}_m K$, $\acute{\Psi}_{44} = -2\bar{\eta}_s \bar{d}_m Z K$.

With the saturation problem, four cases considered are as follows

Case 1: $S_d(\dot{q}_m) \leq \bar{S}_{dm}$ and $S_d(\dot{q}_s) \leq \bar{S}_{ds}$

We have $\bar{\varphi}_m = 1$, and with the definition of $\bar{\eta}_m$, we have $\bar{\eta}_m \leq 1$. Therefore, the LMI is as inequality (13.23).

Case 2: $S_d(\dot{q}_m) > \bar{S}_{dm}$ and $S_d(\dot{q}_s) > \bar{S}_{ds}$

$$\bar{\varphi}_m < \frac{\sum_{j=1}^{n} \left|q_{m_j}\right| \bar{S}_{dm}}{D_m \left\|\dot{q}_m\right\|^2}, \bar{\eta}_m < \frac{\sum_{j=1}^{n} \left|q_{m_j}\right| \bar{S}_{Pm}}{K \left\|\dot{q}_m\right\|^2} \tag{13.36}$$

Then the LMI is as inequality (13.24).

Case 3: $S_d(\dot{q}_m) \leq \bar{S}_{dm}$ and $S_d(\dot{q}_s) > \bar{S}_{ds}$

Then the LMI is as inequality (13.25).

Case 4: $S_d(\dot{q}_m) > \bar{S}_{dm}$ and $S_d(\dot{q}_s) \leq \bar{S}_{ds}$

Then the controller parameters and the saturation upper bound should suit the LMI (13.26).

With the choice of the controller parameters and the system parameters, we have that $V \geq 0$ and $\dot{V} \leq 0$. Therefore, $\dot{q}_m, \dot{q}_s, q_m - q_s \in L_\infty$ and the proof is completed.

Theorem 13.2 *In the bilateral teleoperation system (13.1) with the controller (13.19), the absolute values of the velocities* $|\dot{q}_m|$ *and* $|\dot{q}_s|$ *and the position error* $|q_m - q_s|$ *tend to zero asymptotically in free motion if all the conditions in theorem I are satisfied.*

Proof of Theorem 13.2. From the inequality (13.34), it is possible to see that $\dot{q}_m, \dot{q}_s \in L_2$. Based on the results of Theorem 13.1, we have $\dot{q}_m, \dot{q}_s$ and $q_m - q_s \in L_\infty$. Using the fact that $q_m - q_s(t - d_s(t)) = q_m - q_s + \int_{t-d_s(t)}^{t} \dot{q}_s(t)dt$ and $\int_{t-d_s(t)}^{t} \dot{q}_s(t)dt \in L_\infty$, we have $q_m - q_s(t-d_s(t)) \in L_\infty$, we have $q_s - q_m(t - d_m(t)) \in L_\infty$.

Based on the model of master-slave manipulators we can get

$$\begin{aligned} \ddot{q}_m &= M_m^{-1}(q_m)[\tau_m - C_m(q_m, \dot{q}_m)\dot{q}_m - G_m(q_m)] \\ \ddot{q}_s &= M_s^{-1}(q_s)[\tau_s - C_s(q_s, \dot{q}_s)\dot{q}_s - G_s(q_s)] \end{aligned} \tag{13.37}$$

With the boundedness of $\tau_m, \tau_s, \dot{q}_m, \dot{q}_s, q_m - q_s(t - d_s(t))$ and $q_s - q_m(t - d_m(t))$, it is obvious that $\ddot{q}_m$ and $\ddot{q}_s$ are also bounded. Therefore, we have $\dot{q}_m \to 0$ and $\dot{q}_s \to 0$ as $t \to \infty$ based on Barbalat' lemma. Next we prove that the tracking error $q_m - q_s$ converges to zero asymptotically. If we can prove $\ddot{q}_m \to 0$ and $\ddot{q}_s \to 0$, the result of tracking error converging to zero is straightforward.

Now we need to show that the time derivative of $\ddot{q}_m$ and $\ddot{q}_s$ are bounded. By differentiation of the two former equations, we have

$$\begin{aligned} \dddot{q}_m &= \frac{dM_m^{-1}(q_m)}{dt}[\tau_m - C_m(q_m, \dot{q}_m)\dot{q}_m - G_m(q_m)] \\ &\quad + M_m^{-1}(q_m)[-C_m(q_m, \dot{q}_m)\ddot{q}_m - \frac{dC_m(q_m, \dot{q}_m)\dot{q}_m}{dt} \\ &\quad + \frac{d\tau_m}{dt} - \frac{dG_m(q_m)}{dt}] \\ \dddot{q}_s &= \frac{dM_s^{-1}(q_s)}{dt}[\tau_s - C_s(q_s, \dot{q}_s)\dot{q}_s - G_s(q_s)] \\ &\quad + M_s^{-1}(q_s)[\frac{d\tau_s}{dt} - C_s(q_s, \dot{q}_s)\ddot{q}_s \\ &\quad - \frac{dC_s(q_s, \dot{q}_s)\dot{q}_s}{dt} - \frac{dG_s(q_s)}{dt}] \end{aligned} \tag{13.38}$$

Properties 1.1, 1.2, 1.3 and 1.4, with the boundedness of $\dot{q}_m, \dot{q}_s, q_m - q_s(t - d_s(t))$, $q_m(t - d_m(t)) - q_s$, we have $\ddot{q}_m \in L_\infty$ and $\ddot{q}_s \in L_\infty$. By using Barbalet's lemma, we have $\ddot{q}_m \to 0$ and $\ddot{q}_s \to 0$ as $t \to \infty$. Hence

$$\begin{aligned} \lim_{t\to\infty} |q_m - q_s(t - d_s(t))| &= 0 \\ \lim_{t\to\infty} |q_s - q_m(t - d_m(t))| &= 0 \\ \lim_{t\to\infty} |q_m - q_s| &= 0 \end{aligned} \tag{13.39}$$

Remark 13.8 To deal with the stability and synchronization problem existing in the teleoperation system with input saturation and time-varying delay, an effective control scheme was proposed in [10] with state feedback. Unlike [10], in this chapter the output feedback control problem is considered for teleoperation system under bounded input and time-varying delay. A new SP+Sd-type controller is designed by using the estimated velocity information. The new controller is better than SP+d-type at dealing with the input saturation problem with the unknown velocity. Moreover, the delay-dependent stability criteria are derived. These criteria can give a standard for the choice of the controller parameters in real applications.

13.5 Simulation and Experiment Results

In this section, the simulation and experimental results for the proposed finite time velocity observer and the controller are given to show the effectiveness of the proposed control approaches.

13.5.1 Simulation Results

In simulation, the master and the slave manipulators are considered to be a pair of 2-DOF planar robots with revolute joints. The saturation functions are chosen as $S_P(x) \triangleq sign(x)\min\{|x|, \bar{S}_P\}$, $S_d(x) \triangleq sign(x)\min\{|x|, \bar{S}_d\}$ and $S_{am}(x) = S_{as}(x) \triangleq sign(x)\min\{|x|, \bar{S}_a\}$. $K = diag(20, 20)$, $D_m = D_s = diag(50, 25)$, $\bar{S}_p = diag(10, 1)$, $\bar{S}_d = diag(50, 25)$, $\bar{S}_a = diag(15, 10)$. With the controller parameters and the allowable maximum time delays are $d_m(t) = 0.5 + 0.1\sin t$ and $d_s(t) = 0.4 + 0.2\sin t$. The human insert force is shown in Fig. 13.2. Figures 13.3, 13.4, 13.5 and Fig. 13.6 show the actual position, velocity signals and the estimated position, velocity signals for the master and the slave, respectively. The position synchronization errors are shown in Fig. 13.7. In Fig. 13.8, the control torques are given for the master and the slave.

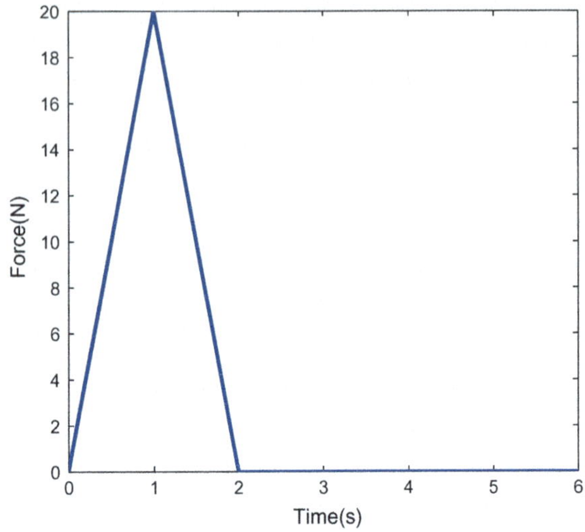

Fig. 13.2 Human-force input F

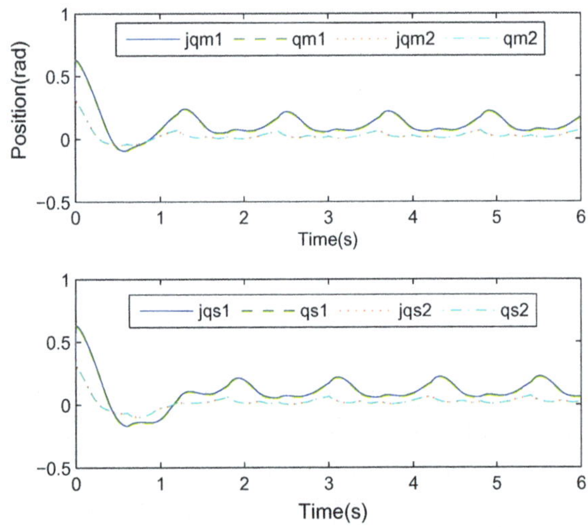

Fig. 13.3 The actual position and the estimated position for the master

From these figures, we can see that, the velocity observer proposed in this chapter can estimate the velocity in a high speed. Moreover, with the estimated velocity information and the bounded inputs, the teleoperation system can get a good work performance.

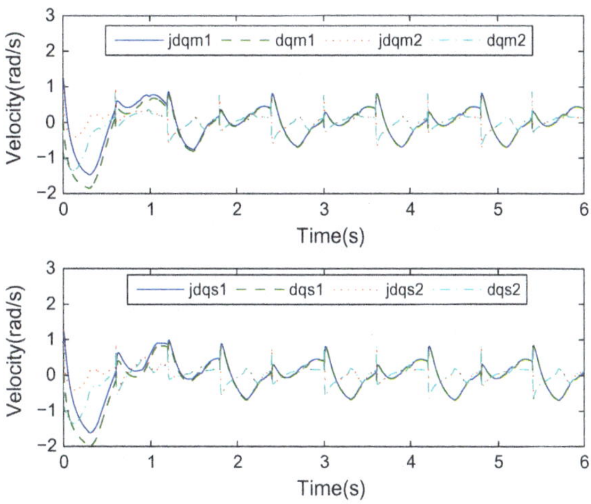

Fig. 13.4 The actual velocity and the estimated velocity for the master

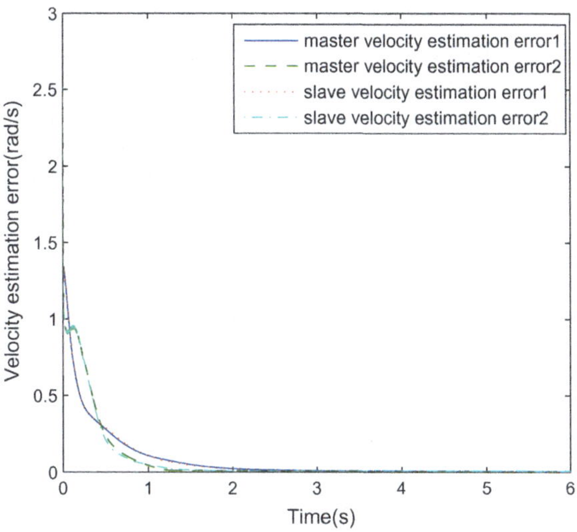

Fig. 13.5 The actual position and the estimated position for the slave

13.5.2 Experimental Results

In this subsection, the experiments on a teleoperated pair of 3-DOF PHANToM manipulators are also performed. To demonstrate the effectiveness of the proposed finite time velocity observer and the SP+Sd-type control scheme, three different time delays are considered: (i) $d_m(t) = 0.2 + 0.1\sin(t)$ and $d_s(t) = 0.2 + 0.1\sin(t)$ (ii)

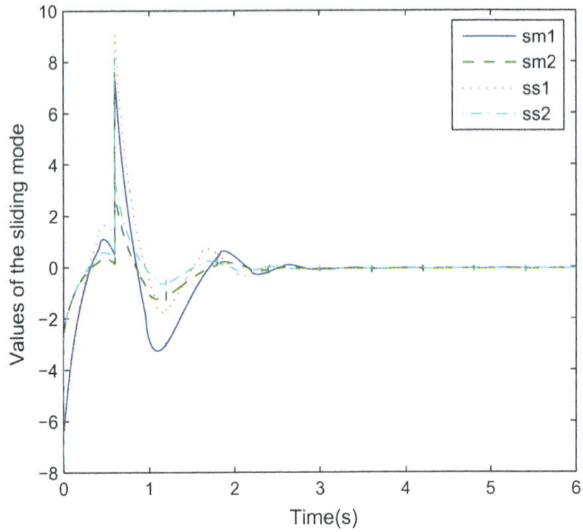

Fig. 13.6 The actual velocity and the estimated velocity for the slave

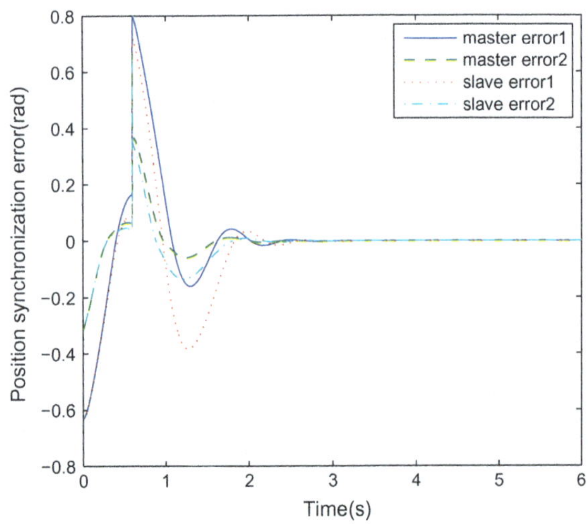

Fig. 13.7 The position synchronization errors at the master side and the slave side

$d_m(t) = 0.5 + 0.1\sin(t)$ and $d_s(t) = 0.5 + 0.1\sin(t)$ (iii) $d_m(t) = 0.5 + 0.2\sin(t)$ and $d_s(t) = 0.5 + 0.2\sin(t)$. It should be notice that in Figs. 13.10, 13.11, 13.12, 13.13, 13.14, 13.15, 13.16, 13.17 and 13.18, the sub-figures denote the the first joint, the second joint and the third joint, respectively.

The upper bounds of the input are set as $U_1 = 0.23$, $U_2 = 0.3$ and $U_3 = 0.3$, respectively. $\bar{g} = 0.07$, $\bar{S}_P = 0.15$, thus $\bar{S}_d = 0.08$. Under case (i), the actual velocity and the estimated velocity are shown in Fig. 13.9. The velocity estimation errors are shown in Fig. 13.10. The positions of the master and the slave are shown in Fig. 13.11. Under case (ii), the actual velocity and the estimated velocity are shown in

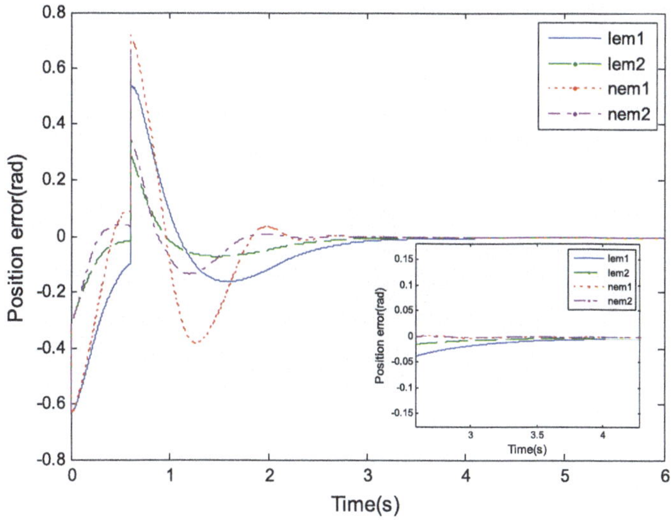

Fig. 13.8 The control torques for the master and the slave

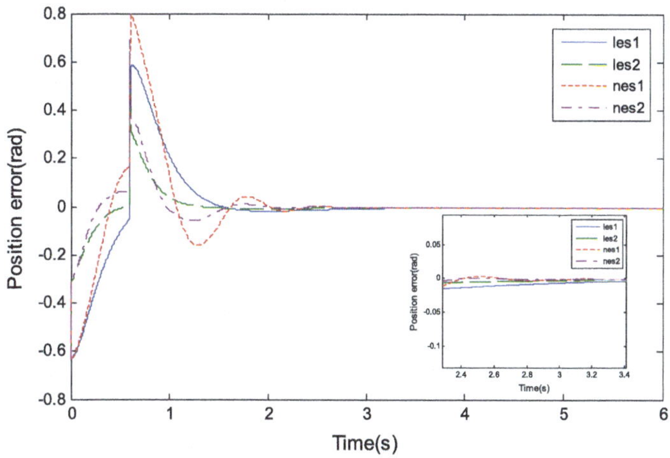

Fig. 13.9 Teleoperation system

Fig. 13.12. The velocity estimation errors are shown in Fig. 13.13. The positions of the master and the slave are shown in Fig. 13.14. Under case (iii), the actual velocity and the estimated velocity are shown in Fig. 13.15. The velocity estimation errors are shown in Fig. 13.16. The positions of the master and the slave are shown in Fig. 13.17.

Fig. 13.10 The actual velocity and the estimated velocity

Fig. 13.11 The velocity estimation error

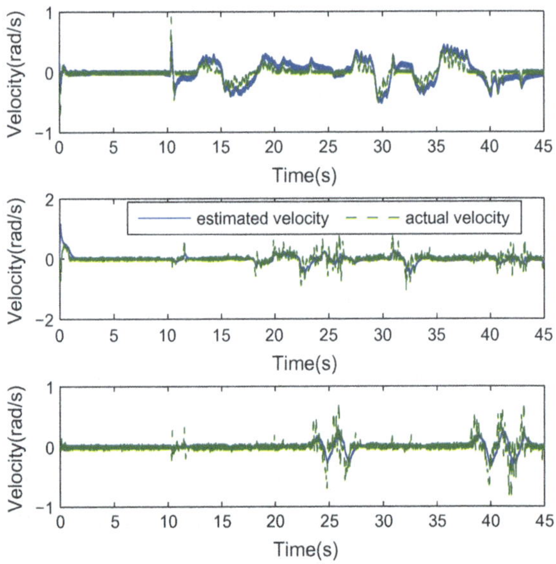

As we can see in above figures, it is obvious that the finite velocity observer can provide fast estimation rate and high estimation precision. Under different time delays, the slave manipulator can follow the trajectory of the master manipulator with the new control scheme.

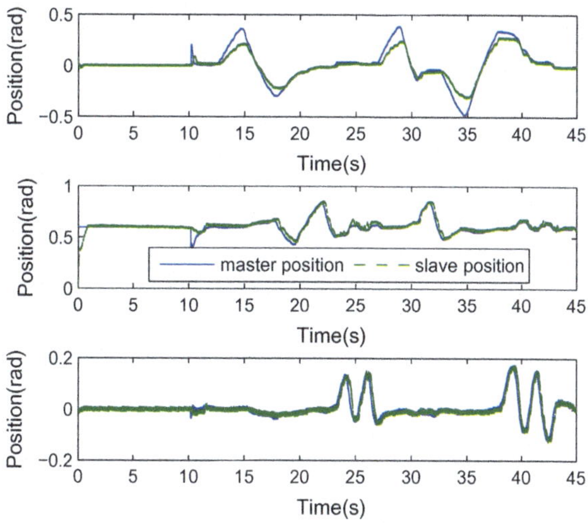

Fig. 13.12 The position of the master and the slave

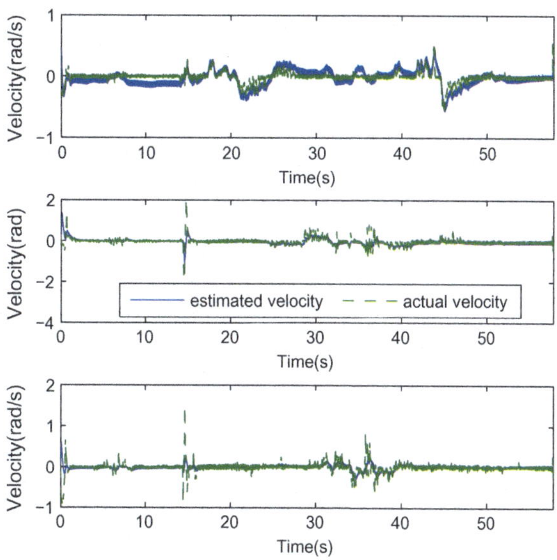

Fig. 13.13 The actual velocity and the estimated velocity

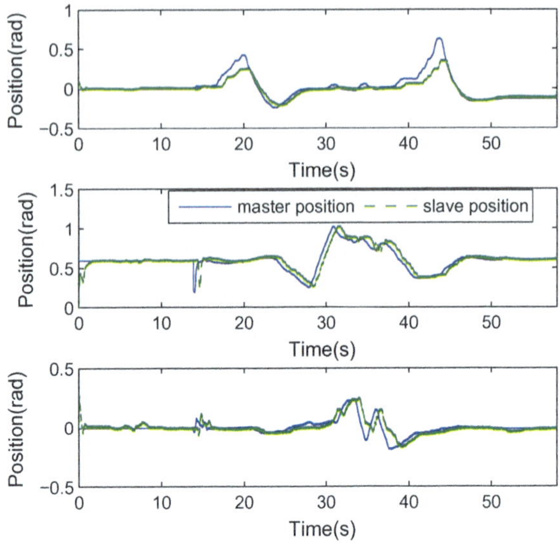

Fig. 13.14 The velocity estimation error

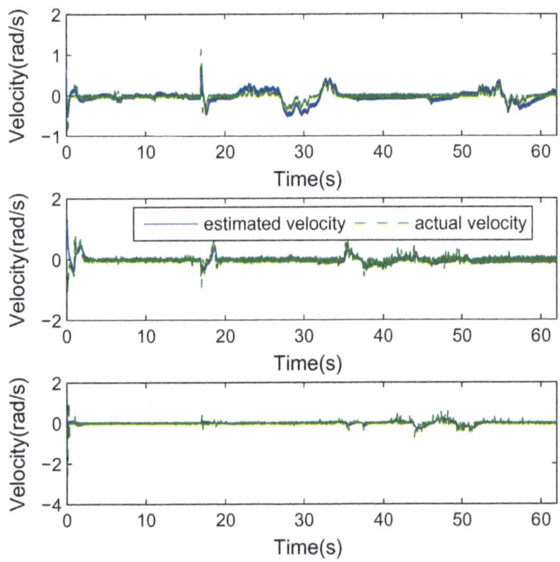

Fig. 13.15 The position of the master and the slave

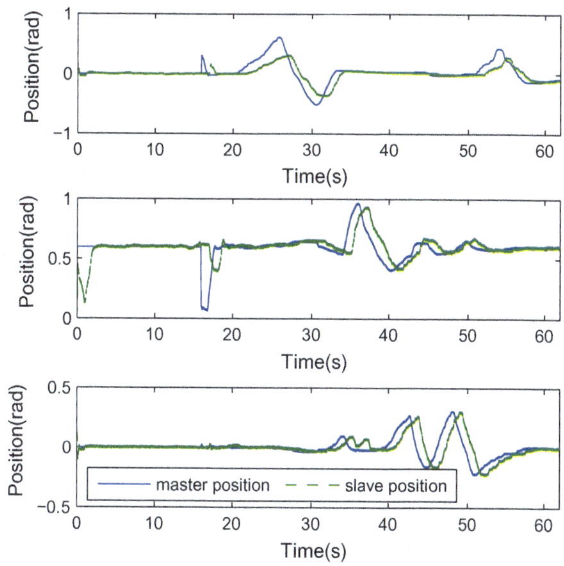

Fig. 13.16 The actual velocity and the estimated velocity

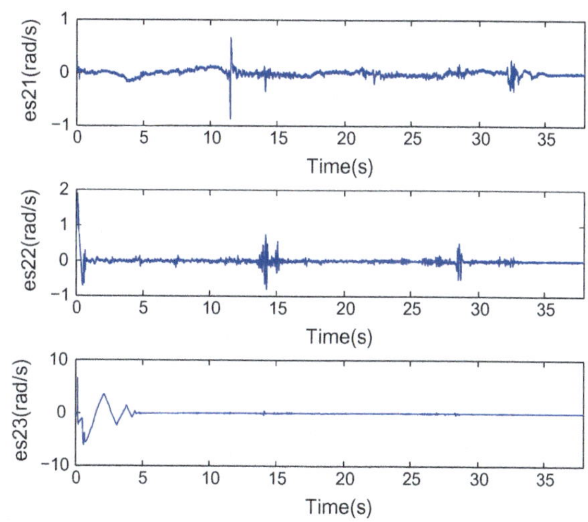

Fig. 13.17 The velocity estimation error

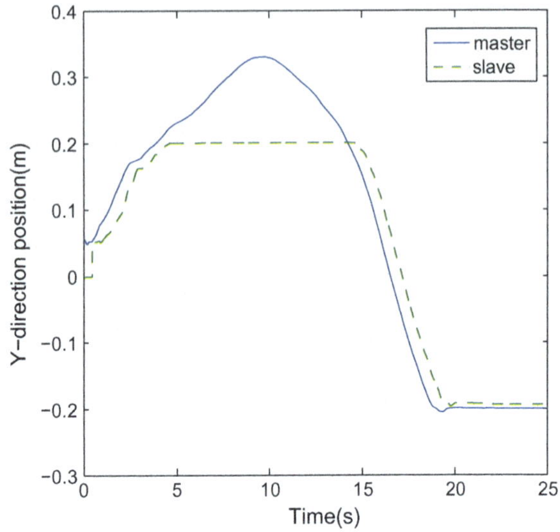

Fig. 13.18 The position of the master and the slave

13.6 Summary

In this chapter, a new output-feedback based adaptive SP+Sd-type controller is proposed. Without the velocity information, a new fast terminal sliding mode velocity observer is designed to estimate the unknown velocity signals. Furthermore, the adaptive method is used to approximate the unknown gravity term, thus the corresponding output-feedback based adaptive SP+Sd-type controllers are developed for the teleoperation system with time-varying delay and bounded inputs. By applying the properties of the saturation function and the proper Lyapunov functions, the delay-dependent stability conditions are deduced. The relationships are built among the controller parameters, the maximum allowable time delays and the upper bound of the saturation functions. Both the simulation and experiment are performed and the results show the effectiveness of the proposed control method. In the future, the system uncertainties and external disturbances will be considered and the force feedback will be applied to improve the transparency of the teleoperation system.

References

1. D. Lee, M.W. Spong, Passive bilateral teleoperation with constant time delay. IEEE Trans. Robot. Autom. **22**(2), 269–281 (2006)
2. E. Nuño, R. Ortega, N. Barabanov, L. Basañez, A globally stable pd controller for bilateral teleoperators. IEEE Trans. Robot. **24**(3), 753–758 (2008)
3. C.C. Hua, X.P. Liu, Delay-dependent stability criteria of teleoperation systems with asymmetric time-varying delays. IEEE Trans. Robot. **26**(5), 925–932 (2010)

4. X. Yang, C.C. Hua, J. Yan et al., Synchronization analysis for nonlinear bilateral teleoperator with interval time-varying delay. Int. J. Robust Nonlinear Control **25**(3), 2142–2161 (2014)
5. I.G. Polushin, P.X. Liu, C.H. Lung, A force reflection algorithm for improved transparency in bilateral teleoperation with communication delay. IEEE/ASME Trans. Mechatron. **12**(3), 361–374 (2007)
6. Y.Q. Ye, Y.J. Pan, T. Hilliard, Bilateral teleoperation with time-varying delay: a communication channel passification approach. IEEE/ASME Trans. Mechatron. **18**(4), 1431–1434 (2013)
7. A. Zavala-Rio, V. Santibanez, Simple extensions of the pd with gravity compensation control law for robot manipulators with bounded inputs. IEEE Trans. Control Syst. Technol. **14**(5), 958–965 (2006)
8. E. Aguinaga-Ruiz, A. Zavala-Rio, V. Santibanez, Global trajectory tracking through static feedback for robot manipulators with bounded inputs. IEEE Trans. Control Syst. Technol. **17**(4), 934–943 (2009)
9. D.J. Lopez-Araujo, A. Zavala-Rio, V. Santibanez, F. Reyes, Output feedbck adaptive sp-sd-type control with an extended continuous adaptation algorithm for the global regulation of robot manipulators with bounded inputs. Int. J. Adv. Robot. Syst. **10**(17), 1–13 (2013)
10. F. Hashemzadeh, I. Hassanzadeh, M. Tavakoli, Teleoperation in the presence of varying time delays and sandwich linearity in actuators. Automatica **49**(9), 2813–2821 (2013)
11. C.C. Hua, X.P. Liu, Teleoperation over the internet with/without velocity signal. IEEE Trans. Instrumen. Measurement **60**, 4–13 (2011)
12. I. Sarras, E. Nuño, M. Kinnaert, L. Basañez, Output-feedback control of nonlinear bilateral teleoperators, in *American Control Conference (ACC)*, pp. 27–29 (2012)
13. D.Y. Zhao, S.Y. Li, Q.M. Zhu, Output feedback terminal sliding mode control for a class of second order nonlinear systems. Asian J. Control **15**(1), 237–247 (2013)
14. Q.L. Hu, B. Xiao, D.W. Wang, Spacecraft attitude fault tolerant control with terminal sliding mode observer. J. Aerosp. Eng. **15**(1), 1–15 (2013)
15. A.M. Zou, Finite-time output feedback attitude tracking control for rigid spacecraft. IEEE Trans. Control Syst. Technol. **22**(1), 338–345 (2014)
16. C.P. Tan, X.H. Yu, Z.H. Man, Terminal sliding mode observers for a class of nonlinear systems. Automatica **46**(8), 1401–1404 (2010)
17. C.C. Hua, X.P. Liu, A new coordinated slave torque feedback control algorithm for network-based teleoperation systems. IEEE/ASME Trans. Mechatron. **18**(2), 764–774 (2013)

The manufacturer's authorised representative in the EU is Springer Nature Customer Service Centre GmbH, Europaplatz 3, 69115 Heidelberg, Germany. If you have any concerns regarding our products, please contact ProductSafety@springernature.com

Printed and bound by CPI Group (UK) Ltd, Croydon, CR0 4YY
29/07/2026
02180610-0004